Extraction/Exclusion

GEOPOLITICAL BODIES, MATERIAL WORLDS

Series Editors: Jason Dittmer, Ian Klinke, and Jennifer L. Fluri

This series publishes studies that originate in a range of different fields that are nonetheless linked through their common foundation: a belief that the macro-scale of geopolitics is composed of trans-local relations between bodies and materials that are only understandable through empirical examination of those relations. It is the interaction of these elements that produces the forces that shape global politics, often with outcomes that differ from the predictions of macro-scaled theories. This world poses questions: how do materialities such as the built environment and the body reproduce global power structures, how are they caught up in violent transformations and how do they become sites of resistance? How do assemblages of human and non-human elements both fortify and transform political space? What possibilities for political change are latent within the present?

Titles in the Series

The Geopolitics of Real Estate: Reconfiguring Property, Capital and Rights, Dallas Rogers

Choreographies of Resistance: Mobilities, Bodies and Politics, Tarja Väyrynen, Eeva Puumala, Samu Pehkonen, Tiina Vaittinen, and Anitta Kynsilehto

Life Adrift: Climate Change, Migration, Critique, edited by Andrew Baldwin and Giovanni Bettini

Territory Beyond Terra, edited by Kimberley Peters, Philip Steinberg, and Elaine Stratford

Chemical Bodies: The Techno-Politics of Control, edited by Alex Mankoo and Brian Rappert

Camps Revisited: Multifaceted Spatialities of a Modern Political Technology, edited by Irit Katz, Diana Martin, and Claudio Minca

Undersea Geopolitics: Sealab, Science, and the Cold War, Rachael Squire

Extraction/Exclusion: Beyond Binaries of Exclusion and Inclusion in Natural Resource Extraction, edited by Stephanie Postar, Negar Elodie Behzadi, and Nina Nikola Doering

Architectures of Security: Design, Control, Mobility, edited by Benjamin J. Muller and Can E. Mutlu

Earth Futures: On Anthropocenes and the Ends of Geopolitics, Deborah Dixon

Extraction/Exclusion

Beyond Binaries of Exclusion and Inclusion in Natural Resource Extraction

Edited by
Stephanie Postar, Negar Elodie Behzadi, and
Nina Nikola Doering

ROWMAN & LITTLEFIELD
Lanham • Boulder • New York • London

Rowman & Littlefield
Bloomsbury Publishing Inc, 1359 Broadway, New York, NY 10018, USA
Bloomsbury Publishing Plc, 50 Bedford Square, London, WC1B 3DP, UK
Bloomsbury Publishing Ireland, 29 Earlsfort Terrace, Dublin 2, D02 AY28, Ireland
www.bloomsbury.com

86-90 Paul Street, London EC2A 4NE

First published in the United States of America 2024
Paperback edition published 2026

British Library Cataloguing in Publication Information available

Library of Congress Cataloging-in-Publication Data

Names: Postar, Stephanie B. (Stephanie Beth), 1988- editor. | Behzadi, Negar Elodie, 1984- editor. | Doering, Nina Nikola, 1987- editor.
Title: Extraction/exclusion : beyond binaries of exclusion and inclusion in natural resource extraction / edited by Stephanie Postar, Negar Elodie Behzadi, and Nina Nikola Doering.
Description: Lanham : Rowman & Littlefield, [2024] | Series: Geopolitical bodies, material worlds | Includes bibliographical references.
Identifiers: LCCN 2023041371 (print) | LCCN 2023041372 (ebook) | ISBN 9781786615367 (hardcover) | ISBN 9781786615374 (epub)
Subjects: LCSH: Mineral industries—Social aspects—Case studies.
Classification: LCC HD9506.A2 E95 2024 (print) | LCC HD9506.A2 (ebook) | DDC 338.2—dc23/eng/20231018
LC record available at https://lccn.loc.gov/2023041371
LC ebook record available at https://lccn.loc.gov/2023041372

For product safety related questions contact productsafety@bloomsbury.com

Contents

Acknowledgments

The theoretical framing for this edited collection grew from the two-day Interdisciplinary Conference on Extraction and Exclusion held in 2017 at the University of Oxford. Many of the chapters in this collection were first presented there. Since the conference, the COVID-19 pandemic and its ongoing impacts indelibly impacted the book you read today, particularly the timeline to publication. Thank you to all of the authors for persevering with this project despite incredible challenges.

Prof. Patricia Daley and Dr. Ian Klinke provided encouragement for the conference and critical early support for this edited collection. Financial support for the conference that inspired this book came from St. Antony's College, University of Oxford, the School of Geography and Environment, University of Oxford, and the Economic and Social Research Council (ESRC). Support for indexing came from the Ciriacy-Wantrup Postdoctoral Fellowship at the University of California, Berkeley.

Negar Behzadi would like to thank her students of the unit Critical Political Ecologies of Extraction and Conservation (2020/2021 and 2022/2023), at the School of Geographical Sciences at the University of Bristol, for their early engagement with ideas and concepts developed in this collection. She would also like to thank Forough and Shahrokh Behzadi for their constant faith and support.

Nina Doering would like to thank Stephanie and Negar for the joint work on this book and for carrying it to the finish line when circumstances forced her to take a step back for a while. This project began with an ESRC doctoral fellowship. Time and space was provided by the Research Institute for Sustainability - Helmholtz Centre Potsdam (RIFS) to complete the edited collection. Nina is grateful for the support of Max Zott and her family.

Stephanie Postar is grateful to Nina and Negar for collaborating on this incredible project. The drafting of this book took place with the support of the Ciriacy-Wantrup Postdoctoral Fellowship at the University of California, Berkeley. The Land Lab at the University of California, Berkeley, provided a forum for the development of ideas for this edited collection. Revisions to the collection took place with the support of a British Academy Postdoctoral Fellowship. Thanks also to Adeen and Michael Postar.

Introduction

Challenging Inclusion as a Solution to Exclusion in Natural Resource Extraction

Negar Elodie Behzadi, Stephanie Postar, and Nina Nikola Doering

In October 2018, 50,000 people converged on the Hambach Forest to protest logging to make way for the expansion of Germany's largest coal mine, the Tagebau Hambach. Local residents were joined by climate activists furious about Germany's continued reliance on electricity from lignite, the least efficient kind of coal. Following years of protests, the demonstration successfully blocked the advance of the mine. Coal can be linked to rising German power in the nineteenth century, as well as two horrific world wars and postwar regeneration. Today, German coal-power supporters argue that renewable energy cannot reliably power urban areas and steel manufacturing and that any transition away from coal needs to occur slowly to protect regions dependent on mining. In 2007, a multistakeholder group negotiated the phase-out of subsidies for domestic hard coal mining.[1] Then, in 2011, the German Government's *Energiewende* (energy transition) committed to a shift toward renewable energy (wind, solar). Hard coal mining ended in 2018 and in 2019, a multistakeholder Coal Exit Commission proposed ending all domestic coal-fired power generation by 2038 or sooner.[2] These cross-constituency domestic agreements converged with international obligations from the 2016 Paris Climate Accords, and Germany pledged to reduce carbon emissions and limit the rise in global temperatures. As Germany missed its emissions targets, coalitions of protesters remained at the Hambach Forest and other coal mines across Germany. Activists have used direct action, mass demonstrations, and court cases to highlight Germany's failure to comply with climate policies in time to protect future generations from the impending climate disaster.

On the other side of the world, in northern Chile's Atacama Desert, another story of resource extraction unfolds. The rise of lithium mining in the Atacama Salt Flats seems disconnected from processes that have developed in Germany. Yet, the same global decarbonization campaign influences this part of the "lithium triangle."[3] Global calls for a transition to renewable energy in the context of the climate crisis have fuelled interest in lithium, a critical mineral for electric vehicle batteries and other green technologies. Chile has recently become the second-largest producer of the mineral globally.[4] In the salt flats, miners from two multinational companies pump brine—salty water containing lithium—into large ponds, where it evaporates to precipitate the precious mineral.[5] The process strains already scarce ground water supplies, harming biodiversity and threatening agriculture, the main livelihood of the Atacameño Indigenous communities. To mitigate these impacts, both companies paid taxes to the government and cash compensations to local residents. In response to the negative impacts of mining, public resistance multiplied through roadblocks, hunger strikes, and lawsuits. At times, protestors invoked the state's failure to respect Indigenous peoples' right to "Free, Prior, and Informed Consent" together with their Andean worldviews. At other times, Indigenous communities have decided to engage with what lithium mining companies framed as participation or benefits—for instance by contributing to environmental monitoring, or by accepting jobs and compensatory payments.

The two cases introduced above point to the core theme of this book: contemporary natural resource extraction. They speak of extraction as a physical process, of surface coal mining in Germany and lithium mining using ground water in Chile. As such, these stories focus on the resource itself—a hydrocarbon (coal) and a mineral (lithium) and how it transforms landscapes. Resource extraction, however, involves more than the physical processes of removing natural resources from their context. It is also inherently geopolitical, encompassing and exceeding broader social, cultural, environmental, and politico-economic processes and formations that are entangled with physical extraction. From both the German and Chilean cases, we can see that they each contextualize relationships of power that materialize at multiple scales, from global negotiations and (trans)national strategies to the everyday lives of miners, activists, human, and/or nonhuman actors. As shown in these cases, we see how certain groups, voices, people, and bodies are included in, and excluded from, extraction in various ways.

Following a long tradition of politico-ecological approaches this book aims to unpack these power dynamics in nature/society relations in resource extraction.[6] Our approach recognizes the ways that the macro-scale geopolitics of socio-environmental issues are deeply entangled with and formed by local, everyday lives, experiences, and bodies. In particular, this edited collection focuses on how the seemingly opposing forces of inclusion and exclusion play out across scales in extractive contexts. Exclusion of individuals and communities from access to land, decision-making, and planning, as well as monetary and nonmonetary benefits accompanies mining and energy projects around the world, as illustrated by the two examples of the complexities of resource extraction that start this introduction. From the German case, people are excluded from ever-expanding surface mining areas. Meanwhile, in Chile, Indigenous peoples have regularly been excluded from decision-making and their rights have been violated in the development of mining projects. International financial institutions, private companies, some states, and their allies promise, however, that sustainable extractive-led development can be an engine to stimulate inclusive global economic growth. In Chile, for instance, inclusionary mechanisms in the form of compensations aim to alleviate the negative impacts of mining on people's livelihoods. Yet, in both cases and beyond, material experiences show that inequality and breaches of rights (human rights, Indigenous land rights, and others), not prosperity, seem to be the reliable outcome of resource practices, as we will consider in depth later in this introduction.

The tension between exclusion and inclusion that we identified in the Chilean and German examples allows us to define the core argument of this book. Our approach takes the predominant narrative of inclusive extractive-led development as a starting point for the underlying critique of this collection: that extraction is predicated on exclusions (practical, epistemic, and ontological), that inclusions (rhetorical and practical) often result in further exclusions, and these in turn reproduce violent exploitative processes on the ground. We find the narratives of the successes of inclusionary extractive-led development to be deceptive because the language of inclusion obscures unchanged systems of domination, dispossession, and violence that remain essentially intact beneath the veneer of inclusivity. The dualistic epistemology of inclusion/exclusion is, we show, rooted in a longer tradition of dualistic thinking that has dominated Eurocentric, colonial perspectives of nature/society relations. Binary thinking, despite having been resisted by many actors, continues to marginalize alternative understandings of and approaches to the uses of natural resources, reproducing the very exclusions it claims to address, as we demonstrate in this collection.

This book brings together twelve empirical chapters written by scholars and activists who participated in a conference hosted by St Antony's College at the University of Oxford in 2017. Each chapter starts with the local, every day, and embodied aspects of the geopolitics of resource extraction by examining the ways in which various forms of inclusion and exclusion are produced, experienced, and contested in mining, oil, gas, and other extractive operations. Together, the chapters build a multidisciplinary approach to critiquing the inclusion narrative through our understanding of natural resources, and what we conceptualize in this introduction as "resource exclusions" and "resource inclusions." "Resource exclusions," and "inclusions," we argue, are entangled, plural, and political and must be read in the context of larger systems of power, including those of race, class, gender, Indigeneity, and others. They also occur at all points in the life-cycle of resource projects (exploration, construction, operation, and beyond) and can be read at multiple scales (from the global to the body). New extractive projects are often built on exclusions determined by legacies of sometimes forgotten extractive activities, including those of colonialism, and reproduced through forms of (neo) colonial presences. By highlighting the complexity and interconnectedness of the sociocultural, historical, economic, and political dynamics that underlie resource operations, we illustrate the problematic impacts of the simple binaries that have come to define both dominant thinking about and attempts to address experiences of exclusions and inclusions. Instead, we refine our understanding of power in resource extraction including, for instance, by challenging various essentialist views of local communities (as illustrated by the Chilean case) as systematically against resource extraction and/or states and extractive actors as monolithic.

This collection frames the twelve empirical chapters around six "naturalized" binaries in resource extraction in order to recognize and challenge them: Haves/Have-Nots, Oppressors/Oppressed, Human/Nonhuman, Static Materials/Dynamic Materials, Large-Scale/Small-Scale, and finally, Inclusion/Exclusion. As the editors, we recognize that framing the book in this way in itself reinforces binary thinking. We, however, believe that making visible these binaries and their reifications is a starting point to challenge them. Each of the six parts also has a short theoretical introduction that retraces intellectual debates around the binaries and will support readers in contextualizing the empirical chapters in a broader genealogy of thought.

This introduction is organized into four sections. We start by teasing out the complexities of exclusions and inclusions in the two resource extraction cases that start this introduction to foreground some of the key questions and contributions at the heart of this collection. Section two reviews the current politico-ecological moment in resource extraction, by grounding the specificities of contemporary inclusions and exclusions of extraction in a longer history. In the third section, we situate our approach among academic debates on natural resources, as well as notions of exclusion and inclusion. We detail our approach and theoretical framework to unsettling this dualistic epistemology and bring together previously siloed critical literature to build new cartographies of resistance to resource exclusions. In the fourth section, we outline the empirical chapters and sections that form this edited collection.

We hope that this edited collection will support intellectual discussions among different categories of readers and invite them to approach both the introduction to this collection and the overall book flexibly based on their previous knowledge of extractivism. Some of the sections in this introduction might feel familiar to critical resource scholars and other political ecologists. We hope that such readers will find value in the rich empirical material that this book provides and how chapters develop nuanced perspectives on the geopolitics of resource extraction, and on resource exclusions and inclusions beyond binary understandings. For readers less familiar with the history and main debates in resource extraction scholarship, the introduction to this book will retrace some key concepts and binaries central to our understanding of extractivism. Extensive footnotes throughout this collection's introduction will also drive to further readings—on political ecology and others—that will contextualize the more theoretical and empirical contributions of this book. The introduction to each of the six parts of the book will also provide readers with a bridge to explore theoretical orientations that might be unfamiliar.

KEY QUESTIONS AND CONTRIBUTIONS OF THIS BOOK

The Chilean and German cases that start this introduction point to a series of key questions and contributions of this book. First, the two examples raise questions about how resource extraction produces various forms of exclusion. In Germany, Chile, and elsewhere, individuals and communities (human and nonhuman) are excluded from land, from the benefits and revenues of extraction, and/or from decision-making and planning. Who and what gets excluded on the ground before, during, and after resource exploitation? Is exclusion in contemporary resource extraction different from how it might have happened in the past? And how are these types of exclusion produced, reproduced, and contested geographically (across places and scales), historically (throughout time), onto-epistemologically (through ways of being and knowing)? At the same time, the variety of sites, modes of resource extraction, temporalities, and places covered by these cases reflect both the diversity that this book also aims to cover as well as similarities and linkages between seemingly disparate geographies. This raises another important question: How are exclusions that occur in distinct places and which involve different resources, linked? To what extent are the forms of exclusion that accompany the mining of minerals and hydrocarbons unique or similar to other types of extraction (hydropower, tourism, forestry, etc.)? Making visible and understanding what we will call "resource exclusions" is one of the core objectives of this book. Exclusions in extraction are, as mentioned earlier, plural, and they take various forms that we aim to explore in this collection. We consider that understanding the ways in which these exclusions happen on the ground and at different scales is key to a critical reading of the geopolitics of resource extraction.

Second, the Chilean and German examples illustrate the contrast between the violence of diverse types of exclusion and the paradoxical mildness inherent in the discourse of "extractive inclusion." The idea that "inclusion" can negate the diverse forms of exclusions in resource extraction without causing additional harm is challenged empirically. Still, inclusionary rhetoric, central to the unfolding of the macro-scale geopolitics of resource extraction, promises that resource extraction can be an engine to stimulate inclusive economic growth. Aligned with inclusionary discourses, international organizations, governments, and private companies introduce sets of managerial and practical responses including state-level regulations and domestic and international agreements (e.g., Aarhus Convention 1998; ILO Convention 169). Free, Prior, and Informed Consent (FPIC discussed below), participation, and corporate social responsibility projects are among the managerial solutions to the mounting problems corporations face when their extractive activities are blocked by dissatisfied local populations or threatened by international or local movements. On the surface, the widespread adoption of inclusionary managerial solutions suggests that extractive industries have shifted to operate based on a model of including more people in their prosperity.

Yet, empirical data suggest that inclusion is ambivalent—exclusion often continues despite the rhetoric otherwise. "Resource exclusions" take new forms in the context of inclusionary practices, as illustrated by the two examples that start this introduction. In Chile, lithium extraction companies' contribution to national taxes feeds into the government's argument that, despite negative local impacts and protests, mining can be inclusive, and a vehicle for national development. The diverse positions within Indigenous communities also illustrate the complexity of inclusion where segments of negatively impacted communities decide for various reasons to be included in potential benefits of lithium mining operations, leading to tensions within communities. In Germany, a multistakeholder group negotiated the end of hard coal mining, reflecting how participation in decision-making about resource extraction can be deployed to include the very people whose livelihoods will be eliminated. The inclusion of coal miners in the Coal Exit Commission also appears to set miners' livelihoods today against the prosperity of future generations threatened by unrelenting carbon emissions. These ambivalent inclusions urge us to consider several questions: Where does the inclusionary discourse come from? How is it articulated on the ground? To what extent have these participatory and other so-called "inclusionary tools" alleviated the negative impacts of resource extraction? And what have been their other effects? Raising these questions engages with the second core objective of this book: to unpack what we refer to as "resource inclusions" and their role in producing, reproducing, and exacerbating different forms of exclusion. Examining the local politics and practices of inclusion sheds light on the way broader macro-scale discourses are also reworked when confronted with the materiality of extraction.

Finally, the two examples above point to the central contributions of this book, which entails questioning the politics of resource knowledge. Our first contribution lies in interrogating how binary framings, which often overlap with each other, marginalize other practices, and forms of knowledge. At the same time, we, as researchers/scholars/activists, determine who and what gets included, and therefore, epistemically excluded from our understanding of resource extraction. Which places, scales, topics, actors, and notions (of exclusion, inclusion, and resistance) are foregrounded and obscured in publications (like this book), discussions, and teachings? The extractive nature of research has long been critiqued, particularly by Indigenous scholars and activists (discussed further below). As editors educated and working in Western research institutions, we recognize that social science research on resource extraction, including critical resource scholarship, has produced broad epistemic exclusions that need challenging. We are aware that by producing this book we have also participated in forms of knowledge extraction and exclusion. By opening ourselves up to this interrogation we see the flaws in our own research, writing, and editing. Nonetheless, we see this book as an effort to produce knowledge that challenges simplistic understandings of inclusions and exclusions in resource extraction. This is a step toward unsettling dominant literature on sustainable extractive-led development and building new cartographies of resistance to adverse "resource exclusions" and "inclusions."

EXCLUSION AND INCLUSION IN CONTEMPORARY RESOURCE EXTRACTION

What characterizes contemporary resource extraction? How do notions of exclusion and inclusion help explain this "moment" of extraction? In this section, we unpack the politico-ecological forces that have sedimented throughout history and led to the consolidation of the binary of exclusion/inclusion.

Neoliberal Extraction as Exclusion

In this book, we see "contemporary" forms of resource extraction as coinciding with the period of late global capitalism that started with the neoliberal turn of the 1980s. Contemporary forms of extraction are not new but predicated on a longer history that we trace back to modern colonialism and earlier forms of privatized globalizations. The neoliberal ideology espoused during the 1980s advanced deregulation and privatization around the world, with support from international organizations and agreements. Starting in full force after the economic crises of the late 1970s, countries of the "Global

North" and "South" restructured their economies and reduced social welfare. In the Global South, structural adjustment policies pushed countries to attract foreign industries and investors—particularly in the form of extractive industries—in exchange for loans from the World Bank and the International Monetary Fund (IMF) (Bebbington et al. 2008). Foundational to neoliberal ideology was the idea that wealth created in free-market capitalism would "trickle-down" from the rich to the poor. This "inclusive" ideology, however, did not achieve its promises, and resulted in uneven forms of development and exclusions (Smith 2010). These include inequalities in wealth between sites of extraction and sites of expansion of value (i.e., through production, commodification, and consumption) as well as uneven distribution of power among actors involved in resource extraction (i.e., multinational companies, states, mining communities, among others).

Neoliberal forms of extraction build upon an existing extractivist system. Extractivism is a mode of accumulation based on the removal and exploitation of raw natural resources in (so-called) peripheral places, and the dislocation of those resources for conversion into commodities to fuel industrial development (Acosta 2013; Gomez-Barris 2017). Contemporary plunder and appropriation can be traced back to the birth of colonialism and capitalism, with the exploitation of raw materials essential to the development of the Global North (Acosta 2013, 62, 63). Resource extraction and its inherent exclusions were one of the main pillars of colonial, racist, capitalist, (hetero) patriarchal expansion and consolidation (Himley et al. 2021). Historically, European and other empires built their prosperity on the violent extraction of a range of resources from occupied lands (minerals, like gold, copper, silver, and diamonds, as well as wildlife, cash crops, timber, water, and others) (Scott 2008; Acosta 2013; Lajus 2011; Morris-Suzuki 2013). The foundations of extractivism emerged from multiple layers of exclusion, for instance, from the ownership and benefits of resources and of local communities from their lands and resources. Exclusions took the form of dispossession, socio-spatial and cultural displacements, disenfranchisement, environmental degradation, and direct violence, particularly toward Indigenous peoples (Kobayashi and de Leeuw 2010; Cameron 2015; Samson and Gigoux 2017). Other violent extractions, including of labor (of miners and others) and the de-humanization of bodies through systems of slavery and exploitation, made and continue to make extractivism possible (Bezuidenhout and Buhlungu 2011).

The exclusions and violence created by sedimented histories of colonial extractivism remain central to neoliberal and neocolonial modes of appropriation (Preston 2017; Murrey and Jackson 2019; Wise 2021). However, neoliberal extraction constitutes a period of material intensification (Bridge 2009)[7] marked by specificities, starting with the attributes of global demand and the forms of exclusion that it shapes. Dramatic increases in global population simultaneously drive demand for electricity and consumer goods (Degani et al. 2020; Sherman et al. 2022). The center of gravity of the global economy has also changed, with the rise of new powers (including China, India, Brazil, and others). Global demand drives resource extraction for hydrocarbons (coal production, for instance, has not decreased despite calls for decarbonization, with China leading production and consumption (Wang et al. 2020) and for other minerals and/or sediments (Bisht 2022; Krausmann et al. 2018).[8] At the same time, in response to the climate crisis, the global push for decarbonization leads to the increase in the mining of specific minerals—lithium, nickel, copper, and others (Grandell et al. 2016).[9] The decrease in demand at the beginning of the COVID-19 pandemic fuelled hopes for a global green revolution based on the phasing out of hydrocarbons. High levels of demand have, however, been maintained, spurring continued extraction during the pandemic to the time of publishing this book.

As a result of such changes in demand, new mining landscapes, property arrangements, and miners have emerged. Novel extractive technologies render accessible previously inaccessible sites or unprofitable resources.[10] Technological advances, together with distinct forms of financialization and territorialization, created "new" extractive frontiers (among many: Peluso and Lund 2011; Peluso 2017; Klinger 2017; Rasmussen and Lund 2018). Specific to contemporary extraction, multinational companies have consolidated power in multiple ways, including through joint ventures (for instance, public–private partnerships). In these arrangements, "new" actors are increasingly prominent—Chinese, Brazilian, and other companies engage in these "South-South" agreements (Cher 2016; Lee 2018; Mawdsley 2018).[11] The interconnectedness of markets and communication technologies have also changed how miners themselves operate—for instance, they can be more reactive to boom and bust cycles, as we can see with flexible artisanal miners who navigate multiple livelihoods (Zhu and Peluso 2021).

In response to the unevenness of extractive development and its layers of exclusions, global movements, states, and local communities organized various forms of resistance. Across scales, people refused to accept the failures of trickle-down economics and exploitative industrial practices, in many ways continuous with resistance movements of the past (Jalbert et al. 2017; Willow 2018; McNeish and Shapiro 2021).[12] At the local level, people resisted the negative environmental and social impacts of resource extraction through violent and nonviolent protest, legal action, and other forms of resistance (Caretta and Zaragocin 2020; Jenkins 2017; among many). Indigenous scholars and rights-holders, for instance, have led resistance movements (e.g., Estes 2019) and foregrounded counterstories to resist colonial narratives that enable land dispossession and extraction (Fjellheim 2020), thereby also opposing simplistic narratives of Indigenous peoples as "victims" (McGregor et al.

2020, 37) and Indigenous lands as "toxic places and spaces of exploitation, extraction, and suffering" (Aldred et al. 2021, 90). At times, international solidarity movements joined local protesters, facilitated by wider access to communication technologies.[13] In some countries, recent nationalizations of domestic extractive industries aimed to revert some of the damage caused by privatization measures.[14] While commentators have at times denounced the neoextractivist nature of these forms of resource nationalisms, countries that engaged in such reforms were often motivated by an anti-neoliberal ethos. Crossing scales, various global environmental movements coalesced in response to the violence of the neoliberal model and its insistence on framing industrial pollution and carbon emissions as externalities.

The Inclusionary Model of Extraction

The various forms of resistance discussed above denounced the failures of neoliberal extraction. Rather than leading to systemic change, however, critiques of the violence and inequalities produced by resource-extractive projects brought about what we refer to in this book as "inclusionary" rhetoric and practices. Ideals of inclusivity in growth and just distribution have deep historical roots, but we trace the contemporary narrative of inclusion in resource extraction back to the notions of the "trickle-down" benefits espoused by free-market capitalism. In this framing, supposedly "efficient" private companies distribute the benefits of extraction, thanks to the nominal inclusion of the wider public in lucrative private enterprises. The exclusions of contemporary extraction set the stage to advance the inclusionary narrative beyond trickle-down economics to new models of development (sustainable development, "green economic growth") as well as new inclusive practices (corporate tools, participatory mechanisms, and systems of accountability).

The 1987 Brundtland Commission Report marked a key milestone in recognizing the exclusions of neoliberal capitalism, instead promoting "sustainable development."[15] Critiqued for its underlying contradictions and inefficiencies, the logics of "sustainable development" further consolidated power in corporations by encouraging "corporate oxymorons" (Kirsch 2010, 87) such as "sustainable mining" and extractive-led development (Yakovleva and Nickless 2022; Lodhia 2018). The idea of sustainable development evolved to respond to mounting climate concerns, continued inequalities, and lasting poverty, to embrace the concept of the "green economy" and "inclusive green growth." The green growth narrative refutes the idea of the neoliberal global economic system as a barrier to environmental sustainability and poverty reduction but instead advances continued growth and market-based mechanisms as the solution (Goodman and Salleh 2013). From the outset, this strategy has met resistance from diverse actors who critique the "neoliberalization of nature" through commodification, marketization, and privatization (Heynen and Robbin 2005).[16] Critics challenge the implications of a "green economy" built on continued extraction (and the violence of involuntary and adverse inclusion through pollution, forced labor, and sociopolitical disruption) and exploitation (Goodman and Salleh 2013).

Recognizing the shifting rhetoric, extractive corporations have responded with a set of managerial tools glossed as inclusionary. Multinational hydrocarbon and mining companies from the 1980s spearheaded the global rise in Corporate Social Responsibility (CSR) programs. Corporations attempted to improve their reputations by contributing to Western understandings of development and adopting narratives of "sustainability" (Gilberthorpe et al. 2016; Hilson 2012). In the Global South, some CSR programs assumed functions of government left void by structural adjustment policies and austerity measures (i.e., health care provision, infrastructure) (Hilson 2012). Experiences from around the world, however, illustrate that CSR programs regularly fail to create lasting positive impact. Rather, they further consolidate power in corporations, strengthen social inequalities, and create novel exclusions (Gilberthorpe and Banks 2012; Frynas 2005). This is in part because CSR is vague conceptually and ill-defined in practice. For instance, it is often difficult to delineate voluntary CSR programs from legally mandated activities such as Environmental Impact Assessments (EIA) and Social Impact Assessments (SIA), which are pre-extraction assessments of anticipated local and regional effects of resource extraction.[17]

Core to this inclusionary model is the necessary participation of affected communities in identifying impacts, in decision-making and management, and in benefiting from extractive activities. Grassroots resistance demanded participation in alignment with critiques of top-down development of the 1990s (Cornwall 2011). Over time, powerful actors adopted participation as a key concept of the inclusive model of resource extraction. The IMF, the World Bank, and other funders made EIAs, SIAs, and monitoring conditional for loans, effectively mandating certain forms of participation in the extractive projects they funded. Free, Prior, and Informed Consent (FPIC, see UNDRIP) for projects affecting Indigenous lands and waters has become ubiquitous in discussions of Indigenous participation.[18] From this requirement for participation, extractive industries (sometimes alongside governments) started negotiating legally binding agreements to obtain consent from impacted communities (Impact and Benefit Agreements, or IBAs). For instance, companies agreed to measures (job creation, training, payments, business and infrastructure programs, and environmental protection) aimed at mitigating the sociocultural and environmental impacts of extractive projects and ensuring benefit-sharing (O'Faircheallaigh and Corbett 2005). Participation in the financial benefits of extraction also at times took the form of sovereign wealth funds set up by resource-rich countries. These funds aim to manage existing (or anticipated) resource wealth, reflect aspirations for economic diversification, and ensure intergenerational fairness.[19]

Alongside the expansion of these participatory processes, skeptical voices emerged. Critics highlighted how participation conceals and reinforces unequal power dynamics, asserts limited or no influence on project development, excludes groups and knowledges, and fosters distrust (Cooke and Kothari 2001). EIAs and SIAs and consultation requirements in other formats as part of licensing processes, for instance, continue to disregard Indigenous and local knowledge and, by design, do not extend beyond the project development phase (Lane et al. 2003; Lawrence and Larsen 2017; Baker and Westman 2018; Fjellheim 2023). Indigenous and non-Indigenous scholars and rights-holders have also denounced the practical shortcomings and violations of FPIC (Schilling-Vacaflor 2016; Mahanty and McDermott 2013; Mitchell et al. 2019). These critiques show how the industry's participatory measures with Indigenous and other local communities often amount to little more than window dressing, forms of what Murrey and Jackson (2019) call "localwashing." Analyses of existing IBAs show that the great majority fall short of their potential for a more just distribution of benefits and costs, at times reproducing inequalities and hindering participation (O'Faircheallaigh and Corbett 2005).[20] Extractive industries and their allies responded to these critiques by focusing on improving methods and instruments and measuring their success in terms of greater inclusion (i.e., a larger number of people drawn into essentially the same flawed participation procedures) (Cooke and Kothari 2001).

Another pillar in the inclusive model of contemporary extractive development is the rise of accountability mechanisms. At the local level, domestic regulations and laws made it possible to seek inclusion (in benefits, ownership, compensation for environmental damage, etc.) through lawsuits and by changing laws to favor participation (Sawyer 2004).[21] Lawyers representing impacted communities also took these cases to the national and international stage via a range of court systems (Sawyer 2015; Watts 2005). At the international level, legally binding protocols and agreements set limits for carbon emissions (i.e., The Kyoto Protocol and the Paris Climate Agreement) and capped pollution (i.e., Minamata Convention on Mercury). Multistakeholder coalitions formed of representatives of governments, industries, donors, civil society, and financial institutions also set standards for their industries.[22] Like other inclusionary mechanisms, these accountability procedures have been critiqued for failing to reach their objectives, maintaining the status quo, and creating novel exclusions (Kutney 2014; Allan 2019; Bieri 2016; Sovacool et al. 2016; Spiegel et al. 2015).

Dismissing these inclusionary actions and instruments univocally or seeing them as inherently exclusionary would nevertheless be too simplistic. Despite the limitations of inclusionary practices, we recognize that, for communities impacted by extraction (Indigenous and otherwise) and activists, these inclusionary tools were hard won, and still have the potential to work after revision or better implementation (see also Mitchell et al. 2019). Powerful examples include Indigenous self-developed FPIC protocols (Lightfoot 2020). Still, these inclusionary tools have made inroads into a system fundamentally based on continued violent colonial and racist exclusion and exploitation (see also McGregor et al. 2020), which endures by re-inventing itself to stay the same. "Including" impacted communities is also complicated as Indigenous and local communities are rarely unified in their opposition to or support of extractive projects, not only because impacts differ among social groups but also because individuals and groups have different needs and expectations (Anthias 2018). In addition, Indigenous peoples may also choose to engage with inclusionary instruments and spaces for various and complex strategic reasons (Leifsen et al. 2017; see also Fjellheim 2023).

RESOURCE EXCLUSIONS AND INCLUSIONS: OUR APPROACH BEYOND BINARIES

Scholars of resource extraction responded in different ways to the tensions between the consolidation of inclusionary rhetoric and practices and the exclusions of contemporary extractivism. In this section, we ask: How has resource scholarship engaged with these tensions? How has this scholarship relied on and reproduced the binaries that this book aims to challenge? And how can we unsettle this dualistic epistemology to build new cartographies of resistance to resource exclusion? By engaging with these questions, we define the approach that we take in this book.

Binaries in Resource Scholarship

The pervasive question of why resource-rich countries in the Global South seem to remain poor drove much of the resource literature since the 1980s (Badeeb, Lean, and Clark 2017). Mainstream geopolitics and economics scholarship on "resource wars" (Klare 2001), "Dutch disease,"[23] and the resource curse[24] (Auty 1993; Humphreys, Sachs, and Stiglitz 2007; Haber and Menaldo 2011) focused on studying the "failed" (largely "Southern") state governance of resource wealth and associated poverty, resource conflicts, and violence. While this literature highlighted some of the challenges facing resource-rich states, it fed into the same inclusionary rhetoric underpinning sustainable development (Kirsch 2010). For these scholars, the problem lay at sites of resource extraction (i.e., with domestic institutional barriers in "Southern" states, local conflicts, technocratic failures of governance, corruption), rather than in the exclusionary structure of neoliberal forms of contemporary extraction (Gilberthorpe and Rajak 2017). We see such perspectives as apolitical as they conceal the systemic nature of exclusions in the neoliberal era, leaving us with scattered understandings of the nuances of exclusions and inequalities beyond economic and statist perspectives (Bannon and Collier 2003; Le Billon 2008; De Soysa and Neumayer 2007; Gilberthorpe and Rajak 2017; Lahiri-Dutt 2006). In particular, these views reproduce a series of binaries and hierarchies: for example, that resource wealth was "well-managed" by certain states (generally: "Global North") and mismanaged in others (generally: "Global South"),[25] and that "natural" materials should be exploited for the economic and broader development of societies.

Critical resource scholarship emerged partly in response to this apolitical statist and economistic resource curse literature (Himley et al. 2021; Watts 2004; de Oliveira 2007). Scholars from across the social sciences highlighted the diverse forms of exclusion common in extractive landscapes—exclusions from jobs, from resources, from decision-making, and from understandings of resources themselves. Critical resource scholars do not always directly engage with the fields of critical geopolitics (Agnew 2001; Dodds and Sidaway 1994; Dodds 2001; Koopman et al. 2021), feminist geopolitics (Dixon and Marston 2013; Dowler and Sharp 2001; Hyndman 2001; Massaro and Williams 2013), or political ecology as they come from a variety of disciplines that draw from beyond these three fields. Nevertheless, they use concepts, ideas, and approaches that resonate with the core principles of these theoretical bodies of work and at times directly engage with notions of geopolitics. Critical resource scholars have, for example, switched attention away from monolithic perspectives of the state or global institutions as the main sites of politics. They have turned instead to empirical perspectives and on-the-ground experiences of resource extraction, revealing how zooming in opens new windows to understand the (geo)politics of extraction (Childs 2018; Squire and Dodds 2019; Zalik 2018). This includes a challenge to grand narratives and discourses on the geopolitics and economics of extraction written from the perspective of powerful institutions and political/economic actors as well as a recognition of often overlooked scales (such as the mundane, the everyday, or the body) and actors (such as Indigenous people, local activists, migrants, etc.) as agents of (geo)politics. Central to critical resource scholarship is also a commitment to material, grounded, and empirical understandings that this book follows (Postar and Behzadi 2022; Behzadi 2019; Caretta and Zaragocin 2020; Hauserman et al. 2020; Jenkins

2017). At the heart of this body of literature also lies a series of criticisms that explicitly or implicitly challenge dualistic epistemologies, including the binary of exclusion/inclusion in the violence of resource extraction, the conceptualization of resources as "natural" entities separate from society, and the opposition of North and South. Among the critical literature, Indigenous scholars, for instance, offer counternarratives which make visible complex epistemologies and ontologies often excluded or marginalized (e.g., Fjellheim 2020; Todd 2018).

As critical resource scholars, and as the editors of this collection, we recognize, however, that critical resource scholarship still (re)produces and relies upon binaries and epistemic exclusions. While we do not aim to draw an exhaustive list here, we see some of these limitations as particularly salient. Regionalized literature has, for example, produced richly detailed empirical material; yet, it has left us with a limited understanding of how processes of extraction connect across different sites.[26] Certain resources, like oil or gold, with robust bodies of literature also tend to cite internally and remain siloed away from other forms of extraction, resources, and materials. The spectacular nature of large-scale extractive operations also continues to overshadow and dissociate from small-scale and artisanal extraction processes (Barrett and Worden 2014). While critical resource scholarship aims to give voice to those excluded from resource operations (Cameron 2015; Cameron and Levitan 2014; Anthias 2018; Murrey 2016; Lahiri-Dutt 2012; Behzadi 2019; Postar and Behzadi 2022), it has sometimes focused on one axis of identity (race, gender, age, class, Indigeneity) to the detriment of more holistic understandings. Indigenous scholars are making crucial contributions to critical resource scholarship (e.g., Altanmirano-Jimenez 2021; Deer and Nagle 2017; Estes 2019; Fjellheim 2023; among many others). Nevertheless, Indigenous knowledges and voices continue to be underrepresented and marginalized in the Western academic system (Todd 2016) and policy-making. Indigenous scholars have referred to this exclusion of knowledge as "epistemic ignorance" (Kuokkanen 2008) and "a form of environmental injustice" (McGregor et al. 2020, 37). These limitations lead critical resource scholarship to too often remain dispersed by region and siloed by scale or discipline. Like in much of academia, more work remains to elevate marginalized voices to positions of epistemic power. These epistemic exclusions—among others—have prevented the emergence of a cohesive block that has the ability to unsettle dominant literature on "sustainable" and inclusive extractive-led development and the binaries and dualistic epistemologies that this narrative draws upon. As editors, and as we will highlight in the next section, we see our book as joining a rich tradition of work that has committed to challenging some of these limitations, while recognizing also limitations and exclusions in our own approach.

Beyond Binaries

In this book, we aim to go beyond the binaries that drive our understanding of contemporary natural resource extraction. We particularly show that the dualistic epistemology of inclusion/exclusion marginalizes alternative understandings of and approaches to the uses of natural resources, reproducing the very exclusions it claims to address. The paragraphs that follow sketch the origins of binary thinking and critiques as well as discuss some of their effects on understandings of resource extraction. We also give a broad overview of literature we rely on to counter binary thinking in resource extraction in this introduction—namely feminist, decolonial, and Indigenous scholarship.

Dualistic thinking is deeply ingrained in Western philosophy and lies at the foundation of modern colonial logics, framing predominant ways of knowing in academia and beyond. Often, the origins of binary thinking in Western philosophy are traced back to Ancient Greece and the philosophy of Plato and Aristotle who argued for the separation of the material body and immaterial soul (Gerson 1986; Heinaman 1990). The oppositional element of binary thinking, particularly with respect to humans (i.e., culture) separated from nature (i.e., matter) was sedimented in modern epistemological categories. Most famously, Descartes's (1996[1641]) perspectives consolidated the formulation of the mind (thinking, immaterial, indivisible) as a substance distinct from the body (matter, unthinking). Early seventeenth- and eighteenth-century critics of binary thinking advanced more relational worldviews by challenging framings separating mind from body and human from nonhuman, among others (Nadler 2019).[27] Still, binaries continued to frame theory and practice. In the nineteenth century, Hegel's (2001[1837]) conceptualization of the human, for example, as the key historical agent similarly frames a world divided between (white, Western, male) humans and all others, underscored by Eurocentric racism.[28] This kind of philosophy emerged during and endorsed the colonial project and its attending exclusions and extractions.

While opposites occur in diverse ways of knowing and relating with the world, this book focuses mainly on how the legacies of Western thinking still dominate knowledge production. These binaries inform contemporary life and economic systems including that of extractivism, exemplified by our earlier discussion of neoliberal capitalism and its associated exclusions. Binaries also shaped the development of academic disciplines, driving their interest in studying (and hence exoticizing and dehumanizing) colonized peoples (Smith 2012). Through positioning the researcher as expert and the "researched" as other, Western academics excluded (and continue to exclude) other ways of knowing as subordinate (i.e., as folklore, tradition, subjective). Contemporary Western academic institutions continue to reflect and, through their teaching and research, reproduce these foundational binaries. This is particularly true of the way scholars, practitioners, and others have come to write about, understand, and practice resource extraction. The very definition of "natural resources" as physical entities to be removed from their initial location for production and consumption is predicated upon the idea of a separation between nature/society and human/nonhuman (McGregor et al. 2020).

In this introduction, we draw from three contemporary bodies of work committed to challenging binary thinking: feminist, decolonial, and Indigenous scholarship.[29] Each body of literature is complex, and this introduction does not claim to cover its breadth. However, we interpret each of these literatures as deriving from and/or offering critiques of a foundational dualistic epistemology engrained in Western thinking, with significant overlaps. Feminist studies spread from an elementary critique of the male/female binary. Decolonial studies emerged from a world marked by colonial presences and absences. Indigenous scholarship problematizes the Western opposition between nature and society, which was deployed to dehumanize and essentialize Indigenous peoples. Each of these bodies of work have also been increasingly in conversation with posthumanist studies.[30] Our decision to rely on these literatures stems first from the nature of contemporary resource extraction. In their exclusions and inclusions, extractive zones are fractured along multiple axes including gender and Indigeneity. As we discussed above, contemporary extraction builds upon colonial exclusions, both practical and epistemic. The continuity of colonialism permeating contemporary resource extraction also links to places and peoples beyond the boundaries of historical empires. Second, our decision reflects the prominence of these literatures in critical resource scholarship, providing valuable theoretical tools to analyze empirical material. In the paragraphs that follow, we highlight several key subthemes addressed by this scholarship in their deconstruction of binaries and that are relevant to resource extraction.

First, these literatures share a commitment to critiquing hierarchies of knowledge. Initially challenging the way women were excluded from knowledge production, more recent feminist scholarship has focused on diverse forms of subjugated knowledges beyond those of women (Sells 1993). This perspective relates to the feminist critique of objectivity in scientific knowledge production which opposes the detached objective knower to conceptualize knowledge as situated, partial, and co-constructed (Haraway 1988). Both critiques of valuable knowledge/subjugated knowledge and objectivity/subjectivity resonate with decolonial scholars' analysis that the naturalized hierarchies of colonialism are "the product of the inferiority of certain peoples and not the cause of their so-called inferiority" (de Sousa Santos 2016, 18). From this stance, decolonial scholars explored hierarchies in the production of the problematic categories of endogenous versus exogenous knowledge (Arowosegbe 2016; Mignolo 2009). Along similar lines, the systemic devaluation of Indigenous worldviews and marginalization of Indigenous intellectual work has been central to Indigenous critique (Smith 2012; Todd 2016; Chiblow 2019; Kuokkanen 2008). The binary of modernity versus nonmodernity frames colonized and marginalized populations as backward and to be developed. Knowledge systems, serving as a mirror for these hierarchies, assign value to certain natural resources and sanction the use of resources in particular ways, as this book highlights.

Second, hierarchies of knowledge are connected to binaries around space and scale. Decolonial scholars challenge the ideas of the core/periphery and of the North/South (de Sousa Santos 2016; Mignolo 2009). Relatedly, the disproportionate violence of extractivism around the world reflects the scalar violence against Indigenous peoples and territories, framed by extractive projects and their allies to be remote, peripheral, inferior, and "other" and therefore ripe for resource exploitation (Aldred et al. 2020; Mollett 2006; Gomez-Barris 2017; Black 2018). Feminist activists and theorists have shown how spaces of production were artificially opposed to spaces of reproduction and, respectively, coded as male and female in capitalist systems (Federici 2004; Kuhn and Wolpe 2013; Gimenez 2018). They have also shown how local, everyday, mundane, intimate, and personal lives were inherently political and entangled with other broader scales, including the global (Marchand and Runyan 2011; Nagar et al. 2002; Massaro and Williams 2013). These literatures privilege speaking from the standpoint of the marginalized and/or the subaltern and foregrounding everyday lives (Harding 1991; hooks 2014; Grosfoguel 2007; Jazeel and Legg 2019). This is particularly relevant in the context of work on resource extraction where the voices of the excluded have for too long remained silent or side-lined.

A focus on subjugated knowledges and grounded voices drives this book and connects to a third set of binaries around identity. Feminists explored how identities are not singular, but fluid, plural, made at the intersection of axes of difference such as gender, race, age, Indigeneity, and class (Crenshaw 1989). Thinking about identity challenges the mind/body dualism which, these literatures have shown, devalues embodied, emotional, and spiritual knowledges (Smith 2012). Scholars have also critiqued essentialisms and the way they obscure the complexity of people's positions in the world (Chakrabarty 2000; Spivak 1988). In the context of resource extraction, for instance, thinking about identity highlights the ambivalence of Indigenous people's positions (Anthias 2014; Cameron 2015; Penfield 2019). It develops understandings of how resource extraction fragments landscapes along multiple axes of difference (Murrey and Jackson 2019; Lahiri-Dutt 2022; Mollett 2006; Elmhirst 2011; Behzadi 2019; Postar and Behzadi 2022).

These naturalized positions subjugating certain people are related to a fourth set of binaries that these literatures challenge—nature/society and nature/culture. Separating people from nature creates the category of "natural" as if it is universal. Scholars of decoloniality and Indigeneity critique this framing for failing to account for ontological differences (among many: Blaser 2013; de la Cadena 2010; Todd 2016) and for reflecting a colonial mentality wherein Indigenous people were lumped into the category of native/savage/uncivilized (Nadasdy 2005; Smith 2012). In this sense, the nature/society nature/culture binaries influence the separation of human from nonhuman/ other than human life (Watts 2013; Smith 2012). Rather than approaching people as separate from the natural world, scholars have shown how identities, subjectivities, and bodies are in constant flux and entangled in mutually constituting relationships with their environments (Elmhirst 2011; Sundberg 2017; Todd 2018). In an even broader sense, this dualistic thinking opposes not only human and nonhuman but also life and death/nonlife. Indigenous scholars, decolonial scholars, new materialists, and posthumanists all explore ways in which what is considered to be "nonlife" in Western thinking (e.g., water) is dynamic and animated (alive) as well as intimately involved in the making of people and culture, challenging representations of nonlife as static and held apart from people (Chiblow 2019; TallBear 2017; Mbembe 2003; Todd 2016; Povinelli 2016; Chen 2012; Bennett 2010). In our engagement with "managerial solutions" to resource exclusions, we are inspired by Indigenous scholars who have noted that the failure to conceive of the "Earth as alive and imbued with spirit" has resulted in "false solutions" to the world's ecological crisis (McGregor et al. 2020, 35–36). Overall, we see challenging the nature/society and nature/culture binaries in the context of resource extraction as looking at how resources (extracted or not) are made and shape people and places (Bakker and Bridge 2006; Richardson and Weszkalny 2014; Ferry and Limbert 2008).[31]

Unsettling Definitions

As we mentioned earlier in this introduction, this book aims to recognize and challenge the binary narratives of inclusion and exclusion in resource extraction. We argue that extraction is predicated on exclusions (practical, epistemic, ontological), that inclusions (rhetorical and practical) often result in further exclusions, and these in turn reproduce violent exploitative processes. The rhetoric of inclusionary extractive-led development and associated inclusionary tools are not the solution but part of the problem. Inclusionary language and practices conceal unchanged systems of domination, dispossession, and violence. This book builds a critique of the binary of exclusion/inclusion through our definitions of natural resources, "resource exclusions," and "resource inclusions" as entangled, plural, and political.

Our politico-ecological view on natural resources is concerned with understanding how power mobilizes raw materials (Bridge 2017) in broader extractivist systems (Gómez-Barris 2017). We challenge the binary of nature/society by going beyond natural resources' material forms to recognize their capacity to span scales and locations. Resources are more than simply valuable, useful substances, or "gifts of nature" to be extracted from a specific place (Bridge 2017; Chiblow 2019; McGregor et al. 2020). They are made (Zimmerman 1933), spatially and temporally distributed as well as energetic (Ferry and Limbert 2008; Richardson and Weszkalnys 2014; Fent and Kojola 2020); and marked by the labor needed for their exploitation (Ferry and Limbert 2008). Defining natural resources requires thinking about scarcity and depletion in terms of social relations (Ferry and Limbert 2008), not only for diminishing resources (fossil fuels) but also for "renewable resources" (plantations, fisheries, and wildlife). As such, defining natural resources from a politico-ecological perspective implies thinking about the ability of a certain group to access, control, and own places of extraction, and the political and economic capital to exploit materials and access markets (Bridge 2017). Natural resources are also assigned values by technology and culture; because of this, the process of labeling something as a resource tells us more about the society than the substance itself (Bridge 2009). Resources are paradoxical: being close in proximity to a resource does not necessarily correlate to access to and control over the material and having access to refined natural resources does not mean you know the origin of that substance (Bridge 2009). For life-giving resources, like water, there is a possibility of (and potentially discomfort around) privatizing and profiting off exclusionary access (Sultana 2011; Strang 2014). This book focuses primarily on minerals, oil, and gas, non renewable materials that can be commodified. However, we do not limit our conceptualization of resources to those narrow categories, and consider, for example, the extractivism of tourism and hydropower alongside resource commodities. We see resources as relational—interconnected and fostering the need to develop further resources of the same and different kinds.

As stated earlier, resource exclusions are nearly always components of resource extraction. Exclusion refers to both the process of excluding and the state of being excluded. We situate our conceptualization of exclusion amid the rhetoric around "social exclusion," widely adopted by policy circles since the 1970s, which was concerned with the growing "underclass" disconnected from or nonparticipatory in mainstream society (Davies 2005; Byrne 2005). Our approach to resource exclusions as plural, however, diverges as it reflects both how exclusions emerge across scales of politico-ecological analysis of extractivism, as well as at the practical, epistemic, and ontological levels. Exclusions can be defined at broader scales (the global, national) as highlighted in our discussion of neoliberal extraction as exclusion, but also co-constituted, experienced, and resisted at more nested scales (the home, the community, the body) (Nagar et al. 2002; Marchand and Runyan 2011).[32] At the practical level, extractive concessions—from the largest offshore oil fields to seasonal small-scale enterprises—are often exceptional spaces, defined by exclusive rights to access and exploit the riches within (Perreault 2013; Lesutis 2019; Cote and Korf 2018). While extraction itself might be associated with defined, exceptional spaces of violent exclusion, the infrastructure moving resources are different. Pipelines criss-crossing national borders and international shipping channels embed extraction in larger networks moving commodities in and out of spaces of exception, while remaining exceptional and defined by exclusionary rights (Barry 2013; Appel 2012; Murrey 2015; Adunbi 2011). Thinking through practical forms of exclusions also entails reflecting on who is excluded from access to resources and their benefits according to individual and group identities (class, gender, age, ability, Indigeneity, and race) as part of intersectional analyses of exclusions (Jenkins 2014; Lahiri-Dutt 2022; Behzadi 2019; Anthias 2018; Cameron 2015; Chiblow 2019). Gender inequity and the oppression, marginalization, and exclusions of women emerge as a result of capitalist and neocolonial modes of production (Gimenez 2005; Lugones 2007). Like others before us, we recognize that these exclusions are linked and built into racial capitalism (Melamed 2015; Hall, Hirsch, and Li 2011). These systemic inequities are informed by and shape epistemic exclusions—whose knowledge and whose voices count in resource-extractive practices and scholarship. Epistemic exclusions are also connected to their ontological forms. As mentioned earlier, the very definition of natural resources is predicated upon a specific ontology that finds its roots in many dominant forms of Western philosophy, law, and economic governance. The idea of resources as extractible materials for economic use emerges from and re-iterates a separation of nature from society which excludes alternative ways of being.

Our understanding of resource inclusions emerges from empirical encounters with inclusionary tools, like CSR and IBAs, and rhetoric, like sustainable mining for development. Inclusion is often defined in policy and within the social exclusion literature as both the state of including or being included, leaving social inclusion undertheorized and defined only negatively—that is, as "not social exclusion" (Cameron 2006, 396). Our politico-ecological conceptualization of resource inclusions departs from definitions that emphasize access, inclusion, or integration as the opposite of exclusion (in natural resource management, social policy, etc.) by contesting the underlying binary thinking that the solution to exclusion is its opposite. Like other critical scholars, we agree that the focus on inclusion/access as the solution to exclusion fails to appreciate or disrupt the fundamentally unequal power relations underpinning systemic exclusions (Ribot and Peluso 2003; Peluso and Ribot 2020; Slee 2013; Williamson 2015; Kafer 2013). Our plural approach makes visible ways in which inclusionary narratives often hide systems and structures of exclusion, highlights the ambivalent nature of inclusions, and recognizes how inclusionary rhetoric and practices create novel exclusions. Inclusions in resource extraction can refer to economic and political involvement through jobs, resource wealth, participation in resource management, and decision-making. The results of such inclusionary mechanisms, however, are often complex, and tend to function within the same exclusionary system, creating what we refer to as ambivalent and precarious inclusions. Resource inclusions can indeed harm materially, through occupational illness from extractive work, environmental health damage from pollution, and involvement in violent dissent and suppression. Additionally, the kinds of seduction and manipulation that companies deploy through their rhetoric and practices of inclusion to "engineer consent" (Brock and Dunlap 2018) for extractive projects often fracture communities (Frederiksen and Himley 2020). People accept or demand their "inclusion" in resource-extractive projects or operations (jobs, cash compensation, etc.) despite recognizing these tensions, personally experiencing their immediate consequences, or worrying about the longer term (Weszkalnys 2016). By pluralizing our conceptualization of inclusion, our approach troubles the binary thinking that assumes that apolitical narratives and tools of inclusion will resolve resource exclusions.

As mentioned before, we are aware that this book is not free of epistemological and ontological exclusions. From the topics, regions, and scholars funded for study, to the citational politics and language barriers in publications, power, and privilege echo in the words of this book. Those inequalities limited and excluded people from participating in this book project from its start. For instance, a key shortcoming of this book is the lack of Indigenous authorship. This collection was also produced in part during the COVID-19 pandemic. The global inequalities and exclusions predating and also exacerbated by the pandemic (Habtom and Scribe 2020) reverberate in this text: chapters are missing from this book and others were added as a result of the unequal impacts of this global health crisis, in terms of physical and mental health as well as its ramifications in terms of academic and reproductive labor. We have learned important lessons from working on this book, including the need to find ways to tackle the exclusionary effects of our own academic practice in the future. Naming and examining exclusions and adverse inclusions in this text takes a step toward linking struggles across sites of extraction, from natural resources to academic research.

STRUCTURE OF THE BOOK

This book is structured around challenging the binary framing of exclusion/inclusion and what we see as a series of corollary binaries in resource extraction. Though the binaries are diverse, they each boil down to inclusionary and exclusionary practices and knowledge. By bringing together empirical material across geographical sites, scales, and disciplines, this edited collection contributes to unsettling these dualistic epistemologies. It is our hope that the multidisciplinary approach of this edited collection will also contribute to further multi-, inter-, and transdisciplinary research in the future. Each of the six binary parts begins with an introductory theoretical framing. Two empirical chapters engage with each binary, although, it should come as no surprise that there is significant overlap, as exclusions beget exclusions.

Part I, "Haves/Have-Nots," unsettles the opposition between powerful actors (multinational companies and their allies) and weaker actors (local communities or Indigenous people). This section links corporations, states, local communities, and other actors, highlighting how they come together to (re)produce exclusions and attempt (new) forms of inclusions (Bridge 2004; Bridge 2008; Bebbington 2009; Conde and Le Billon 2017). Philippe Le Billon draws on his political ecology expertise in resource extraction to write the introduction to this part together with Stephanie Postar and Negar Elodie Behzadi. Asel Doolotkeldieva's chapter on Kyrgyz gold mining challenges conventional North/South and core/periphery binaries by considering inequalities beyond these narrow geographical divisions. She studies the legacies of Soviet rule in the discourses of corporations, the state, and local consultants, and their role in delegitimizing complaints. Teresa Bornschlegl challenges binaries of state/companies/institutions versus local and Indigenous communities. She shows how constitutional changes that include Indigenous ontologies in Ecuador did not transform the logics of extractivism. Via legal actions against the oil industry, she concludes that institutional barriers often maintain the status quo despite new mechanisms allowing environmental and civil complaints.

Drawing on feminist, critical race theory, and post/decolonial theories, part II, "Oppressors/Oppressed," explores the politics of resource exclusion at different scales—from the global to the most intimate scales of the body—at the intersection of gender, race, class, ethnicity, ability, sexuality, Indigeneity, and citizenship (Rocheleau et al. 1996; Elmhirst 2011; Mollett and Faria 2013; Behzadi 2019). It also invites consideration of the ways in which an engagement with feminist post and decolonial theories and methodologies provide tools to resist the subjugation and exclusion of resource knowledges (Murrey 2016, 2017; Rocheleau and Nirmal 2015). Rebecca Elmhirst brings her expertise in feminism and environments and in particular feminist political ecology, to write the introduction to this part. Sharlene Mollett and Amber Murrey investigate the ways racialized sexualized violence emerges from the practices and logics of extractivism with empirical material from a Panamanian tourism enclave and along the Chad-Cameroon oil pipeline. Through the gendered threats of exclusion of young female informal coal miners in Tajikistan, Negar Elodie Behzadi highlights the parallels between the dialectical productive exhaustion of land and resources and the reproductive exhaustion of women's bodies to form an intersectional extractive violence.

Part III focuses on the borderlands of anthropocentred thought via the binary "Human/Nonhuman," considering debates on the "more-than-human," Indigenous ontologies and their role in decision-making processes, and the boundary between life and nonlife (Povinelli 2016; Tallbear 2017). It questions how resource extraction projects both exclude and sometimes operationalize more-than-human entities and Indigenous knowledge in their procedures to claim resources (Biersack 1999; Mauss 2016[1925]; Nash 1979). It also reflects on possible ways to make visible these alternative resource ontologies. Emilie Cameron draws on her expertise on Indigeneity and resources to introduce this part. Drawing from extensive ethnographic research in northwest Greenland, Mark Nuttall raises questions about the inclusion of Indigenous knowledge and the voices of the nonhuman in participatory processes in the context of Arctic oil exploration. By focusing on rumors of a sickness linked to a delayed uranium mining project in Tanzania, Stephanie Postar considers the ways resources have power even before they are extracted.

Part IV, "Static Materials/Dynamic Materials," focuses on resource materialities, temporalities, and affect in the process of resource making and "becoming" (Zimmermann 1933). It particularly highlights the spatio-temporal processes around resource projects and reveals how exploration of resource anticipations and futures shed light on the affective politics around individuals' and communities' experiences of exclusion (Richardson and Weszkalnys 2014; Ferry and Limbert 2008). Gisa Weszkalnys brings her critical perspective as an economic anthropologist to write the introduction for this part. Anna Willow considers different reactions to a major Canadian hydroelectric project, exploring how these diverse responses reflect the individuals' interpretations of themselves as temporal actors in the ongoing story of citizen activism and contesting different visions for the future. Robin West and Isabel Crowhurst unpick the sense of uncertainty felt in two Italian alpine valleys in which local communities ambivalently await the resumption of commercial extraction while maintaining (and expecting the mining companies to support) deep cultural attachment to the local histories of extraction.

Part V, "Large-Scale/Small-Scale," investigates artisanal and small-scale mining (ASM) as activities that destabilize the conceptual and institutional frameworks of resources (Lahiri-Dutt and Macintyre 2016; Bryceson and Jønsson 2010). It draws attention to the often-overlooked experiences of ASM beyond binaries of artisanal/industrial mining, inclusion/exclusion, informal/formal, and illegal/legal, shedding light on the creative, precarious, and illegal practices that often emerge at the margins of larger industrial projects (Fisher 2007; Hilson 2010; Huggins 2016; Lahiri-Dutt 2012). Furthermore, the part highlights how resistance to broader forms of resource development may also be articulated through local populations' involvement in ASM (Geenen 2014; Cuvelier 2014). Kuntala Lahiri-Dutt draws on her expertise in ASM to write the introduction for this part. Through their study of the politics of the formalization of artisanal oil production in East Java, Indonesia, Nanang Kuniawan, Päivi Lujala, and Ståle Angen Rye challenge the binary between inclusion and exclusion in artisanal mining. Gavin Hilson, John Owen, Titus Sauerwein, Massaran Traore, and Eleonore Lebre investigate the possibility of autonomous coexistence between artisanal and industrial mining in sub-Saharan Africa, focusing on the case of Mali.

Part VI, "Inclusion/Exclusion," highlights the precarious resource inclusions that have emerged as a result of new industry practices such as FPIC or revenue-sharing tools (O'Faircheallaigh 2013; Segal 2011; Schilling-Vacaflor 2017). While international agreements stipulate that affected communities, including Indigenous peoples, should be at the heart of decision-making processes, actual practices often fall short of established best practices and produce new forms of exclusions that reinforce neocolonial power asymmetries (Kirsch 2007; Sawyer 2004). Penda Diallo introduces this chapter, drawing on her academic and practical expertise regarding community relations with extractive projects. Leah Horowitz critiques the limits of "consent" through the development of an Impact Benefit Agreement (IBA), which guaranteed financial compensation for impacted communities in New Caledonia without requiring any technical changes to environmentally destructive operations by a multinational nickel mining company. Despite the goal of public consultation for extraction projects in Greenland, Nina Doering highlights the complexities of nonparticipation in resource exploration and production planning, beyond simplistic representations of lack of capacity or indifference.

Gavin Bridge concludes the book with a provocation for readers to consider whether the concept of extraction has become overexposed as scholars increasingly instrumentalize the notion of extraction as a general organizing concept, for instance, as a synonym for capitalism. Bridge argues that beyond its uses in the material contexts of mining, extraction continues to be a valuable concept because of its capacity to engage with the specifics of social ordering and displacement both in the primary sector and in novel sites beyond "the mine," with rich possibilities for thinking beyond binaries including inclusion/exclusion and action/resistance.

NOTES

1. The group included the Federal Government, the German states home to coal mining, the hard coal mining company, and the miners' union.

2. Germany's position against Russia's war in Ukraine initiated a series of knock-on impacts on German energy policies and management. In July 2022, German Parliament approved measures to bring coal-fired power plants back online as part of temporary emergency steps responding to reduced access to Russian gas (Kinkartz 2022). The *Energiewende* increased Germany's reliance on imported gas, including from Russia. EU sales and import bans on Russian gas (in effect at the end of 2022) and Russia pre-emptively shutting down the Nordstream 1 gas pipeline to Germany in September 2022 (Lawson 2022) appear to be forcing Germany, at least temporarily, to rely to a greater extent on coal-powered electricity generation. At the publication of this introduction, however, German Greens party politicians remain committed to ending coal-power generation by 2030.

3. Region of Andean South America in Argentina, Bolivia, and Chile rich in lithium.

4. Batteries are produced elsewhere for electric cars in foreign markets (China, Europe, USA). This boom is the latest chapter in a long Chilean history of extraction—gold mining in the precolonial and colonial periods, and silver, coal, and copper after independence. Today, progressives also want to rewrite Chile's Constitution, to revert historical inequalities, in particular those of the far-right dictatorship era.

5. The two multinational companies are Albermale Corporation (headquarters in Charlotte, North Carolina, USA) and SQM (Sociedad Química y Minera, Chilean company purchased at 23.77 percent by the Chinese company Tianqi Lithium Corporation).

6. Political ecology is an interdisciplinary field of study across anthropology, geography, development studies, and others that emerged in Anglophone academia in the 1980s and aims at challenging apolitical views on the environment. For a genealogy of political ecology, see Perreault et al. (2015); Peet and Watts (1996), Peet et al. (2011), Robbins (2020).

7. Bridge (2009) frames the current period of material intensification as distinct but also one of many in history. The period of industrialization of the late nineteenth century in the Global North being one of them.

8. See Chilean case at the beginning of this introduction for Chinese investments in lithium.

9. As illustrated in the Chilean case at the beginning of this introduction, lithium mining responds to demand for batteries for electric cars and other green technologies for markets situated elsewhere. Renewable energy systems including solar panels and wind turbines similarly drive demand for a range of other resources.

10. This can refer to accessing novel places or sites of extraction (i.e., the Arctic), or resources from sites that were accessible but unprofitable (i.e., deepwater drilling and mining). This is particularly relevant for "unconventional resources" including oil shales and shale gas (Kama 2020).

11. In the Chilean case that starts this introduction, two multinational companies mine lithium in the Atacama Salt flats. In 2018, 23.8 percent share of one of these two companies, SQM, was bought by a Chinese company called Tianqi Lithium Corp.

12. Resistance, strikes, and revolts against exploitative conditions led to large-scale, often violent conflicts throughout history including the 1884 Anzin miners' strike in France (inspiring Emile Zola's Germinal) (Hayat 2017), the 1890–1933 American "mine wars" (Andrews 2008), the 1926 miners' lockout in Britain (Barron 2009), and the 1946 Black miners' strike in apartheid South Africa (Moodie 1986), among many other examples.

13. As in the German coal mine case at the beginning of this introduction, international solidarity impacts resistance to extraction around the world (Hopke 2016).

14. We can understand this "resource nationalism" (Koch and Perreault 2018), as both a political discourse about state distribution of wealth, and as a rejection of private/foreign control of resource wealth in favor of benefiting a collective, like the "nation" or "people." In some cases, this entailed reverting the impacts of decades of neoliberal policies through the legally required public–private partnerships in extractive projects and the outright nationalization of resources, as was the case in several Latin American contexts (Burchardt and Dietz 2014; Lyall and Valdivia 2019), as well as in Central and Inner Asia (Fumagalli 2015; Myadar and Jackson 2019). In different contexts, these kinds of responses to neocolonial extractive interests could be seen across a variety of political landscapes on the left (Riofrancos 2020) as well as the right (Kojola 2019) and in between (Jacob and Pederson 2018). This most recent wave of resource nationalism, as enacted by populist and authoritarian regimes, shares some similarities with earlier waves, for instance in post-independence African countries in the 1960s and 1970s, which were coded as rent-seeking behaviors (Caramento and Saunders 2019).

15. The Brundtland Report defined sustainable development as "development that meets the needs of the present without compromising the ability of future generations to meet their own needs" and led to the concept passing into policy discourse (WCED 1987, 8).

16. Payments for ecosystem services, as seen in major international programs like REDD+, are an example of these market-driven interventions under the banner of green growth (Asiyanbi 2016; Gómez-Baggethun et al. 2010).

17. EIAs and SIAs are used across a range of development projects, including infrastructure like roads and bridges, special economic zones, among others. In some places, EIAs and SIAs are legally mandated, depending on the project (Glasson and Therivel 2012; Vanclay and Esteves 2011).

18. FPIC is a key principle protected by human rights standards according to which all peoples have "the right to self-determination" and to "freely pursue their economic, social and cultural development." It is supported by a series of international

conventions aiming at defending the rights of Indigenous peoples—the Convention on Biological Diversity, the International Labour Organization Convention 169, and the United Nations Declaration on the Rights of Indigenous Peoples (UNDRIP) (FAO 2016).

19. Sovereign wealth funds are generally state-owned funds investing in financial assets such as bonds, real estate, stocks, precious metals, or private funds and equity. They are usually funded by revenues from exports, including from natural resources and have involved ethical questions on how money from these funds should be spent (Armstrong 2013). Longstanding examples can be seen in the Kuwait Investment Authority and the Alaska Permanent Fund, both saving a portion of oil revenues.

20. This is particularly the case for Indigenous people (see, for example, the work of Cameron and Levitan (2014) in Northern Canada or chapter 11 (part VI) from Leah Horowitz in this volume).

21. In the Chilean case, this includes a rewriting of the Constitution to include Indigenous people's worldview, as well as local opposition through lawsuits. This reflects changes in many locations in the world. See also chapter 2 (part I) in this volume from Teresa Bornschlegl on Ecuador.

22. These certifications include reducing the exchange of conflict minerals (Kimberley Process), requiring transparency in governance of extractive industries (EITI), assessing the sustainability of hydropower projects (Hydropower Sustainability Assessment Protocol), and others.

23. The "Dutch disease" describes the economic conditions when a country experiences negative consequences from relying on a single commodity, often natural resources. Resource booms exhibiting Dutch disease cause an increase in commodity prices and an associated increase in wages (and exchange rates), which ultimately damages non-resource sectors by making them less competitive (Ross 1999).

24. The resource curse refers to the failure of resource-rich countries to produce and maintain wealth from their natural resources. This theory shows that resource-rich countries tend to be more authoritarian, have more conflicts, and lower economic growth, compared with other countries (Badeeb et al. 2017).

25. With overlapping binaries framing who and where institutions are perceived to be incapable of so-called "correct management."

26. Some insightful exceptions to this trend can be found in comparative work (Caretta et al. 2020; Arboleda 2020).

27. Inspired by and departing from Cartesian philosophy, Benedict de Spinoza posited a unity of mind and body, where neither is superior to the other (Nadler 2019; Navaro-Yashin 2009; Singh 2013). Like Spinoza, many other scholars critiqued the binary framings posed by Descartes from a relational approach. Margaret Cavendish, a Descartes contemporary and marginalized scholar, opposed the Cartesian distinction and superiority of humans over animals, arguing instead for the equality of all ways of knowing by human and nonhuman animals (Cooney 2019). Similarly, Leibniz supposed that animals must have souls, challenging the Cartesian thesis that the immaterial soul was a purely human trait, with animals as mere automatons (Anfray 2019).

28. Hegel's depiction of the master–slave dialectic, however, complicates the dualistic thinking as it recognizes the (white, male) master (a collective figure) as incapable of historical agency because of its dependence on the slave (Buck-Morss 2000).

29. We recognize that the very broad categories of scholarship here in practice represent diverse scholars who have multiple scholarly identities, but for clarity, we are organizing them into these three groups. For more on decolonial scholarship, see Mignolo and Walsh (2018). For more on feminist scholarship, see Hesse-Biber (2012). For more on Indigenous scholarship see Hokowhitu et al. (2020).

30. Decolonial and Indigenous thinkers have critiqued the failure of posthumanist scholarship for marginalizing and subordinating alternative ways of knowing the nonhuman world (Todd 2016; Watts 2013; Sundberg 2014).

31. See this volume, part VI Introduction, Static materials/Dynamic materials by Giza Weszkalnys.

32. For a discussion of exclusions in resource extraction at more nested scales, see Caretta, Zaragocin, Turley, and Orellana (2020), Postar and Behzadi (2022), Behzadi (2019), or Murrey (2016).

REFERENCES

Acosta, Alberto. 2013. "Extractivism and neoextractivism: Two sides of the same curse." In *Beyond Development: Alternative Visions from Latin America*, edited by M. Lang and D. Mocrani, 61–86. Amsterdam: Transnational Institute.

Adunbi, Omolade. 2011. "Oil and the production of competing subjectivities in Nigeria: 'Platforms of possibilities' and 'pipelines of conflict'." *African Studies Review* 54 (3): 101–120. https://doi: 10.1353/arw.2011.0064

Agnew, John. 2013. "The origins of critical geopolitics." In *The Ashgate Research Companion to Critical Geopolitics*, edited by Klauss Dodds and Merje Kuss. London: Routledge.

Aldred, Terri-Leigh, Charis Alderfer-Mumma, Sarah de Leeuw, May Farrales, Margo Greenwood, Dawn Hoogeveen, Ryan O'Toole, Margot W. Parkes, and Vanessa Sloan Morgan. 2021. "Mining sick: Creatively unsettling normative narratives about industry, environment, extraction, and the health geographies of rural, remote, northern, and Indigenous communities in British Columbia." *The Canadian Geographer / Le Géographe canadien* 65 (1): 82–96.

Allan, Jen Iris. 2019. "Dangerous incrementalism of the Paris Agreement." *Global Environmental Politics* 19 (1): 4–11.

Altamirano-Jiménez, Isabel. 2021. "Indigenous women refusing the violence of resource extraction in Oaxaca." *AlterNative: An International Journal of Indigenous Peoples* 17 (2): 215–223. https://doi/10.1177/11771801211015316.

Andrews, Thomas G. 2008. *Killing for Coal: America's Deadliest Labor War*. Cambridge: Harvard University Press.

Anfray, Jean-Pascal. 2019. "Leibniz and Descartes." In *The Oxford Handbook of Descartes and Cartesianism*, edited by Steven Nadler, Tad M. Schmaltz, and

Delphine Antoine-Mahut, 720–737. Oxford: Oxford University Press. https://doi/10.1093/oxfordhb/9780198796909.013.45.

Anthias, Penelope. 2018. *Limits to Decolonization: Indigeneity, Territory, and Hydrocarbon Politics in the Bolivian Chaco*. Ithaca: Cornell University Press.

Appel, Hannah C. 2012. "Walls and white elephants: Oil extraction, responsibility, and infrastructural violence in Equatorial Guinea." *Ethnography* 13 (4): 439–465. https://doi: 10.1177/1466138111435741.

Arboleda, Martín. 2020. *Planetary Mine: Territories of Extraction Under Late Capitalism*. London and Brooklyn: Verso.

Armstrong, Chris. 2013. "Sovereign wealth funds and global justice." *Ethics and International Affairs* 27 (4): 413–428. https://doi: 10.1017/S0892679413000361.

Arowosegbe, Jeremiah O. 2016. "Endogenous knowledge and the development question in Africa." *Cambridge Review of International Affairs* 29 (2): 611–635. https://doi: 10.1080/09557571.2015.1060692.

Asiyanbi, Adeniyi P. 2016. "A political ecology of REDD+: Property rights, militarised protectionism, and carbonised exclusion in Cross River." *Geoforum* 77: 146–156. https://doi: 10.1016/j.geoforum.2016.10.016.

Auty, Richard M. 1993. *Sustaining Development in Mineral Economies: The Resource Curse Thesis*. London: Routledge.

Badeeb, Ramez Abubakr, Hooi Hooi Lean, and Jeremy Clark. 2017. "The evolution of the natural resource curse thesis: A critical literature survey." *Resources Policy* 51: 123–134.

Baker, Janelle Marie, and Clinton N. Westman. 2018. "Extracting knowledge: Social science, environmental impact assessment, and indigenous consultation in the oil sands of Alberta, Canada." *The Extractive Industries and Society* 5 (1): 144–153.

Bakker, Karen, and Gavin Bridge. 2006. "Material worlds? Resource geographies and the 'matter of nature'." *Progress in Human Geography* 30 (1): 5–27. https://doi: 10.1191/0309132506ph588oa.

Bannon, Ian, and Paul Collier, eds. 2003. *Natural Resources and Conflict: What We Can Do, Natural Resources and Violent Conflict. Options and Actions*. Washington, DC: The World Bank. https://doi: 10.1596/0-8213-5503-1.

Barron, Hester. 2009. *The 1926 Miners' Lockout: Meanings of Community in the Durham Coalfield*. Oxford: Oxford University Press.

Barry, Andrew. 2013. *Material Politics: Disputes Along the Pipeline*. Chichester: Wiley-Blackwell.

Bebbington, Anthony. 2009. "Latin America: Contesting extraction, producing geographies." *Singapore Journal of Tropical Geography* 30 (1): 7–12. https://doi: 10.1111/j.1467-9493.2008.00349.x.

Bebbington, Anthony, Leonith Hinojosa, Denise Humphreys Bebbington, Maria Luisa Burneo, and Ximena Warnaars. 2008. "Contention and ambiguity: Mining and the possibilities of development." *Development and Change* 39 (6): 887–914. https://doi: 10.1111/j.1467-7660.2008.00517.x.

Behzadi, Negar Elodie. 2019. "Women miner's exclusion and Muslim masculinities in Tajikistan: A feminist political ecology of honor and shame." *Geoforum* 100: 144–152.

Bennett, Jane. 2010. *Vibrant Matter a Political Ecology of Things*. Durham: Duke University Press.

Bezuidenhout, Andries, and Sakhela Buhlungu. 2011. "From compounded to fragmented labour: Mineworkers and the demise of compounds in South Africa." *Antipode* 43 (2): 237–263. https://doi: 10.1111/j.1467-8330.2010.00758.x.

Bieri, Franziska. 2016. *From Blood Diamonds to the Kimberley Process: How NGOs Cleaned Up the Global Diamond Industry*. London: Routledge.

Biersack, Aletta. 1999. "The Mount Kare Python and his gold: Totemism and ecology in the Papua New Guinea highlands." *American Anthropologist* 101 (1): 68–87. https://doi: 10.1093/bioinformatics/btp629.

Bisht, Arpita. 2022. "Sand futures: Post-growth alternatives for mineral aggregate consumption and distribution in the global south." *Ecological Economics* 191 (April 2021): 107233. https://doi: 10.1016/j.ecolecon.2021.107233.

Black, Megan. 2018. *The Global Interior: Mineral Frontiers and American Power.* Cambridge: Harvard University Press.

Blaser, Mario. 2013. "Notes towards a political ontology of 'environmental' conflicts." In *Contested Ecologies: Dialogues in the South on Nature and Knowledge*, edited by Lesley Green, 13–27. Cape Town: Human Sciences Research Council Press.

Bridge, Gavin. 2004. "Mapping the bonanza: Geographies of mining investment in an era of neoliberal reform." *Professional Geographer* 56 (3): 406–421. https://doi: 10.1111/j.0033-0124.2004.05603009.x.

Bridge, Gavin. 2008. "Global production networks and the extractive sector: Governing resource-based development." *Journal of Economic Geography* 8 (3): 389–419. https://doi: 10.1093/jeg/lbn009.

Bridge, Gavin. 2009. "Material worlds: Natural resources, resource geography and the material economy." *Geography Compass* 3 (3): 1217–1244. https://doi: 10.1111/j.1749-8198.2009.00233.x.

Bridge, Gavin. 2017. "Resource extraction." In *International Encyclopedia of Geography: People, the Earth, Environment and Technology*, edited by Douglas Richardson, Noel Castree, Michael F. Goodchild, Audrey Kobayashi, Weidong Liu, and Richard A. Marston. Chichester, UK: John Wiley & Sons. https://doi.org/10.1002/9781118786352.wbieg1047.

Brock, Andrea, and Alexander Dunlap. 2018. "Normalising corporate counterinsurgency: Engineering consent, managing resistance and greening destruction around the Hambach coal mine and beyond." *Political Geography* 62: 33–47. https://doi: 10.1016/j.polgeo.2017.09.018.

Bryceson, Deborah Fahy, and Jesper Bosse Jønsson. 2010. "Gold digging careers in rural east Africa: Small-scale miners' livelihood choices." *World Development* 38 (3): 379–392. https://doi: 10.1016/j.worlddev.2009.09.003.

Buck-Morss, Susan. 2000. "Hegel and Haiti." *Critical Inquiry* 26 (4): 821–865.

Burchardt, Hans-Jürgen, and Kristina Dietz. 2014. "(Neo-)extractivism—A new challenge for development theory from Latin America." *Third World Quarterly* 35 (3): 468–486. https://doi: 10.1080/01436597.2014.893488.

Byrne, David, ed. 2005. *Social Exclusion*. Berkshire: Open University Press.

Cameron, Angus. 2006. "Geographies of welfare and exclusion: Social inclusion and exception." *Progress in Human Geography 30* (3): 396–404. https://doi.org/10.1191/0309132506ph614pr.

Cameron, Emilie. 2015. *Far Off Metal River: Inuit Lands, Settler Stories, and the Marking of the Contemporary Arctic*. Vancouver and Toronto: UBC Press.

Cameron, Emilie, and Tyler Levitan. 2014. "Impact and benefit agreements and the neoliberalization of resource governance and indigenous-state relations in Northern Canada." *Studies in Political Economy* 93 (1): 25–52.

Cameron, Emilie, Sarah de Leeuw, and Caroline Desbiens. 2013. "Indigeneity and ontology." *Cultural Geographies* 21 (1): 19–26.

Caramento, Alex, and Richard Saunders. 2019. "Capitalism and resource nationalism in Africa." *Capitalism in Africa* (blog), *Review of African Political Economy*, October 17, 2019. https://roape.net/2019/10/17/capitalism-and-resource-nationalism-in-southern-africa/.

Caretta, Martina Angela, and Sofia Zaragocin. 2020. "Women's resistance against the extractive industry: Embodied and water dimensions." *Human Geography* 13 (1): 3–5. https://doi: 10.1177/1942778620910893.

Caretta, Martina Angela, Sofia Zaragocin, Bethani Turley, and Kamila Torres Orellana. 2020. "Women's organizing against extractivism: Towards a decolonial multi-sited analysis." *Human Geography* 13 (1): 49–59. https://doi: 10.1177/1942778620910898.

Chakrabarty, Dipesh. 2007. *Provincializing Europe: Postcolonial Thought and Historical Difference*. Princeton: Princeton University Press.

Chen, Mel Y. 2012. *Animacies: Biopolitics, Racial Mattering, and Queer Affect*. Durham and London: Duke University Press.

Cheru, Fantu. 2016. "Emerging southern powers and new forms of South–South cooperation: Ethiopia's strategic engagement with China and India." *Third World Quarterly* 37 (4): 592–610. https://doi: 10.1080/01436597.2015.1116368.

Chiblow (Ogamauh annag qwe), Susan. 2019. "Anishinabek Women's Nibi Giikendaaswin (Water Knowledge)." *Water* 11 (2): 209. https://doi.org/10.3390/w11020209.

Childs, John. 2018. "Extraction in four dimensions: Time, space and the emerging geo(-)politics of deep-sea mining." *Geopolitics* 25 (1): 189–213.

Conde, Marta, and Philippe Le Billon. 2017. "Why do some communities resist mining projects while others do not?" *Extractive Industries and Society* 4 (3): 681–697. https://doi: 10.1016/j.exis.2017.04.009.

Cooke, Bill, and Uma Kothari, eds. 2001. *Participation: The New Tyranny?* London and New York: Zed Books.

Cooney, Hadley. 2019. "Cavendish vs. Descartes on mechanism and animal souls." In *The Oxford Handbook of Descartes and Cartesianism*, edited by Steven Nadler, Tad M. Schmaltz, and Delphine Antoine-Mahut, 642–658. Oxford: Oxford University Press. https://doi: 10.1093/oxfordhb/9780198796909.013.40.

Cornwall, Andrea, ed. 2011. *The Participation Reader.* London and New York: Zed Books.

Côte, Muriel, and Benedikt Korf. 2018. "Making concessions: Extractive enclaves, entangled capitalism and regulative pluralism at the gold mining frontier in Burkina Faso." *World Development* 101: 466–476. https://doi: 10.1016/j.worlddev.2016.11.002.

Crenshaw, Kimberle. 1989. "Demarginalizing the intersection of race and sex: A black feminist critique of antidiscrimination doctrine, feminist theory, and antiracist politics." *The University of Chicago Legal Forum* 1989 (140): 139–167.

Cuvelier, Jeroen. 2014. "Work and masculinity in Katanga's artisanal mines." *Africa Spectrum* 49 (2): 3–26.

Davies, Jonathan S. 2005. "The social exclusion debate: Strategies, controversies and dilemmas." *Policy Studies* 26 (1): 3–27. https://doi: 10.1080/01442870500041561.

Deer, Sarah, and Mary Kathryn Nagle. 2017. "The rapidly increasing extraction of oil, and native women, in North Dakota." *The Federal Lawyer* 64 (3): 35–37.

De La Cadena, Marisol. 2010. "Indigenous cosmopolitics in the Andes: Conceptual reflections beyond 'politics'." *Cultural Anthropology* 25 (2): 334–370. https://doi: 10.1111/j.1548-1360.2010.01061.x.

de Oliveira, Ricardo S. 2007. "Business success, Angola-style: Postcolonial politics and the rise and rise of Sonangol." *Journal of Modern African Studies* 45 (4): 595–619. https://doi: 10.1017/S0022278X07002893.

de Sousa Santos, Boaventura. 2016. "Epistemologies of the south and the future." *From the European South* 1: 17–29.

De Soysa, Indra, and Eric Neumayer. 2007. "Resource wealth and the risk of civil war onset: Results from a new dataset of natural resource rents, 1970-1999." *Conflict Management and Peace Science* 24 (3): 201–218. https://doi: 10.1080/07388940701468468.

Degani, Michael, Brenda Chalfin, and Jamie Cross. 2020. "Fuelling capture: Africa's energy frontiers." *The Cambridge Journal of Anthropology* 38 (2): 1–18. https://doi: 10.3167/cja.2020.380202.

Descartes, René. 1996 [1641]. *Meditations on the First Philosophy.* Translated by John Cottingham. Cambridge: Cambridge University Press.

Dixon, Deborah P., and Sallie A. Marston. 2013. *Feminist Geopolitics. At the Sharp End.* New York: Routledge.

Dodds, Klauss. 2001. "Political geography III: Critical geopolitics after ten year." *Progress in Human Geography* 25 (3): 469–484.

Dodds, Klauss-John, and James Derrick Sidaway. 1994. "Locating critical geopolitics." *Environment and Planning D: Society and Space* 12: 515–524.

Dowler, Lorraine, and Joanne Sharp. 2001. "A feminist geopolitics?" *Space and Polity* 5 (3): 165–176.

Elmhirst, Rebecca. 2011. "Introducing new feminist political ecologies." *Geoforum* 42 (2): 129–132. https://doi: 10.1016/j.geoforum.2011.01.006.

Estes, Nick. 2019. *Our History Is the Future: Standing Rock Versus the Dakota Access Pipeline, and the Long Tradition of Indigenous Resistance.* London: Verso.

FAO. 2016. *Free Prior and Informed Consent an Indigenous Peoples' Right and a Good Practice for Local Communities: MANUAL FOR PROJECT PRACTITIONERS,*Rome: FAO.

Federici, Silvia. 2004. *Caliban and the Witch: Women, the Body and Primitive Accumulation*. Brooklyn: Autonomedia.

Fent, Ashley, and Erik Kojola. 2020. "Political ecologies of time and temporality in resource extraction." *Journal of Political Ecology* 27 (1): 819–829. https://doi: 10.2458/V27I1.23252.

Ferry, Elizabeth Emma, and Mandana E. Limbert, eds. 2008. *Timely Assets: The Politics of Resources and Their Temporalities*. Santa Fe: School for Advanced Research Press.

Fisher, Eleanor. 2007. "Occupying the margins: Labour integration and social exclusion in artisanal mining in Tanzania." *Development and Change* 38 (4): 735–760. https://doi/10.1111/j.1467-7660.2007.00431.x.

Fjellheim, Eva Maria. 2020. "Through our stories we resist: Decolonial perspectives on south Saami history, indigeneity and rights." In *Indigenous Knowledges and the Sustainable Development Agenda*, edited by Anders Breidlid and Roy Krøvel. London: Routledge. https://doi/10.4324/9780367853785-12.

Fjellheim, Eva Maria. 2023. "'You can kill us with dialogue:' Critical perspectives on wind energy development in a nordic-Saami green colonial context." *Human Rights Review* 24: 25–51. https://doi.org/10.1007/s12142-023-00678-4

Frederiksen, Tomas, and Matthew Himley. 2020. "Tactics of dispossession: Access, power, and subjectivity at the extractive frontier." *Transactions of the Institute of British Geographers* 45 (1): 50–64. https://doi: 10.1111/tran.12329.

Frynas, Jedrzej George. 2005. "The false development promise of corporate social responsibility: Evidence from multinational companies." *International Affairs* 81 (3): 581–598.

Fumagalli, Matteo. 2015. "The Kumtor gold mine and the rise of resource nationalism in Kyrgyzstan." *Central Asia Economic Paper Series*, 16. Central Asia Program, George Washington University, Washington DC. http://centralasiaprogram.org/archives/8661.

Geenen, Sara. 2014. "Dispossession, displacement and resistance: Artisanal miners in a gold concession in South-Kivu, Democratic Republic of Congo." *Resources Policy* 41 (1): 90–99. https://doi: 10.1016/j.resourpol.2013.03.004.

Gerson, L.P. 1986. "Platonic dualism." *The Monist* 69 (3): 352–369.

Gilberthorpe, Emma, and Glenn Banks. 2012. "Development on whose terms?: CSR discourse and social realities in Papua New Guinea's extractive industries sector." *Resources Policy* 37 (2): 185–193.

Gilberthorpe, Emma, and Dinah Rajak. 2017. "The anthropology of extraction: Critical perspectives on the resource curse." *Journal of Development Studies* 53 (2): 186–204. https://doi: 10.1080/00220388.2016.1160064.

Gilberthorpe, Emma, Dorice Agol, and Thomas Gegg. 2016. "'Sustainable mining'? Corporate social responsibility, migration and livelihood choices in Zambia." *The Journal of Development Studies* 52 (11): 1517–1532. https://doi.org/10.1080/00220388.2016.1189534.

Gimenez, Martha E. 2005. "Capitalism and the oppression of women: Marx revisited." *Science & Society* 69 (1): 11–32.

Gimenez, Martha E. 2018. *Marx, Women and Capitalist Social Reproduction*. Leiden and Boston: Brill.

Glasson, John, and Riki Therivel. 2012. *Introduction to Environmental Impact Assessment*. London: Routledge. https://doi.org/10.4324/9781315881218.

Gómez-Baggethun, Erik, Rudolf de Groot, Pedro L. Lomas, and Carlos Montes. 2010. "The history of ecosystem services in economic theory and practice: From early notions to markets and payment schemes." *Ecological Economics* 69:1209–1218. https://doi:10.1016/j.ecolecon.2009.11.007.

Gómez-Barris, Macarena. 2017. *The Extractive Zone: Social Ecologies and Decolonial Perspectives*. Durham: Duke University Press. https://doi: 10.1215/9780822372561.

Goodman, James, and Ariel Salleh. 2013. "The 'green economy': Class hegemony and counter-hegemony." *Globalizations* 10 (3): 411–424.

Grandell, Leena, Antti Lehtilä, Mari Kivinen, Tiina Koljonen, Susanna Kihlman, and Laura S. Lauri. 2016. "Role of critical metals in the future markets of clean energy technologies." *Renewable Energy* 95: 53–62. https://doi: 10.1016/j.renene.2016.03.102.

Grosfoguel, Ramón. 2007. "The epistemic decolonial turn: Beyond political-economy paradigms." *Cultural Studies* 21 (2–3): 211–223.

Haber, Stephen, and Victor Menaldo. 2011. "Do natural resources fuel authoritarianism? A reappraisal of the resource curse." *American Political Science Review* 105 (1): 1–26. https://doi: 10.1017/s0003055410000584.

Habtom, Sefanit, and Megan Scribe. 2020. "To breathe together: Co-conspirators for decolonial futures." Yellowhead Institute, 2 June 2020. https://yellowheadinstitute.org/2020/06/02/to-breathe-together/.

Hall, Derek, Philip Hirsch, and Tania Murray Li. 2011. *Powers of Exclusion: Land Dilemmas in Southeast Asia*. Honolulu: University of Hawai'i Press.

Haraway, Donna. 1988. "Situated knowledges: The science question in feminism and the privilege of partial perspective." *Feminist Studies* 14 (3): 575–599.

Harding, Sandra. 1991. *Whose Science? Whose Knowledge?: Thinking from Women's Lives*. Ithaca: Cornell University Press.

Hausermann, Heidi, Janet Adomako, and Maya Robles. 2020. "Fried eggs and all-women gangs: The geopolitics of Chinese gold mining in Ghana, bodily vulnerability and resistance." *Human Geography* 13 (1): 60–73.

Hayat, Samuel. 2017. "Minor/miner politics: Managerial order and community in the Nord coal mines in the nineteenth century." *Politix* 119 (3): 123–146.

Hegel, Georg Wilhelm Friedrich. 2001 [1873]. *The Philosophy of History*. Translated by J. Sibree. Ontario: Batoche Books.

Heinaman, Robert. 1990. "Aristotle and the mind-body problem." *Phronesis: A Journal for Ancient Philosophy* 35 (1): 83–102.

Hesse-Biber, Sharlene Nagy, ed. 2012. *Handbook of Feminist Research: Theory and Praxis*. London: SAGE Publications. https://dx.doi.org/10.4135/9781483384740.

Heynen, Nik, and Paul Robbins. 2006. "The neoliberalization: Governance, privatization, enclosure and valuation." *Capitalism Nature Socialism* 16 (1): 5–8.

Hilson, Gavin. 2010. "'Once a miner, always a miner': Poverty and livelihood diversification in Akwatia, Ghana." *Journal of Rural Studies* 26 (3): 296–307. https://doi: 10.1016/j.jrurstud.2010.01.002.

Hilson, Gavin. 2012. "Corporate social responsibility in the extractive industries: Experiences from developing countries." *Resources Policy* 37 (2): 131–137. https://doi:10.1016/j.resourpol.2012.01.002.

Himley, Matthew, Elizabeth Havice, and Gabriela Valdivia, eds. 2021. *The Routledge Handbook of Critical Resource Geography*. London and New York: Routledge.

Hokowhitu, Brendan, Aileen Moreton-Robinson, Linda Tuhiwai-Smith, Chris Andersen, and Steve Larkin, eds. 2020. *Routledge Handbook of Critical Indigenous Studies*. Abingdon: Routledge. https://doi.org/10.4324/9780429440229

hooks, bell. 2014. *Feminist Theory: From Margin to Center*. New York: Routledge.

Hopke, Jill E. 2016. "Translocal anti-fracking activism: An exploration of network structure and tie content." *Environmental Communication* 10 (3): 380–394. https://doi: 10.1080/17524032.2016.1147474.

Huggins, Chris. 2016. "Artisanal and small-scale mining: Critical approaches to property rights and governance." *Third World Thematics* 1 (2): 151–154.

Humphreys, Macartan, Jeffrey Sachs, and Joseph E. Stiglitz. 2007. *Escaping the Resource Curse*. New York: Columbia University Press.

Hyndman, Jennifer. 2001. "Towards a feminist geopolitics." *The Canadian Geographer* 45 (2): 210–222.

International Labour Organization (ILO). 1989. *Indigenous and Tribal Peoples Convention, C. 169*. https://www.ilo.org/dyn/normlex/en/f?p=NORMLEXPUB:55:0::NO::P55_TYPE,P55_LANG,P55_DOCUMENT,P55_NODE:REV,en,C169,/Document.

Jacob, Thabit, and Rasmus Hundsbæk Pedersen. 2018. "New resource nationalism? Continuity and change in Tanzania's extractive industries." *Extractive Industries and Society* 5 (2): 287–292. https://doi: 10.1016/j.exis.2018.02.001.

Jalbert, Kirk, Anna Willow, David Casagrande, and Stephanie Paladino, eds. 2017. *ExtrACTION: Impacts, Engagements, and Alternative Futures*. Abingdon: Routledge.

Jazeel, Tariq, and Stephen Legg, eds. 2019. *Subaltern Geographies*. Athens: University of Georgia Press.

Jenkins, Katy. 2014. "Women, mining and development: An emerging research agenda." *The Extractive Industries and Society* 1 (2): 329–339.

Jenkins, Katy. 2017. "Women anti-mining activists' narratives of everyday resistance in the Andes: Staying put and carrying on in Peru and Ecuador." *Gender, Place and Culture* 24 (10): 1441–1459. https://doi: 10.1080/0966369X.2017.1387102.

Kama, Kärg. 2020. "Resource-making controversies: Knowledge, anticipatory politics and economization of unconventional fossil fuels." *Progress in Human Geography* 44 (2): 333–356. https://doi: 10.1177/0309132519829223.

Kinkartz, Sabine. 2022. "Germany's energy U-turn: Coal instead of gas." *DW*, August 4, 2022. https://www.dw.com/en/germanys-energy-u-turn-coal-instead-of-gas/a-62709160.

Kirsch, Stuart. 2007. "Indigenous movements and the risks of counterglobalization: Tracking the campaign against Papua New Guinea's Ok Tedi mine." *American Ethnologist* 34 (2): 303–321. https://doi: 10.1525/ae.2007.34.2.303.

Kirsch, Stuart. 2010. "Sustainable mining." *Dialectical Anthropology* 34: 87–93.

Klare, Michael T. 2001. *Resource Wars: The New Landscape of Global Conflict*. New York: Metropolitan Books.

Klinger, Julie Michelle. 2017. *Rare Earth Frontiers: From Terrestrial Subsoils to Lunar Landscapes*. Ithaca and London: Cornell University Press.

Kobayashi, Audrey, and Sarah de Leeuw. 2010. "Colonialism and the tensioned landscapes of indigeneity." In *The SAGE Handbook of Social Geographies*, edited by Susan J. Smith, Rachel Pain, Sallie A. Marston, and John Paul Jones, 118–139. London: SAGE Publications Ltd. https://dx.doi.org/10.4135/9780857021113.n5.

Koch, Natalie, and Tom Perreault. 2018. "Resource nationalism." *Progress in Human Geography* 43 (4): 611–631. https://doi: 10.1177/0309132518781497.

Kojola, Erik. 2019. "Bringing back the mines and a way of life: Populism and the politics of extraction." *Annals of the American Association of Geographers* 109 (2): 371–381. https://doi: 10.1080/24694452.2018.1506695.

Koopman, Sara, Simon Dalby, Nick Megoran, Jo Sharp, Gerry Kearns, Rachel Squire, Alex Jeffrey, Vicky Squire, and Gerard Toal. 2021 "*Critical Geopolitics*/critical geopolitics 25 years on." *Political Geography* 90 (21): 2–9.

Krausmann, Fridolin, Christian Lauk, Willi Haas, and Dominik Wiedenhofer. 2018. "From resource extraction to outflows of wastes and emissions: The socioeconomic metabolism of the global economy, 1900–2015." *Global Environmental Change* 52 (December 2017): 131–140. https://doi: 10.1016/j.gloenvcha.2018.07.003.

Kuhn, Annette, and AnnMarie Wolpe, eds. 2013. *Feminism and Materialism: Women and Modes of Production*. London and New York: Routledge.

Kuokkanen, Rauna. 2008. "What is hospitality in the academy? Epistemic ignorance and the (im)possible gift." *The Review of Education, Pedagogy, and Cultural Studies* 30 (1): 60–82. https://doi.org/10.1080/10714410701821297.

Kutney, Gerald. 2014. *Carbon Politics and the Failure of the Kyoto Protocol*. London: Routledge.

Lahiri-Dutt, Kuntala. 2006. "'May God give us chaos, so that we can plunder': A critique of 'resource curse' and conflict theories." *Development* 49: 14–21.

Lahiri-Dutt, Kuntala. 2012. "Digging women: Towards a new agenda for feminist critiques of mining." *Gender, Place and Culture* 19 (2): 193–212.

Lahiri-Dutt, Kuntala. 2022. "New directions in research on women and gender in extractive industries." *The Extractive Industries and Society* 9.

Lahiri-Dutt, Kuntala, and Martha Macintyre, eds. 2006. *Women Miners in Developing Countries: Pit Women and Others*. London: Routledge.

Lajus, Julia. 2011. "Colonization of the Russian north: A frozen frontier." *In Cultivating the Colony: Colonial States and their Environmental Legacies*, edited by Christina Folke Ax, Niels Brimnes, Niklas Thode Jensen, and Karen Oslund. Athens, OH: Ohio University Press.

Lane, Marcus B., Helen Ross, A. P. Dale, and Roy E. Rickson. 2003. "Sacred land, mineral wealth, and biodiversity at Coronation Hill, Northern Australia: Indigenous knowledge and SIA." *Impact Assessment and Project Appraisal* 21 (2): 89-98. https://DOI:10.3152/147154603781766374.

Lawrence, Rebecca, and Rasmus Kløcker Larsen. 2017. "The politics of planning: Assessing the impacts of mining on Sami lands." *Third World Quarterly* 38 (5): 1164–1180. DOI: 10.1080/01436597.2016.1257909.

Lawson, Alex. 2022. "Nord stream 1: Gazprom announces indefinite shutdown of pipeline." *The Guardian*, September 2, 2022. https://www.theguardian.com/business/2022/sep/02/nord-stream-1-gazprom-announces-indefinite-shutdown-of-pipeline.

Le Billon, Philippe. 2008. "Diamond wars? Conflict diamonds and geographies of resource wars." *Annals of the Association of American Geographers* 98 (2): 345–372. https://doi: 10.1080/00045600801922422.

Lee, Ching Kwan. 2018. *The Specter of Global China: Politics, Labor, and Foreign Investment in Africa*. Chicago: University of Chicago Press.

Leifsen, E., Gustafsson, M. T., Guzmán-Gallegos, M. A., and Schilling-Vacaflor, A. 2017. "New mechanisms of participation in extractive governance: Between technologies of governance and resistance work." *Third World Quarterly* 38: 1043–1057. https://doi.org/10.1080/01436597.2017.1302329

Lesutis, Gediminas. 2019. "Spaces of extraction and suffering: Neoliberal enclave and dispossession in Tete, Mozambique." *Geoforum* 102: 116–125. https://doi: 10.1016/j.geoforum.2019.04.002.

Lightfoot, Sherly. 2020. "Indigenous laws and governance in indigenous self-developed FPIC protocols." https://fnemc.ca/wp-content/uploads/2015/07/Sheryl-Lightfoot-Expert-Paper-Indigenous-Legal-Framework.pdf

Lodhia, Sumit K., ed. 2018. *Mining and Sustainable Development: Current Issues*. Abingdon: Routledge.

Lugones, María. 2007. "Heterosexualism and the colonial/modern gender system." *Hypathia* 22 (1): 186–219.

Lyall, Angus, and Gabriela Valdivia. 2019. "The speculative petro-state: Volatile oil prices and resource populism in Ecuador." *Annals of the American Association of Geographers* 109 (2): 349–360. https://doi: 10.1080/24694452.2018.1531690.

Mahanty, Sango, and Constance L. McDermott. 2013. "How does 'free, prior, informed consent' (FPIC) impact social equity? Lessons for REDD+." *Land Use Policy* 35: 406–416.

Marchand, Marianne H., and Anne Sisson Runyan, eds. 2011. *Gender and Global Restructuring: Sightings, Sites and Resistances*. 2nd Edition. London: Routledge.

Massaro, Vanessa A., and Jill Williams. 2013. "Feminist geopolitics." *Geography Compass* 7 (8): 567–577.

Mauss, Marcel, and Jane I. Guyer, transl. 2016 [1925]. *The Gift*. Chicago: HAU Books.

Mawdsley, Emma. 2018. "The 'southernisation' of development?" *Asia Pacific Viewpoint* 59 (2): 173–185. https://doi: 10.1111/apv.12192.

Mbembe, Achille. 2003. "Necropolitics." *Public Culture* 15 (1): 11–40.

McGregor, Deborah, Steven Withaker, and Mahisha Sritharan. 2020. "Indigenous environmental justice and sustainability." *Current Opinion in Environmental Sustainability* 43: 35–40. https://doi.org/10.1016/j.cosust.2020.01.007.

Melamed, Jodi. 2015. "Racial capitalism." *Critical Ethnic Studies* 1 (1): 76–85.

Mignolo, Walter D. 2009. "Epistemic disobedience, independent thought and decolonial freedom." *Theory, Culture and Society* 26 (8): 159–181. https://doi: 10.1177/0263276409349275.

Mignolo, Walter, and Catherine E. Walsh. 2018. *On Decoloniality: Concepts, Analytics, and Praxis*. Durham: Duke University Press.

Mitchell, Terry, Courtney Arsenau, Darren Thomas, and Peggy Smith. 2019. "Towards an indigenous-informed relational approach to free, prior, and informed consent (FPIC)." *International Indigenous Policy Journal* 10 (4). https://doi.org/10.18584/iipj.2019.10.4.8372.

Mollett, Sharlene. 2006. "Race and natural resource conflicts in Honduras: The Miskito and Garifuna struggle for Lasa Pulan." *Latin American Research Review* 41 (1): 76–101.

Mollett, Sharlene, and Caroline Faria. 2013. "Messing with gender in feminist political ecology." *Geoforum* 45: 116–125. https://doi: 10.1016/j.geoforum.2012.10.009.

Moodie, T. Dunbar. 1986. "The moral economy of the black miners' strike of 1946." *Journal of Southern African Studies* 13 (1): 1–35. https://doi: 10.1080/03057078608708130.

Morris-Suzuki, Tessa. 2013. "The nature of empire: Forest ecology, colonialism and survival politics in Japan's imperial order." *Japanese Studies* 33 (3): 225–242.

Murrey, Amber. 2015. "Invisible power, visible dispossession: The witchcraft of a subterranean pipeline." *Political Geography* 47: 64–76. https://doi: 10.1016/j.polgeo.2015.04.004.

Murrey, Amber. 2016. "Slow dissent and the emotional geographies of resistance." *Singapore Journal of Tropical Geography* 37 (2): 224–248. https://doi: 10.1111/sjtg.12147.

Murrey, Amber. 2017. "Decolonising the imagined geographies of 'witchcraft'." *Third World Thematics: A TWQ Journal* 2 (2–3): 157–179. https://doi: 10.1080/23802014.2017.1338535.

Murrey, Amber, and Nicholas A. Jackson. 2019. "A decolonial critique of the racialized 'localwashing' of extraction in Central Africa." *Annals of the American Association of Geographers* 110 (3): 1–24. https://doi: 10.1080/24694452.2019.1638752.

Myadar, Orhon, and Sara Jackson. 2019. "Contradictions of populism and resource extraction: Examining the intersection of resource nationalism and accumulation by dispossession in Mongolia." *Annals of the American Association of Geographers* 109 (2): 361–370. https://doi: 10.1080/24694452.2018.1500233.

Nadasdy, Paul. 2005. "Transcending the debate over the ecologically noble Indian: Indigenous peoples and environmentalism." *Ethnohistory* 52 (2): 291–331. https://doi: 10.1215/00141801-52-2-291.

Nadler, Steven. 2019. "Spinoza, Descartes, and the 'Stupid Cartesians'." In *The Oxford Handbook of Descartes and Cartesianism*, edited by Steven Nadler, Tad

M. Schmaltz, and Delphine Antoine-Mahut, 658–677. Oxford: Oxford University Press. https://doi: 10.1093/oxfordhb/9780198796909.013.41.

Nagar, Richa, Victoria Lawson, Linda McDowell, and Susan Hanson. 2002. "Locating globalization: Feminist (re)readings of the subjects and spaces of globalization." *Economic Geography* 78 (3): 257–284.

Nash, June C. 1979. *We Eat the Mines and the Mines Eat Us: Dependency and Exploitation in Bolivian Tin Mines*. New York: Columbia University Press.

Navaro-Yashin, Yael. 2009. "Affective spaces, melancholic objects: Ruination and the production of anthropological knowledge." *Journal of the Royal Anthropological Institute* 15 (1): 1–18. https://doi: 10.1111/j.1467-9655.2008.01527.x.

O'Faircheallaigh, Ciaran. 2013. "Community development agreements in the mining industry: An emerging global phenomenon." *Community Development* 44 (2): 222–238. https://doi: 10.1080/15575330.2012.705872.

O'Faircheallaigh, Ciaran, and Tony Corbett. 2005. "Indigenous participation in environmental management of mining projects: The role of negotiated agreements." *Environmental Politics* 14 (5): 629–647. https://doi: 10.1080/09644010500257912.

Peet, Richard, Paul Robbins, and Michael Watts, eds. 2011. *Global Political Ecology*. Abingdon and New York: Routledge.

Peet, Richard, and Michael Watts, eds. 1996. *Liberation Ecologies: Environment, Development and Social Movements*. London and New York: Routledge.

Peluso, Nancy Lee. 2017. "Plantations and mines: Resource frontiers and the politics of the smallholder slot." *Journal of Peasant Studies* 44 (4): 954–989. https://doi: 10.1080/03066150.2017.1339692.

Peluso, Nancy Lee, and Christian Lund. 2011. "New frontiers of land control: Introduction." *Journal of Peasant Studies* 38 (4): 667–681. https://doi: 10.1080/03066150.2011.607692.

Peluso, Nancy Lee, and Jesse Ribot. 2020. "Postscript: A theory of access revisited." *Society and Natural Resources* 33 (2): 300–306. https://doi: 10.1080/08941920.2019.1709929.

Penfield, Amy. 2019. "Extractive pluralities: The intersection of oil wealth and informal gold mining in Venezuelan Amazonia." In *Indigenous Life Projects and Extractivism: Ethnographies from South America*, edited by Cecilie Vindal Ødegaard and Juan Javier Rivera Andía, 75–93. Cham: Springer Verlag.

Perreault, Tom. 2013. "Dispossession by accumulation? Mining, water and the nature of enclosure on the Bolivian Altiplano." *Antipod* 45 (5): 1050–1069. https://doi: 10.1111/anti.12005.

Perreault, Tom, Gavin Bridge, and James McCarthy, eds. 2015. *The Routledge Handbook of Political Ecology*. London and New York: Routledge. https://doi.org/10.4324/9781315759289.

Postar, Stephanie, and Negar Elodie Behzadi. 2022 (in press). "'Extractive bodies': A feminist counter-topography of two extractive landscapes." *Geoforum*. https://doi.org/10.1016/j.geoforum.2022.09.005.

Povinelli, Elizabeth A. 2016. *Geontologies: A Requiem to Late Liberalism*. Durham and London: Duke University Press.

Preston, Jen. 2017. "Racial extractivism and white settler colonialism: An examination of the Canadian Tar Sands mega-projects." *Cultural Studies* 31 (2–3): 353–375. https://doi: 10.1080/09502386.2017.1303432

Rasmussen, Mattias Borg, and Christian Lund. 2018. "Reconfiguring frontier spaces: The territorialization of resource control." *World Development* 101: 388–399. https://doi: 10.1016/j.worlddev.2017.01.018.

Ribot, Jesse C., and Nancy Lee Peluso. 2003. "A theory of access." *Rural Sociology* 68 (2): 153–181. https://doi:10.1111/j.1549- 0831.2003.tb00133.x.

Richardson, Tanya, and Gisa Weszkalnys. 2014. "Resource materialities." *Anthropological Quarterly* 87 (1): 5–30.

Riofrancos, Thea. 2020. *Resource Radicals: From Petro-Nationalism to Post-Extractivism in Ecuador*. Durham and London: Duke University Press.

Rocheleau, Dianne E., and Padini Nirmal. 2015. "Feminist political ecologies: Grounded, networked and rooted on Earth." In *The Oxford Handbook of Transnational Feminist Movements*, edited by Rawwida Baksh and Wendy Harcourt, 793–814. Oxford: Oxford University Press. https://doi: 10.1093/oxfordhb/9780199943494.013.032.

Rocheleau, Dianne E., Barbara P. Thomas-Slayter, and Esther Wangari, eds. 1996. *Feminist Political Ecology: Global Issues and Local Experiences*. London: Routledge.

Ross, Michael L. 1999. "The political economy of the resource curse." *World Politics* 51 (2): 297–322.

Samson, Colin, and Carlos Gigoux. 2017. *Indigenous Peoples and Colonialism: Global Perspectives*. Cambridge, UK: Polity.

Sawyer, Suzana. 2004. *Crude Chronicles: Indigenous Politics, Multinational Oil, and Neoliberalism in Ecuador*. Durham and London: Duke University Press.

Sawyer, Suzana. 2015, "Crude contamination: Law, science, and indeterminacy in Ecuador and beyond." In *Subterranean Estates: Life Worlds of Oil and Gas*, edited by Hannah Appel, Arthur Mason, and Michael Watts, 126–146. Ithaca and London: Cornell University Press.

Schilling-Vacaflor, Almut. 2016. "Who controls the territory and the resources? Free, prior, informed consent as a contested human rights practice in Bolivia." *Third World Quarterly* 38 (5): 1058–1074. https://DOI: 10.1080/01436597.2016.1238761.

Scott, Heidi V. 2008. "Colonialism, landscape and the subterranean." *Geography Compass* 2 (6): 1853–1869. https://doi: 10.1111/j.1749-8198.2008.00164.x.

Segal, Paul. 2011. "Resource rents, redistribution, and halving global poverty: The resource dividend." *World Development* 39 (4): 475–489. https://doi: 10.1016/j.worlddev.2010.08.013.

Sells, Laura. 1993. "Feminist epistemology: Rethinking the dualisms of atomic knowledge." *Hypathia* 8 (3): 202–210.

Shapiro, Judith, and John-Andrew Mcneish, eds. 2021. *Our Extractive Age: Expressions of Violence and Resistance*. Abingdon: Routledge.

Sherman, Peter, Haiyang Lin, and Michael McElroy. 2022. "Projected global demand for air conditioning associated with extreme heat and implications for electricity grids in poorer countries." *Energy and Buildings* 268: 112198. https://doi: 10.1016/j.enbuild.2022.112198.

Singh, Neera M. 2013. "The affective labor of growing forests and the becoming of environmental subjects: Rethinking environmentality in Odisha, India." *Geoforum* 47: 189–198. https://doi: 10.1016/j.geoforum.2013.01.010.

Slee, Roger. 2013. "How do we make inclusive education happen when exclusion is a political predisposition?" *International Journal of Inclusive Education* 17 (8): 895–907. https://doi: 10.1080/13603116.2011.602534.

Smith, Linda Tuhiwai. 1999. *Decolonizing Methodologies: Research and Indigenous Peoples*. London: Zed Books.

Smith, Neil. 2010. *Uneven Development: Nature, Capital, and the Production of Space*. 3rd edition. London: Verso.

Sovacool, Benjamin K., Götz Walter, Thijs Van de Graaf, and Nathan Andrews. 2016. "Energy governance, transnational rules, and the resource curse: Exploring the effectiveness of the extractive industries transparency initiative (EITI)." *World Development* 83: 179–192.

Spiegel, Samuel, Susan Keane, Steve Metcalf, and Marcello Veiga. 2015. "Implications of the Minamata convention on mercury for informal gold mining in Sub-Saharan Africa: From global policy debates to grassroots implementation?" *Environment, Development and Sustainability* 17 (4): 765–785. https://doi: 10.1007/s10668-014-9574-1.

Spivak, Gayatri Chakravorty. 1988. "Can the Subaltern speak?" In *Marxism and the Interpretation of Culture*, edited by Cary Nelson and Lawrence Grossberg, 271–313. Urbana: University of Illinois Press.

Squire, Rachel, and Klauss Dodds. 2019. "Introduction to the special issue: Subterranean geopolitics." *Geopolitics* 25 (1): 4–16.

Strang, Veronica. 2014. "Fluid consistencies. Material relationality in human engagements with water." *Archaeological Dialogues* 21 (2): 133–150. https://doi: 10.1017/S1380203814000130.

Sultana, Farhana. 2011. "Suffering for water, suffering from water: Emotional geographies of resource access, control and conflict." *Geoforum* 42 (2): 163–172. https://doi: 10.1016/j.geoforum.2010.12.002.

Sundberg, Juanita. 2014. "Decolonizing posthumanist geographies." *Cultural Geographies* 21 (1): 33–47. https://doi: 10.1177/1474474013486067.

Sundberg, Juanita. 2017. "Feminist political ecology." In *International Encyclopedia of Geography: People, the Earth, Environment and Technology*, edited by Douglas Richardson, Noel Castree, Michael F. Goodchild, Audrey Kobayashi, Weidong Liu, and Richard A. Marston. Chichester, UK: John Wiley & Sons. https://doi.org/10.1002/9781118786352.wbieg0804.

Tallbear, Kim. 2017. "Beyond the life / not-life binary: A feminist-Indigenous reading of cryopreservation, interspecies thinking, and the new materialisms." In *Cryopolitics: Life in a Melting World*, edited by Joanna Radin and Emma Kowal, 179–202. Cambridge, Massachusetts: The MIT Press.

Todd, Zoe. 2016. "An Indigenous feminist's take on the ontological turn: 'Ontology' is just another word for colonialism." *Journal of Historical Sociology* 29 (1): 4–22. https://doi: 10.1111/johs.12124.

Todd, Zoe. 2018. "Refracting the state through human-fish relations: Fishing, Indigenous legal orders and colonialism in north/western Canada." *Decolonization: Indigeneity, Education & Society* 7 (1): 60–75.

United Nations. 1998. *Aarhus Convention on Access to Information, Public Participation in Decision-Making and Access to Justice in Environmental Matters.* https://unece.org/environment-policy/public-participation/aarhus-convention/introduction.

Vanclay, Frank, and Ana Maria Esteves, eds. 2011. *New Directions in Social Impact Assessment: Conceptual and Methodological Advances.* Cheltenham: Edward Elgar.

Wang, Qiang, Xiaoxin Song, and Yi Liu. 2020. "China's coal consumption in a globalizing world: Insights from multi-regional input-output and structural decomposition analysis." *Science of the Total Environment* 711: 134790. https://doi:10.1016/j.scitotenv.2019.134790.

Watts, Michael. 2004. "Resource curse? Governmentality, oil and power in the Niger Delta, Nigeria." *Geopolitics* 9 (1): 50–80. https://doi:10.1080/14650040412331307832.

Watts, Michael. 2005. "Righteous oil? Human rights, the oil complex, and corporate social responsibility." *Annual Review of Environmental Resources* 30: 373–407. https://doi: 10.1146/annurev.energy.30.050504.144456.

Watts, Vanessa. 2013. "Indigenous place-thought & agency amongst humans and non-humans (first woman and sky woman go on a European world tour!)." *Decolonization: Indigeneity, Education & Society* 2 (1): 20–34.

WCED. 1987. *Our Common Future: The Brundtland Report 1987.* Oxford: Oxford University Press.

Weszkalnys, Gisa. 2016. "A doubtful hope: Resource affect in a future oil economy." *Journal of the Royal Anthropological Institute* 22 (S1): 127–146. https://doi:10.1111/1467-9655.12397.

Williamson, Bess. 2015. "Access." In *Keywords for Disability Studies*, edited by Rachel Adams, Benjamin Reiss, and David Serlin, 14–17. New York: NYU Press.

Willow, Anna J. 2018. *Understanding ExtrACTIVISM: Culture and Power in Natural Resource Disputes.* Abingdon: Routledge.

Wise, Louise. 2021. "The genocide-ecocide nexus in Sudan: Violent 'development' and the racial-spatial dynamics of (neo)colonial-capitalist extraction." *Journal of Genocide Research* 23 (2): 189–211.

Yakovleva, Natalia, and Edmund Nickless, eds. 2022. *Routledge Handbook of the Extractive Industries and Sustainable Development.* Abingdon: Routledge.

Zhu, Annah, and Nancy Lee Peluso. 2021. "From gold to rosewood: Agrarian change, high-value resources, and the flexible frontier-makers of the twenty-first century." In *The Routledge Handbook of Critical Resource Geography*, edited by Matthew Himley, Elizabeth Havice, and Gabriela Valdivia, 345–357. London and New York: Routledge. https://doi: 10.4324/9780429434136-29.

Zimmermann, Erich W. 1933. *World Resources and Industries: A Functional Appraisal of the Availability of Agricultural and Industrial Resources.* New York: Harper and Row.

Part I

HAVES/HAVE-NOTS

Unsettling the Political Ecologies of Extraction

Philippe Le Billon, Stephanie Postar, and Negar Elodie Behzadi

Political ecologies of extraction have played a seminal role in exposing the exploitation and subjectification within contemporary logics and practices of extractive development (Acosta 2013; Junka-Aikio and Cortes-Severino 2017; Veltmeyer 2016). Analyzing relations between local communities, multinational companies, the state, and environments, these studies have exposed the uneven power relations, various forms of violence, and exclusions/inclusions associated with extraction and extractivism (Bridge 2008; Bebbington 2009, 2015). In doing so, political ecology studies can help recognize some of the binaries frequently framing extraction, particularly those opposing powerful and weak "actors." In this introduction, we historically trace ideas of haves/have-nots to politico-economic approaches that interpret ownership and control over resources as the driver of economic development.[1] We recognize that this pattern has shaped binaries of North/South, core/periphery, and developed/underdeveloped. At the same time, our political ecology approach helps us problematize these binaries by complexifying our understanding of power. We particularly identify three levels of "unsettling" the haves/have-nots binary that we illustrate through a review of the two chapters that form this subsection.

The first level of "unsettling" refers to the broader scale analysis of uneven development shaped by and shaping resource extraction. Definitions of extractivism often rely on the idea of a division between North/South and the exploitation and plunder of the South for the benefit of the industrialization of the North (Acosta 2013). Recognizing these foundational inequalities is important to historicize the violence of extractivism. Yet, recent patterns of development and associated politico-ecological analyses have challenged this binary in multiple ways. "New" actors (China, Brazil, etc.) playing an increasingly prominent role in extractive-led development challenge assumptions about a North/South binary by making the landscape of development more "polycentric" (Mawdsley 2018) (e.g., "South-South" developments). Furthermore, scholars have examined notions of neoextractivism by these "new" actors and their role in reproducing systems of extractivism, for instance, by critiquing global rhetoric that homogenizes Chinese-led extractive development projects, investors, and capital (Lee 2018). Central to the logics of neoextractivism are also forms of resource nationalisms, that is, a "political discourse [. . .] about how a state and its populations should derive and distribute profits derived from natural resources" and a "collective belonging expressed through the idiom of natural resources" (Koch and Perreault 2018, 611–612). Progressive left governments' resource nationalisms often focus on reverting past wrongs of neoliberal extractive-led development (see for instance, the cases of Ecuador or Bolivia); while in other places, neoextractivist models can be motivated by desires to integrate into the global neoliberal development model.

Political ecologists have also looked at forms of development that reproduce ideas of core/periphery but do not fall within the North/South binary frame. This includes studies of forms of internal colonizations that happen in both "Southern" and "Northern" countries through resource development (Dunlap 2020; Hall 2012; Hernandez and Newell 2022). It also encompasses analyses of nature/society relations in (post-)Soviet countries and the ways in which Soviet modernization, post-Soviet environmental trajectories, and their socio-environmental effects both compare and differ from Western (or Northern) development (Fleming 2017; Davidov 2013; Wooden 2014). In studies of resource extraction, scholars have theorized the specificities of post-Soviet politico-ecological changes and their impacts on both communities and the environment (Artman 2020; Kumagalli 2015; Behzadi 2019; Doolotkelieva 2015). In her chapter in this section "Gold as a Frictious Fiction: Managing gold extraction through technocratic discourse in postsocialist Kyrgyzstan," Asel Doolotkeldieva speaks to these subthemes. Her ethnographic work challenges ideas of North/South and Soviet/post-Soviet. She shows how technocratic discourses use Soviet legacies and tropes to cast mining protesters as social parasites in the context of Kyrgyzstan's mining conflicts. She also highlights ways in which these discourses contribute to the expansion of mining frontiers internally, and materialize through multiple levels of violent exploitations and exclusions. This expansion is motivated by a neoextractivist narrative that frames domestic resource extraction as central to economic development and nation-building in the post-Soviet era.

The second level of "unsettling" problematizes the designation of who "has" or "has-not." This framing contests the assumption that there are naturalized categories determining the distribution of wealth and power by recovering the agency of the "have-nots." Rather than homogenizing the "have-nots" as weak, passive, or backward recipients of development, a broad array of scholarship clarifies how those who are usually considered as the "have-nots" act as agents in the face of the effects of resource extraction. People in this category resist and refuse extraction in a number of ways, from broad-scale anti-extractivism (Riofrancos 2020) to intimate and emotional forms of dissent (Murrey 2016; González-Hidalgo and Zografos 2019; Postar and Behzadi 2022). Resistance can take the form of community-level consent refusal during consultation processes (Leifsen et al. 2017; Middeldorp and Le Billon 2020), plebiscites, or *consultas* seeking to demonstrate popular local opposition to specific extraction projects (Laplate and Nolin 2014; Walter and Leire 2017; McNeish 2017), and broader struggles for territorial autonomy (Anthias 2018). To stop ongoing extractive projects, especially during their construction or extension phases, the "have-nots" also create physical disruptions. This can take the form of street protests, blockades, or sabotage, with Environmental Justice Atlas-based studies demonstrating the diversity of conflicts, resistance processes, and outcomes (see Temper et al. 2015). These often-material manifestations of resistance reflect both the value of their instrumental and symbolic effects for communities facing the coloniality of a dominant society reproduced through systematic economic marginalization, political paternalism, and racism (Blomley 1996). Asel Doolotkeldieva's chapter highlights some of these resistance strategies in the context of Kyrgyzstan, revealing communities' frustration with what she refers to as the "fictitious" character of extractive-led economic development. Unmet promises around the economic benefits of gold extraction have spurred "friction" in the form of anti-mining protests and at times violent conflicts that have marked the country's history over the past ten years. In her chapter, "Civil complaints and environmental justice in post-neoliberal Ecuador," Teresa Bornschlegl embeds her institutional ethnography in a broader Ecuadorian context where anti-extraction and Indigenous resistances demanded and attained constitutional changes.

Political ecologists have explored such examples that speak to outward resistance strategies. However, those who are usually considered to be the have nots do not only deploy agency through direct forms of resistance. They also, at times, actively choose to participate in and negotiate the terms of their inclusions, sometimes leading to further exclusions. Far fewer political ecology studies have examined these aspects of tacit assent or resigned consent for extraction (Bebbington et al. 2008). Local communities and their allies have sought to stall, compensate, or remediate extractive activities, thereby seeking to exclude themselves or at least positively alter their inclusion into extractivist logics and assemblages (Condé and Le Billon 2017; O'Faircheallaigh and Corbett 2005). Despite consent rights, especially for Indigenous communities, prior consultation processes very rarely forestall extractive projects, and mostly serve in negotiating project benefits (Zaremberg and Torres-Wong 2018). Studies of Impact Benefit Agreements (IBAs) have pointed to some of their ambivalent effects, such as enabling "communities" to negotiate their own terms of inclusion. However, these agreements are often on highly unequal terms—between extractive companies and communities as well as within communities. Furthermore, IBAs allow the state to jettison its statutory duty while serving as the enforcer of private (extractivist) contracts (Peterson St-Laurent and Le Billon 2015). Often narrowly understood as environmental clean-up, remediation is a continuation of extractive processes, reshaping local landscapes and economies, benefiting or harming local communities, and requiring greater public participation and attention to justice concerns (Beckett and Keeling 2018; Eklund et al. 2019). In this subsection, Teresa Bornschlegl highlights how Ecuadorian constitutional changes that include Indigenous ontologies in Ecuador did not transform the exclusionary and exploitative logics of extractivism. In her chapter, Bornschlegl shows how marginalized communities use civil complaints to challenge the exclusions produced particularly by companies' failures to prevent or remediate oil spills and other instances of environmental harm caused by extraction. At the same time, she highlights the limitations of these legal mechanisms, revealing the inconsequential nature of some of these complaints in a broader system of power imbalances.

At the third level, political ecology perspectives contend with *how* power operates rather than *who* holds power. By examining rhetorical tactics and discourses, as this volume does with the "inclusionary narrative," political ecologists have traced the development and evolution of relationships of power structuring natural resource management (Fortmann 1995; Leach and Fairhead 2000). Rhetorical strategies to resist extraction are often countered by extractive companies and their allies (in governments and elsewhere) through adjusting extractivist narratives to protesters' claims, changing contexts, and different audiences (Dietz 2019), dismissing lack of consent, and coercively imposing extractive activities (Frederiksen and Himley 2020; Brock and Dunlap 2018). In this absorption of opposition narratives, political ecologists have mapped the ambiguous dialectical shifts between the "haves" and the "have-nots," highlighting how power re-establishes itself by co-opting its rivals.[2] In this section, Doolotkeldieva's chapter focuses on discursive power—she shows how technocratic discourses, articulated by governmental and corporate specialists and reproduced by mass media, are used to discard certain grievances and delegitimize others. Doolotkeldieva highlights how these discourses "represent anti-mining resistance as an irrational and criminal conduct, and in this way strip them from the larger societal support and clear the ground for the advance of corporate extractivism" (chapter 1, this volume). Bornschlegl's chapter starts with the recognition that the Ecuadorian Constitution requires the state to "integrally repair" social and environmental damages, as informed by the concept of *Buen Vivir*—a way of life grounded in Indigenous cosmovisions that aims to respect all dimensions of society, collective communities, and the dignity of people. She concludes that despite the limited practical successes of this rhetorical commitment, "[t]he progressive constitutional provisions and the institutional adjustments in the environmental governance they entailed are still better than no environmental laws at all" (chapter 2, this volume).

Political ecologists have also critiqued monolithic notions of governments as homogenous, coordinated, singular entities perpetrating extractive exclusions and violence and therefore necessarily allied with extractive corporations. Instead, they look explicitly or implicitly at how power operates through assemblages or networks of actors including governments and other entities. At times, these actors reinforce one another's positions with regard to resource development, at others; however, they hold distinct, contradictory, or unexpected positions. Although most governments are pro-extraction, some have banned specific extractive sectors, often as a result of a combination of grassroots campaigning, bi-partisan political consensus (e.g., over water security), and low resource dependence (Bebbington et al. 2018; Gaulin and Le Billon 2020). In addition, in some cases, extraction is contradictorily banned by some parts of the government and permitted by others (Arnold and Neupane 2017). Along these lines, notions of "fragmented sovereignty" (Lund 2011) challenge the very idea that governmental institutions are single-handedly capable of successfully governing natural resources (ruling, controlling), problematizing the artificial opposition of state/community. Doolotkelieva's chapter illustrates this by highlighting the multiplicity of actors beyond the state that come together to advance corporate extractivism. Governments, alongside corporate actors and mass media frame anti-mining protests and protesters as backward, criminal, and opposed to development. In contrast, Teresa Bornschlegl exposes the tensions within the Ecuadorian state itself. Her methodological use of institutional ethnography allows her to look at procedures around civil complaints related to environmental damages as a window into the Ecuadorian state's internal struggles between extractivism and post-extractivism. Even though the Ecuadorian state formalized its post-neoliberal status through changes in its Constitution, "the struggle inside the state continues to be won by the extractivist state project" (chapter 2 this volume), and the "post-neoliberal state continues to be 'neo-extractivist in disguise'" (Swampa 2019, cited in Bornschlegl, chapter 2, this volume). As revealed in the two chapters that follow, problematizing relations of power between actors of extraction leads to a better understanding of the reasons why and the modalities through which resource extraction and its associated exclusions continue in the present moment, despite resistance and knowledge of adverse impacts.

NOTES

1. We could link this to Adam Smith's *The Wealth of Nations*. More recently, neo-evolutionist Julian Steward (1955) similarly theorized how the entanglement of nature and culture stimulates cultural and technological changes.

2. For another example of how this works in practice, see the chapter by Leah Horowitz on the negotiations for an Impact Benefit Agreement in New Caledonia in part VI.

REFERENCES

Acosta, Alberto. 2013. "Extractivism and neoextractivism: Two sides of the same curse." In *Beyond Development: Alternative Visions from Latin America*, edited by Miriam Lang, Lyda Fernando, and Nick Buxton, 61–86. Amsterdam: Transnational Institute.

Anthias, Penelope. 2018. *Limits to Decolonization: Indigeneity, Territory, and Hydrocarbon Politics in the Bolivian Chaco*. Ithaca: Cornell University Press.

Arnold, Gwen, and Kaubin Wosti Neupane. 2017. "Determinants of pro-fracking measure adoption by New York southern tier municipalities." *Review of Policy Research* 34 (2): 208–232. https://doi: 10.1111/ropr.12212.

Artman, Vincent. 2021. "Resource nationalism and spiritual pathways to sustainability in Kyrgyzstan." In *Religion, Sustainability and Place: Moral Geographies of the Anthropocene*, edited by Steven Silvern and Edward H. Davis. Singapore: Palgrave Macmillan. https://doi.org/10.1007/978-981-15-7646-1_11.

Avcı, Duygu, and Consuelo Fernández-Salvador. 2016. "Territorial dynamics and local resistance: Two mining conflicts in Ecuador compared." *The Extractive Industries and Society* 3 (4): 912–921. https://doi.org/10.1016/j.exis.2016.10.007.

Bebbington, Anthony, Benjamin Fash, and John Rogan. 2019. "Socio-environmental conflict, political settlements, and mining governance: A cross-border comparison, El Salvador and Honduras." *Latin American Perspectives* 46 (2): 84–106. https://doi.org/10.1177/0094582X18813567.

Bebbington, Anthony, Leonith Hinojosa, Denise Humphreys Bebbington, Maria Luisa Burneo, and Ximena Warnaars. 2008. "Contention and ambiguity: Mining and the possibilities of development." *Development and Change* 39 (6): 887–914. https://doi.org/10.1111/j.1467-7660.2008.00517.x.

Beckett, Caitlynn, and Arn Keeling. 2019. "Rethinking remediation: Mine reclamation, environmental justice, and relations of care." *Local Environment* 24 (3): 216–230. https://doi.org/10.1080/13549839.2018.1557127.

Blomley, Nicholas. 1996. "'Shut the province down': First nations blockades in British Columbia, 1984-1995." *BC Studies: The British Columbian Quarterly* 111: 5–35. https://doi.org/10.14288/bcs.v0i111.1361.

Bridge, Gavin. 2008. "Global production networks and the extractive sector: Governing resource-based development." *Journal of Economic Geography* 8 (3): 389–419. https://doi.org/10.1093/jeg/lbn009.

Conde, Marta, and Philippe Le Billon. 2017. "Why do some communities resist mining projects while others do not?" *The Extractive Industries and Society* 4 (3): 681–697. https://doi.org/10.1016/j.exis.2017.04.009.

Davidov, Veronica. 2013. "Soviet and post-Soviet inequalities through the lens of political ecology: A foundation for further inequality." *Journal of Globalization Studies* 4 (1): 95–104.

Dietz, Kristina. 2019. "Contesting claims for democracy: The role of narratives in conflicts over resource extraction." *The Extractive Industries and Society* 6 (2): 510–518. https://doi.org/10.1016/j.exis.2019.03.004.

Doolotkeldieva, Asel. 2015. *Social Mobilisations, Politics and Society in Contemporary Kyrgyzstan*. PhD diss in Politics, University of Exeter. http://hdl.handle.net/10871/21665.

Dunlap, Alexander. 2020. "Bureaucratic land grabbing for infrastructural colonization: Renewable energy, L' Amassada, and resistance in southern France." *Human Geography* 13 (2): 109–126. https://doi.org/10.1177/1942778620918041

Eklund, Ruth L., Landon C. Knapp, Paul A. Sandifer, and Rita C. Colwell. 2019. "Oil spills and human health: Contributions of the Gulf of Mexico research initiative." *GeoHealth* 3 (12): 391–406. https://doi.org/10.1029/2019GH000217.

Fleming, Jake. 2017. "Toward vegetal political ecology: Kyrgyzstan's walnut-fruit forest and the politics of graftability." *Geoforum* 79: 26–35. https://doi.org/10.1016/j.geoforum.2016.12.009.

Fortmann, Louise. 1995. "Talking claims: Discursive strategies in contesting property." *World Development* 23 (6): 1053–1063. https://doi.org/10.1016/0305-750X(95)00024-7.

Fumagalli, Matteo. 2015. "The Kumtor gold mine and the rise of resource nationalism in Kyrgyzstan." Central Asia Economic Papers 16: 1–10, August 2015. https://centralasiaprogram.org/kumtor-gold-mine-rise-resource-nationalism-kyrgyzstan.

Gaulin, Nicolas, and Philippe Le Billon. 2020. "Climate change and fossil fuel production cuts: Assessing global supply-side constraints and policy implications." *Climate Policy* 20 (8): 888–901. https://doi.org/10.1080/14693062.2020.1725409.

Gogo, Kato, and Prince Nweke. 2008. "Management, remediation and compensation in cases of crude oil spills in Nigeria: An appraisal." *Journal of Mineral Resources Law* 8 (1): 27–42.

González-Hidalgo, Marien, and Christos Zografos. 2018. "Emotions, power, and environmental conflict: Expanding the 'emotional turn' in political ecology." *Progress in Human Geography* 44 (2): 235–255. https://doi.org/10.1177/0309132518824644.

Greenhough, Beth. 2016. "Vitalist geographies: Life and the more-than-human." In *Taking-Place: Non-Representational Theories and Geography*, edited by Anderson Ben, and Paul Harrison, 51–68. London: Routledge.

Hall, Rebecca. 2012. "Diamond mining in Canada's Northwest territories: A colonial continuity." *Antipode* 45 (2): 376–393. https://doi.org/10.1111/j.1467-8330.2012.01012.

Hernandez, Daniel Sofa, and Peter Newell. 2022. "Oro blanco: Assembling extractivism in the lithium triangle." *The Journal of Peasant Studies* 49 (5): 945–968. https://doi.org/10.1080/03066150.2022.2080061.

Junka-Aikio, Laura, and Catalina Cortes-Severino. 2017. "Cultural studies of extraction." *Cultural Studies* 3 (2–3): 175–184. https://doi.org/10.1080/09502386.2017.1303397.

Kothari, Ashish, Ariel Salleh, Arturo Escobar, Federico Demaria, and Alberto Acosta. 2019, "Crisis as opportunity: Finding pluriversal paths." In *Postdevelopment in Practice*, edited by Klein Elise, and Carlos Eduardo Morreo, 100–116. London: Routledge.

Laplante, J. P., and Catherine Nolin. 2012. "Consultas and socially responsible investing in Guatemala: A case study examining Maya perspectives on the indigenous right to free, prior, and informed consent." *Society & Natural Resources* 27 (3): 231–248. https://doi.org/10.1080/08941920.2013.861554.

Leach, Melissa, and James Fairhead. 2022. "Fashioned forest pasts, occluded histories? International environmental analysis in west African locales." *Development and Change* 31: 35–59. https://doi.org/10.1111/1467-7660.00146.

Leifsen, Esben, Maria-Therese Gustafsson, Maria A. Guzmán-Gallegos, and Almut Schilling-Vacaflor. 2017. "New mechanisms of participation in extractive governance: Between technologies of governance and resistance work." *Third World Quarterly* 38 (5): 1043–1057. https://doi.org/10.1080/01436597.2017.1302329.

Lund, Christian. 2011. "Fragmented sovereignty: Land reform and dispossession in Laos." *Journal of Peasant Studies* 38 (4): 885–905. https://doi.org/10.1080/03066150.2011.607709.

McNeish, John-Andrew. 2017. "A vote to derail extraction: Popular consultation and resource sovereignty in Tolima, Colombia." *Third World Quarterly* 38 (5): 1128–1145. https://doi.org/10.1080/01436597.2017.1283980.

Murrey, Amber. 2016. "Slow dissent and the emotional geographies of resistance." *Singapore Journal of Tropical Geography* 37 (2): 224–248. https://doi.org/10.1111/sjtg.12147.

Mignolo, Walter. 2011. *The Darker Side of Western Modernity: Global Futures, Decolonial Options*. Durham: Duke University Press

O'Faircheallaigh, Ciaran, and Tony Corbett. 2005. "Indigenous participation in environmental management of mining projects: The role of negotiated agreements." *Environmental Politics* 14 (5): 629–647. https://doi: 10.1080/09644010500257912.

Peterson St-Laurent, Guillaume, and Philippe Le Billon. 2015. "Staking claims and shaking hands: Impact and benefit agreements as a technology of government in the mining sector." *The Extractive Industries and Society* 2 (3): 590–602. https://doi.org/10.1016/j.exis.2015.06.001.

Postar, Stephanie, and Negar Elodie Behzadi. 2022 (in press). "'Extractive bodies': A feminist counter-topography of two extractive landscapes." *Geoforum*. https://doi.org/10.1016/j.geoforum.2022.09.005.

Riofrancos, Thea. 2020. *Resource Radicals: From Petro-Nationalism to Post-Extractivism in Ecuador*. Durham and London: Duke University Press.

Salom, Andreas T., and Sonja Kivinen. 2019. "Closed and abandoned mines in Namibia: A critical review of environmental impacts and constraints to rehabilitation." *South African Geographical Journal* 102 (3): 389–405. https://doi.org/10.1080/03736245.2019.1698450.

Sarmiento, Fausto, and Sarah Hitchner, eds. 2017. *Indigeneity and the Sacred: Indigenous Revival and the Conservation of Sacred Natural Sites in the Americas*. Oxford and New York: Berghahn Books.

Schulz, Karsten A. 2017. "Decolonizing political ecology: Ontology, technology and 'critical' enchantment." *Journal of Political Ecology* 24 (1): 125–143. https://doi.org/10.2458/v24i1.20789.

Temper, Leah, Daniela Del Bene, and Joan Martinez-Alier. 2015. "Mapping the frontiers and front lines of global environmental justice: The EJAtlas." *Journal of Political Ecology* 22 (1): 255–278. https://doi.org/10.2458/v22i1.21108.

Veltmeyer, Henry. 2016. "Extractive capital, the state and the resistance in Latin America." *Sociology and Anthropology* 4 (8): 774–784. https://doi.org/10.13189/sa.2016.040812.

Walter, Mariana, and Leire Urkidi. 2017. "Community mining consultations in Latin America (2002–2012): The contested emergence of a hybrid institution for participation." *Geoforum* 84: 265–279. https://doi.org/10.1016/j.geoforum.2015.09.007.

Wong, Marcela Torres. 2018. *Natural Resources, Extraction and Indigenous Rights in Latin America: Exploring the Boundaries of Environmental and State-Corporate Crime in Bolivia, Peru, and Mexico*. Abingdon and New York: Routledge.

Wooden, Amanda. 2014. "Kyrgyzstan's dark ages: Framing and the 2010 hydroelectric revolution." *Central Asian Survey* 33 (4): 463–481. https://doi.org/10.1080/02634937.2014.989755.

Zabbey, Nenibarini, Kabari Sam, and Adaugo Trinitas Onyebuchi. 2017. "Remediation of contaminated lands in the Niger Delta, Nigeria: Prospects and Challenges." *Science of the Total Environment* 586: 952–965. https://doi.org/10.1016/j.scitotenv.2017.02.075.

Zaremberg, Gisela, and Marcela Torres-Wong. 2018. "Participation on the edge: Prior consultation and extractivism in Latin America." *Journal of Politics in Latin America* 10 (3): 29–58. https://doi.org/10.1177/1866802X1801000302.

1

Gold as a Frictious Fiction

Managing Gold Extraction Through Technocratic Discourse in Postsocialist Kyrgyzstan

Asel Doolotkeldieva

Resistance to extraction in Kyrgyzstan exists only in places with no previous history of mining. In the former Soviet mining towns, all is good, people do not protest, they understand the necessity of extraction.

—CEO of a Bishkek-based company that provides scientific and technical assessments of mining projects

All these claims about environmental violations are just a sort of speculation, an attempt to get a compensation and attract attention. Conflicts are created by local mafia.

—Local expert affiliated to the metallurgical trade union

The core problem is a Soviet behavior of social parasitism, still prevalent among locals, whereby people still think that the state should deliver for them. Presently, they expect the same from investors.

—CSR officer of a foreign mining company, operating in southern Kyrgyzstan

For people in extractive areas, the notions of exclusion and inclusion have diverse and sometimes conflicted meanings (Bridge 2009). These notions reflect interests of multiple groups and shape resource struggles differently around environmental justice (Martínez-Alier 1997), in defense of place (Escobar 2008) or against dispossession (Andrews 2018; Lesutis 2019), among others.[1] Yet, the quotes from representatives of mining companies, independent consultants, and government experts that start this chapter reference master frames by which diverse local struggles get reduced to a limited number of culturally recognizable themes. Since Kyrgyzstan's independence from the Soviet Union in 1991, the leadership of this gold-rich Central Asian state has defined gold mining as the driving force of economic growth (Bogdetsky et al. 2001). However, despite years of investments in mining, the country's economy has remained poor, and local populations have often felt excluded from the benefits of these resources (Pomfret 2018). These exclusions have led to multiple anti-mining protests and at times violent conflicts over the past decades (Wooden 2013, 2017, 2018). The technocratic response, articulated by the ruling government, mining corporations, independent experts and relayed by mass media, frames specific perspectives on these conflicts. This chapter proposes investigating the epistemic exclusions produced by expert discourses and their material workings (see also Murrey and Mollett in this collection).

Underlying this research objective is the understanding of exclusionary power inspired by Foucault's concept of governmentality (1991) and its expansion by resource scholars to notions of "resource governmentality" (Watts 2004). Power, as shown by Foucault, goes beyond state sovereign power and coercive power and includes the art of governing that shapes people's lives and behaviors, sometimes in subtly violent ways (1979). Technocratic discourses are one of the "techniques and forms of knowledge" that seek "to shape conduct by working through the desires, aspirations, interests and beliefs of various actors" (Dean 2010, 18). In resource-extractive contexts, discourses produced by organizations, in and outside the state, shape certain truths and knowledge about the industry, actors, territories, and struggles. These discourses have practical effects and participate in controlling and regulating places, people, and their livelihoods.

This chapter argues that the technocratic discourses articulated by governmental and corporate specialists and reproduced by mass media proceed by sorting out and discarding the grievances of local communities and environmentalists that are deemed illegitimate for neoliberal governmentality. The technocratic discourses work on public perceptions to represent anti-mining resistance as an irrational and criminal conduct, and in this way strip them from the larger societal support and clear the ground for the advance of corporate extractivism. By mapping illegitimate behavior in resource areas, discourses produce a possibility for physical violence by way of governmental intervention on the ground. However, beyond these immediate effects, they provide society with a framework of meaning-making and language about linkages between extraction and development, extraction and politics, and extraction and identity in the postsocialist context where not long ago extractivism was once put into the service of modernization.

Within these linkages and consequent constructions of resource zones, I identify three dominant discourses which demonstrate the ways the government came to represent the grievances of anti-mining protesters. In a snapshot, the narratives root the causes of anti-mining resistance in (i) criminalization of society in the post-independence era, (ii) the socialist past interpreted in its double manifestation of strong mining labor, and yet social parasitism, (iii) and finally "traditionalism" as a backward alternative to extraction. Far from intending to draw a homogenous picture of institutions governing natural resource management in Kyrgyzstan, my focus on technocratic discourses reveals a governmentality logic that prioritizes particular descriptions of mining conflicts. By doing so, it excludes other voices shaping what I refer to as epistemic forms of exclusions. The narratives identified above have considerable effects as they organize knowledge and practices of extraction in space and time as I will touch upon in the conclusion of this chapter.

To explore the emergence of these epistemic exclusions, I draw on ethnographic material collected over the past nine years at five mining projects across the country: Bozumchak, Jerooy, Taldy Bulak Levoberejny (Orlovka), Chaarat Gold, Taldy Bulak (Aral). During this research, I undertook semi-structured interviews with different actors, observed local events such as community gatherings and protests, CSR activities, and government actions. To investigate the technocratic discourses, I have identified and interviewed individuals inside and outside of the government who exert influence on decision-making and public opinion regarding gold extraction. The interviews were further complemented by the study of official documents and mass media analysis.[2]

The chapter is structured as follows: I first discuss the political economy of gold in Kyrgyzstan, imagined as the "messy ground," to show the emergence of technocratic responses to it. In the following section, I give examples of technocratic discourses about gold conflicts in their embeddedness in historical, cultural, and geopolitical interpretations. In the conclusion, I reflect on the power and effects of this discursive management of gold extraction.

EMERGENCE OF THE "MESSY GROUND": GOLD AS A FRICTIOUS FICTION

Since the early years of independence from the Soviet Union, the Kyrgyz leadership, under the advice of international financial organizations, has defined gold mining as the driving force of economic growth (Abazov 1999; Ministry of Economy 2014). Gold became a symbol of independent national development: in 2018, Kumtor, the largest gold mine, accounted for nearly 10 percent of the country's GDP (IGF Mining Policy Framework Assessment 2018, 1) and 63 percent of exports together with other minerals and metals

(World Bank 2020, 13). However, gold and "inclusive development" do not equate in Kyrgyzstan, despite three decades of economic liberalization and commodity-based exports. Poverty has declined but remained as high as 31 percent in 2018 (World Bank 2020, 19), and while per-capita GDP is constant, GDP has never caught up with pre-independence levels. Contributing to less than 3 percent of employment, failing to generate spillovers, and exported directly away, extraction of raw materials became "an enclave, an economy within the economy" (World Bank 2020, 13).

While successive governments sought to present mineral extraction as a way out of economic stagnation, residents in mining areas stress the "fictitious" character of gold when they say they do not see any improvement in their life conditions (ethnographic interviews in Talas and Chatkal areas). Yet, despite its fictitious character, gold has been productive in friction,to borrow from Anna Tsing's account of resource extraction in Indonesia (2005). Such perceptions are related to state kleptocracy, growing criminalization, geopolitical encroachment, and political violence under authoritarianism in resource zones. I start my analysis of this messy ground by shortly tracing back successive stages of resource governance and the formation of protracted grievances among local people. This background information will help to better understand the origins of dominant discourses and why they became successful in shaping collective imaginaries about mining conflicts.

The rule of Kyrgyzstan's first president, Askar Akaev (1991–2005), was associated with rapid economic and political liberalization bestowing upon Kyrgyzstan a reputation of the "island of democracy" amid its Central Asian authoritarian neighbors (Anderson 1999). With respect to the extractive industry, according to governmental agencies and independent experts involved, the 1990s were marked by entrepreneurial spirit and wild business. Licenses were obtained in semi-legal ways, territories overrun by oligarchs from all over the post-Soviet and Turkic world, while machinations over land ownership and murky deals flourished. Land speculators, small contractors lobbying investors' interests, would-be environmentalists, and other fortune seekers proliferated. Many Soviet-trained geologists who had converted into corporate specialists during this period recall the 1990s and early 2000s as an era of wild possibilities, far away from central regulation.[3]

However, after Akaev's overthrow in a popular uprising of 2005 and the establishment of President Bakiev's dictatorial rule, mines came under tighter regime control. Some of them were subjected to illegal expropriations involving members of the regime and dubious offshore companies (Doolotkeldieva and Heathershaw 2015). The police and special security forces suppressed local anti-mining protests, allowing the regime's cronies to explore and develop mines at their discretion. Local protesters tell stories of how miners circumvented locals by concluding direct deals with the regime, bribing local elites, and suppressing dissent.[4]

In 2010, in the aftermath of the second popular uprising against President Bakiev, anti-mining conflicts sprung up in every corner of the country, putting in question the previous authoritarian-capitalist contracts (Gullette 2013; NISI 2013).[5] Using a short window of opportunity, local people (sometimes in collaboration with elites, mafia, or rival corporate actors) attacked geological camps, trying to take over ownership and/or negotiate the terms of the conduct of foreign investors. In the recent third popular uprising of October 2020, at least eight mining entities were subjected to attacks by local people.[6] A few cases (e.g., in Chatkal Valley) also exist where local people sought to safeguard the incumbent miners and successfully did so. The reordering of mineral contracts has systematically accompanied anti-regime uprisings, demonstrating a direct relationship between politics and resources.

Multiple grievances and anti-mining claims have emerged, presenting a cacophony of local views, concerns, and interests in relation to mining. This mosaic is further complicated in relation to different life cycles of mining projects. For example, in relation to the biggest and long-serving mine, Kumtor, anti-mining claims have gradually progressed from demands for compensation, to better mining regulation, criticism of CSR, and finally requests for nationalization (Wooden 2017). In younger projects, still in the exploration phase or recently entering the production phase, the claims are hard to discern and categorize. In places I visited, most of the claims identified by Wooden (2017) at Kumtor were present simultaneously and sometimes in contradiction to each other. Some of the demands and concerns were imported from Kumtor to other mining areas due to the exchanges established across mines and the politicization of gold by opposition parties. As Kumtor served as the prime image of environmental degradation and unfair benefit distribution, it fed people's concerns about future risks across the country even where gold projects were in the study phase.

To sum up this section, the decades of corrupt and authoritarian resource governance have led to the building up of protracted concerns and ambiguous claims that are difficult to discern and categorize for people outside of mining places,and sometimes even for local people. The opacity of local conflicts was handy for technocratic actors whose aim was to mobilize wider societal support for extraction-based economy. The rise of anti-mining resistance in resource zones became then a major object of expert discussions in the country in the 2010s. The lack of unified resistance language in resource zones raised questions about the authenticity of grievances and concerns. Government officials, experts, and journalists questioned whether mining-related conflicts are genuine or power-oriented, that is, sponsored by competing political and economic actors. In conversations with government officials, elites, and corporate experts, it was a common thing to denote local activists as backward, mafiosi, parochial, and irrational. Despite relative political liberalization from 2010 and onwards, violence in mining areas continued in its discursive, but not less significant, form. Above, I provided a short overview of the political economy of gold and the emergence of a powerful technocratic response that accompanied the neoliberalization of natural resources. In the next section, I will detail the three dominant narratives designed to manage public expectations related to this important industry despite local criticism.

DISCURSIVE MANAGEMENT OF THE "MESSY GROUND": MAPPING NATURAL RESOURCES AGAINST HISTORICAL, CULTURAL, AND POLITICAL DYNAMICS

As the advance of corporate extraction met resistance, novel technocratic discourses followed as a technique of governmentality designed to demobilize local critics on the one hand, and to facilitate public expectations about resource-based development on the other hand. This section discusses three dominant narratives that emerged in response to recent anti-mining resistance. The narratives advance the following ideas about its providence: (i) criminalization of society in the post-independence era; (ii) the socialist past interpreted in its double manifestation of the strength of mining labor, and yet social parasitism; (iii) and finally "traditionalism" as a backward and parochial alternative to extraction. In the process of unpacking these narratives, I will analyze their origins, wider historical conditions, tensions with other truths, as well as the power relations that they enact. Although I treat the discourses separately, for the sake of analytical clarity, they often circulate simultaneously and at times in contradiction to each other.

Narrative 1: Criminalization of Society in the Post-independence Era

In the aftermath of the post-revolutionary liberalization of 2010, when the attacks on mining operations occurred massively across the country, a powerful technocratic discourse emerged to demonize and depoliticize this resistance. In this subsection, I show how technocratic actors attempted to explain anti-mining resistance as a result of criminalization of society. Employed ever since at every large anti-mining mobilization, the narrative advances a simple but effective idea that (1) anti-mining protests are staged and that (2) local grievances are therefore fake. There is a widespread belief among government officials, the business sector, and independent experts that mafia, elites, or competing mining firms are behind the protests.[7] The discourse invites the public not to trust local claims and look for the enigmatic "third forces" and their particularistic interests as the clue to conflicts. Since the narrative shifts the focus onto the criminalization of rural/mountainous society, it strips local actors of agency and makes exclusions produced by the industry invisible. Furthermore, as the attacks and protests are portrayed as sudden outbursts of criminality, rather than results of protracted tensions, the discourse helps to distance grievances away from mining operations and resource governance. This way, discontinuity is successfully erected between inequalities and exclusions produced by the industry and the episodes of local discontent. In sum, episodes of anti-mining contestations are represented as shallow and illegal acts by criminal and backward elements of society that act on the orders of superiors who themselves stay in the shadow. To enforce this picture, until recently, mass media outlets used to employ the images of young men on top of horses storming modern and expensive facilities of investors' factories. Paradoxically, the preceding history of corrupt and violent resource governance which criminalized and victimized local activists is not taken into account and no effort is made to connect the allegedly sudden insurgencies to this protracted history of exclusions. Instead, the narrative works to portray the investors as victims of opportunist local entrepreneurs who have otherwise a noble mission to bring industrialization, development, and progress in remote Kyrgyzstani Mountains. Consider the answer given by the then head of the State Agency for Promotion and Protection of Investment to my question on the scope of the agency's activities:

> We travel to the countryside and hold educative talks with the locals. Some of these local men who intimidate investors are mere criminals who lack basic education and understanding of industrial processes. Often my job is to put these local aborigines into their places and help the investors to find their way into communities.[8]

Since the anti-mining resistance is explained via the criminalization prism, the discourse normalizes state coercion against local activists. The latter indeed told the author that they were subjected to surveillance and intimidation by the special security forces.[9] Consider the speech of the then-President Almazbek Atambaev addressed to the residents of Talas province after a series of anti-mining demonstrations against the Jeroy project located there:

> In Jerooy, people go against the development of the mine. But why can we develop Kumtor and Bozumchak, and not Jeroy? These people distinguish themselves from the rest of the Kyrgyz nation and put themselves above it. We need to show them their place because we have been babysitting them for five

> years. It is time to show force. If they do not understand nice words, the state should establish discipline in the country by using force.[10]

Thus, Talas province was articulated as an ungovernable place with permanent strongholds of local mafia, with Jeroy, Andash, and Aral deposits enduring this construed reputation since 2010. By looking only at the immediate acts of attacks and lootings at the mines and discarding the historicity of local opposition to mines preceding these events, the technocratic narrative offers a distorted story of the political economy of extraction. Another consequence of the narrative is that environmental claims lose legitimacy. External observers who watch this supposedly backward and criminal behavior at remote mining zones will have no scruples about discarding the authenticity of local grievances and claims,[11] and any environmental collective action is doomed to failure amid such public distrust.

Narrative 2: The Socialist Past Interpreted in Its Double Manifestation of Strong Mining Labor, and Yet Social Parasitism

As several mining projects entered the production phase, local claims and repertoires of collective action have diversified. Labor-based demands and professional strikes have replaced the direct attacks on mines in some former Soviet industrial towns. During the period between 2013 and 2015, a second narrative emerged among experts and government officials to discursively shape the new type of contestations between foreign management and local workers. In this subsection, I demonstrate how this narrative came to offer an explanation for recent labor-based conflicts in connection to the past socialist modernization project. The narrative works to provide a distinction between labor-related conflicts from other forms of contention as a more acceptable behavior at the resource frontier.

Amid massive de-industrialization and de-professionalization of economic sectors following the collapse of the Soviet Union, labor-based conflicts enjoy a more modern symbolism compared to the gangster politics described above. Labor-based conflicts also look "simpler" from the business perspective. Unlike other mining projects in which protracted resistance is enmeshed in complex local structures of subsistence and environmental concerns, labor demands are "readable." From the governmentality standpoint, such conflicts are easier to manage by improving workers' material conditions. For example, the series of violent conflicts that took place during this period at Taldy Bulak Levoberejny mine, co-exploited by a Chinese company and the Kyrgyz government, were framed as labor-based. The nature of conflict between the Chinese management and the local workforce was framed as a struggle for higher wages, better working conditions, and a stronger quota for the Kyrgyz workers vs. Chinese workers. Consider the following opinion of the former state official who was part of the conflict regulation:

> Conflict at Altynken? No, there was none here really. What happened in the beginning was just an inadequate labor politics from the part of the Chinese management. Upon its correction, the conflict ceased to exist. Compared to other mines, we had no problems at all. We are an exemplary mine![12]

Indeed, the Taldy Bulak Levoberejny mine was framed by the business sector and the government as a successful project, where conflicts were resolved at an early stage and operations continued undisturbed.[13] However, by framing conflicts at Altynken narrowly in labor terms, the technocratic narrative silenced again other grievances and concerns such as economic justice, nationalism, and Sinophobia in Orlovka town and neighboring villages where the workers come from. As another low-ranked local official from Orlovka City told me informally, the conflict was allegedly heightened by resource nationalism and Sinophobia. From a governmental standpoint, allowing resource nationalism and Sinophobia to transpire in conflict structures would mean lower rankings in international business indexes.[14]

Valorization of labor also responds to other technocratic strategies of highlighting certain aspects of the socialist past because of its potentiality to offer a path forward to new corporate extractivism. In the past, the Soviet state externalized environmental violations as a necessary cost to pay for modernization (Kalinovsky 2018). Today, previous practices and knowledge of extraction in former mining towns such as Orlovka seem to serve as a bridge to new corporate extraction. Thus, by focusing on the Soviet mining towns, the narrative maps contemporary labor-based conflicts upon extractive territories through the prism of Soviet industrialization, thus drawing on a positive image of modernization.[15] As a CEO of a local company that provides technical expertise to foreign investors told me:

> Workers and residents in old mining towns know that there is no progress without environmental costs. They have accepted it as part of the process. They welcome new investing projects because they mean a renewed life in their towns. But locals in areas with no previous history of mining lack such knowledge and understanding of the industry, and they oppose to it.[16]

The narrative, of course, omits to mention environmental critiques and movements that also existed during the Soviet period (Penati 2019). To continue, however, with the delegitimizing effects of the technocratic discourse, this same Soviet past is simultaneously referenced to criticize residents in rural areas who allegedly develop strong expectations from private companies to provide welfare. According to some government officials and foreign investors, local people are indulged in social parasitism, *ijdivenchestvo* in Russian, an alleged product of the Soviet system. As the opinion of the then head of the Department of Mining Policy at the Ministry of Economy goes:

> The problem of our sector is *ijdivenchestvo* of local people. They got used that the state and state enterprises provide for them. Presently, they expect that private investors do the same for them. But we moved forward to market relations. If we succeed to teach people to take responsibility for themselves and not expect from mining companies, the industry will get a chance for development.[17]

As seen from the quote above, the neoliberal critique of socialist welfare operates by demonizing local people as social parasites and calling their collective expectations of the industry unrealistic, utopian, and irrational. It urges citizens to take control of their welfare, reproducing neoliberalism's central idea that governing starts with responsible citizens. This way community expectations toward mining projects are delegitimized and depoliticized as manifestations of the past ideology rendered obsolete in current market economy conditions. This is well demonstrated by the reflection of a middle-ranked official in the State Agency for Geology and Subsoil to my observation as to why the government ignores the needs and interests of local communities:

> Well, didn't we pass to the market economy relations? If you [the investor] chose to open a business, then it's your responsibility to regulate the relations with local communities. Why should the state meddle in this private business?[18]

Overall, although the narrative shows awareness and acknowledgment of labor-based exclusions in mining areas as the most legitimate, it ends up doing it at the expense of other voices.

Narrative 3: "Traditionalism" as a Backward and Parochial Alternative to Extraction

In this subsection, I will focus on the third narrative that emerged in response to environmental critiques voiced by local people whose subsistence and economic activities were put under question by new mining projects. For decades, as discussed above, the dominant technocratic discourse about criminalization of society reinforced public perceptions of environmental concerns as a smokescreen to cover shadowy economic interests. Both the government and the industry worked tirelessly to delegitimize concrete environmental claims in the past. However, with the rising numbers of mining projects shifting from study to production phase, environmental violations became more visible and harder to silence. Images of destroyed forests, protected areas, and glaciers from Kumtor, tales of polluted rivers and experiences of dust from passing trucks raise awareness of ecological transformations in mining zones.

To be in line with the government's own protracted struggle against the Canadian firm Centerra that exploits Kumtor in which the government recently moved to center its criticism on environmental violations, the new narrative emerged to reflect these changes. The narrative became more sensitive to ongoing ecological transformations and accepts local opposition to mining operations in selected places. Consider the following expert opinion that reflects local environmental concerns:

> In some mining areas, environmental violations are real. And local conflicts occur because of these wrongdoings. If I personally woke up one morning and did not see the mountain in front of my window anymore, I would also protest. The government should work on law enforcement and mining companies should respect ecological regulations if they want to continue working here.[19]

However, what the technocratic narrative conceals and silences is that alternative views of economic development do exist and local environmental concerns have strong variations across the country. Local people do not prioritize extraction rents as the primary source of development but rather seek to diversify economic activities (Tiainen et al. 2014) or preserve mines for a better future (interviews with activists in Talas province). The proponents of environmental agendas are local businessmen, farmers, and herders who are concerned with the transformations and destructions of their ecosystems that serve as traditional models of subsistence, that is, farming and livestock. However, local environmental claims were packed with other claims such as more benefits, more compensations, and better CSR programs (Gullette and Kalybekova 2014) and thus difficult to discern as existing in their own right. But environmental claims have always been part of anti-mining struggles, especially those related to the country's largest mine Kumtor (Wooden 2013; Buxton 2011). Environmental concerns have also been strong in other projects: for example, a local mobilization against Andash project due to the mine's proximity to the village,[20] oppositions to the Jeroy's and Bozumchak's tailing dump plans and others.[21]

Although local NGOs help in politicizing environmental issues, they are slow in gaining prominence in public discussions. In the meantime, most of the expert opinions downgrade local livelihoods as economically unviable and backward. Consider this insight that was shared with me by the former head of the State Agency for Geology and Subsoil:

> We are behind the world economic development. There is no time to catch up and invent a new wheel. We just must get advantage of high prices for gold as far as the world market is favorable to us. But our farmers do not understand that

> they could continue their traditional farming and grazing while allowing miners to mine. These two models are perfectly compatible. So, we as the government should continue our information campaigns aimed at enlightening the locals.[22]

Although the technocratic discourse came to partially accept this environmentalism of the poor (Martínez-Alier 2003; Nixon 2011), it does not grant full legitimacy to proponents of alternative visions of development.

CONCLUSION

This chapter identified several technocratic discourses that emerged in response to local mobilizations against mining operations to shape public discussions about the role of the extractive industry in Kyrgyzstan's post–Soviet development. It is worth noting that these narratives do not embody the diversity of opinions within government agencies and the expert community. Other opinions exist that highlight problems of CSR programs (Doolotkeldieva 2020), environmental and human rights violations, the life-cycle of mines, and other factors that influence local perceptions about extraction. Also, other actors such as environmental NGOs, academic experts, and specialized public foundations resist mainstream technocratic discourses, but their voices are marginal.

One important outcome of the technocratic discourses is that they produce spatial and temporal dynamics of inclusion and exclusion in relation to extractive industry. Spatially, these discourses identify which places belong to the national project of development and which do not. By mapping mining-related conflicts through the industrial/urban/labor lens, they construct resource territories as supportive of the industry and thus owning a proud place in the country's quest for modernization. Where conflicts are categorized through the traditionalist/mountain/rural lens, the territory is identified as backward and reactionary. These discourses also construct specific perspectives on gold in time. They establish connections to the authoritarian and socialist history to exploit consequences of these periods for the advance of new corporate extractivism. By echoing and reproducing the Soviet discourse in Central Asia whereby conquest of nature and a necessary environmental transformation were framed for the sake of development of the USSR's "most backward" regions (Florin 2019), technocratic discourses tend to invisibilize recent violations and exclusions. Thus, the distinction between the new corporate resource development and old state resource development is instrumentalized in both ways: either to blur it or to erect an artificial discontinuity between them depending on the objectives of the momentum.

Technocratic discourses discussed above morph with time to reflect changes in anti-mining dynamics, government, and corporate policies. But some of them are "stickier." For example, the discursive production of resource territories and identities through the lens of criminalization and Soviet mentality creates an image of permanent economies of parasitism: Naryn province was recently shaped as another ungovernable space where old practices of artisanal and state-owned extraction clash with the recent penetration of Chinese and other foreign companies (Mestre 2017; Botoeva 2024). Yet, the mapping is *per se* an exercise of institutional power and exclusion whereby some grievances and concerns related to extraction are muted and discarded, allowing for only simplistic readings of the otherwise complex realities on the ground.

In fact, local anti-mining activists do emphasize the role of mafia, the socialist past, and traditional livelihoods. But they make use of these notions in different directions. For them, postsocialism and authoritarian legacy become productive categories to make sense of the changing nature of post-independence governance. Privatization of natural resources under authoritarian corrupt governments makes them experience a new relationship with the neoliberal state, the one that "outsources" its functions to private foreign actors. These differing meanings and values attached to gold mining underlie multiple ontologies of environmental conflicts (Blaser 2013), and the technocratic discourses render this politics of resistance based on distrust toward the state invisible.

NOTES

1. They are also often struggles over livelihoods (Bebbington et al. 2008), of valuation (Li 2015), against mining capitalism (Kirsch 2014), and resource nationalism (Koch and Perreault 2019).

2. At the time of research, the governmental block included officials of the Ministry of Economy, the State Agency for Promotion and Protection of Investment, the State Agency for Geology and Subsoil, the State Agency for Environment, and the State Inspection for Ecological and Technical Protection. The business block involved representatives of the International Business Council, CSR specialists, local consultancy companies, and specialized law firms. Finally, I have also interviewed members of various specialized professional associations such as the Metallurgical Trade Union, The Association of Miners and Geologists, and geologists.

3. Interviews with Soviet geologists who recycled their knowledge as new CSR officers in private mining companies.

4. Interviews with residents of Talas province, host of several gold projects.

5. Anti-mining protests also occurred in 1990s and 2000s (Buxton 2011), but their dramatic cross-regional upsurge began in 2010 when more projects entered a productive phase and the second popular uprising put an end to authoritarianism.

6. "8 gornodobyvaushih predpriatiy Kyrgyzstana prostaivaut uje mesac. Iz-za mestnih jitelei" [In Kyrgyzstan, 8 mining companies have been halted. Because of

local residents], 2 November 2020. Kaktus Media: https://kaktus.media/doc/425045_8_gornodobyvaushih_predpriiatiy_kyrgyzstana_prostaivaut_yje_mesiac._iz_za_mestnyh_jiteley.html

7. To be fair, some mining companies develop more nuanced understanding of the local political economy of gold. For example, the assumption that is widely shared among mining entities holds that the real instigators are actors with petty interests in local contracts and landownership. According to this belief, local actors with vested interest in the local economy of resource extraction are capable of mobilizing entire communities at their service against the miners. While these evaluations testify to more interest in complex structures of economic investments, greed, corruption, parochialism, and mercantilism are put again forward as the reasons of local discontent (interviews with CSR officers of mining companies in Chatkal valley and Talas province). Such views are typical of many other mining firms around the world, see Welker's discussion of Newmont company in Indonesia (2014).

8. Interview with the then head of the State Agency for Investment Protection and Promotion. The word aborigine was used by the interlocutor in its Russian version—*aborigen*.

9. Interviews with residents in Talas, Chui and Jalal-Abad provinces.

10. "Atambaev: The state should apply force to people who are not interested in the rule of law," 21 July 2013, https://kabarlar.org/news/54081-atambaev-k-lyudyam-kotorye-ne-zainteresovany-v-zakonnosti-gosudarstvo-dolzhno-primenit-silu.html, accessed 5 September 16.

11. Consider for example the following mass media coverage: "Ecologicheskoe prikritie zolotoi lihoradki na Jerue" [Ecological cover-up of the golden rush at Jeroy], 27 December 2017. http://www.msn.kg/showwin.php?type=newsportal&id=18709 or "De-Факто", № 14, 31 мая 2007 г.

12. Interview with a former state official in Kemin district, Chui province.

13. International Business Council. 2019. "Altynken experience is useful for industry, local authorities and communities." *IBC*, 29 August 2019. http://www.ibc.kg/en/news/ibc/4550_altynken_experience_is_useful_for_industry_local_authorities_and_communities

14. Interview with a local official in Orlovka city administration.

15. For more information about Soviet extraction-based modernization project in Central Asia see Wooden (2018), Khazanov (1995), Liu (2018), Kalinovsky and Kamp (2018).

16. Interview with a CEO of a Bishkek-based company providing technical expertise to foreign investors.

17. Interview with a former head of the Department of Mining Policy at the Ministry of Economy. On the role of state enterprises in Soviet industry and modernization, see Collier (2001), Rogers (2015).

18. Interview with a middle-ranked official in the State Agency for Geology and Subsoil.

19. Interview with an environmental expert at a Bishkek-based NGO.

20. "Chleny Pravitelstvennoi komissii po rassmotreniu situacii na mestorojdenii Andash vstretilis s talasscami" [Members of government commission met with Talas

residents to discuss the situation around Andash deposit], 2 February 2007. https://www.for.kg/news-24162-ru.html

21. "Kak talasscy mitingovali protiv razrabotki Djeruya" [How Talas residents protested against the exploitation of Jeroy, 24 July 2015. https://kloop.kg/blog/2015/07/24/video-kak-talastsy-mitingovali-protiv-razrabotki-dzheruya/

22. Interview with the former head of the State Agency for Geology and Subsoil.

REFERENCES

Abazov, Rafis. 1999. "Policy of economic transition in Kyrgyzstan." *Central Asian Survey* 18(2): 197–223. http://doi.org/10.1080/02634939995687.

Anderson, John. 1999. *Kyrgyzstan: Central Asia's Island of Democracy?* Abingdon, UK: OPA.

Andrews, Nathan. 2018. "Land versus livelihoods: Community perspectives on dispossession and marginalization in Ghana's mining sector." *Resource Policy* 58: 240–249. https://doi.org/10.1016/j.resourpol.2018.05.011.

Bebbington, Anthony, Denis Humphreys Bebbington, Jeffrey Bury, Jeannet Lingan, Juan Pablo Muñoz, and Martin Scurrahal. 2008. "Mining and social movements: Struggles over livelihood and rural territorial development in the Andes." *World Development* 36(12): 2888–2905. https://doi.org/10.1016/j.worlddev.2007.11.016.

Blaser, Mario. 2013. "Notes towards a political ontology of 'environmental' conflicts." In *Contested Ecologies: Dialogues in the South on Nature and Knowledge*, edited by Lesley Green, 13–27. Cape Town: Human Sciences Research Council Press.

Bogdetsky, Valentin, Vitaliy Stavinskiy, Emil Shukurov, and Murat Suyunbaev, eds. 2001. *Mining Industry and Sustainable Development in Kyrgyzstan.* www.iied.org. MMSD Working paper N110.

Botoeva, Gulzat. 2024. "Community tensions and small-scale gold mining in Kyrgyzstan: Diverging livelihood strategies of local people." In *Central Asian World*, edited by Madeleine Reeves and Jeanne Féaux de la Croix, 499–514. London: Routledge.

Bridge, Gavin. 2009. "Material worlds: Natural resources, resource geography and the material economy." *Geography Compass* 3(3): 1217–1244. https://doi.org/10.1111/j.1749-8198.2009.00233.x.

Buxton, Charles. 2011. *The Struggle for Civil Society in Central Asia: Crisis and Transformation.* Boulder: Kumarian Press.

Collier, Stephen. 2001. *Post-Soviet Social: Neoliberalism, Social Modernity, Biopolitics.* Princeton: Princeton University Press.

Conde, Marta. 2017. "Resistance to mining: A review." *Ecological Economics* 132: 80–90. https://doi.org/10.1016/j.ecolecon.2016.08.025.

Conde, Marta, and Philippe Le Billon. 2017. "Why do some communities resist mining projects while others do not?" *The Extractive Industries and Society* 4(3): 681–697. https://doi.org/10.1016/j.exis.2017.04.009.

Dean, Mitchell. 2010. *Governmentality: Power and Rule in Modern Society*. 2nd edition. London: Sage.

Doolotkeldieva, Asel. 2020. *Regulating Corporate Social Responsibility (CSR) in the Large-Scale Mining Sector of Kyrgyzstan*. OSCE Academy Policy Brief #60, May 2020. OSCE Academy in Bishkek & Norwegian Institute of International Affairs.

Doolotkeldieva, Asel, and John Heathershaw. 2015. "State as resource, mediator and performer: Understanding the local and global politics of gold mining in Kyrgyzstan." *Central Asian Survey* 34(1): 93–110. https://doi.org/10.1080/02634937.2015.1010853.

Escobar, Arturo. 2008. *Territories of Difference. Place, Movements, Life, Redes*. Durham and London: Duke University Press.

Florin, Moritz. 2019. "Emptying lakes, filling up seas: Hydroelectric dams and the ambivalences of development in late Soviet Central Asia." *Central Asian Survey* 38(2): 237–254. https://doi.org/10.1080/02634937.2019.1584604.

Foucault, Michel. 1991. "Governmentality." In *The Foucault Effect: Studies in Governmentality*, edited by Graham Burchell, Colin Gordon, and Peter Miller, 87–104. Chicago: University of Chicago Press.

Foucault, Michel. 1979. "Power and norms." In *Power, Truth and Strategy*, edited by M. Morris and P. Patton, 59–67. Sydney: Feral.

GIZ. 2016. *Local Content Development in Mining. Opportunities and Challenges in Kyrgyzstan*. Bishkek: GIZ.

Gullette, David. 2013. *Conflict Sensitivity in the Mining Sector of The Kyrgyz Republic*. Bishkek: OSCE Academy.

Gullette, David, and Asel Kalybekova. 2014. *Agreement Under Pressure: Gold Mining and Protests in the Kyrgyz Republic*. Bishkek: FES. http://library.fes.de/pdf-files/id-moe/10927.pdf.

IISD. 2018. *IGF Mining Policy Framework Assessment: Kyrgyzstan*. Winnipeg: ISSD.

Kalinovsky, Artemy M. 2018. *Laboratory of Socialist Development: Cold War Politics and Decolonization in Soviet Tajikistan*. Ithaca and New York: Cornell University Press.

Kalinovsky, Artemy M., and Marianne Kamp. 2018. "From industrialization to extraction: Visions and practices of development in Soviet Central Asia: Introduction to the forum." *Ab Imperio* 2: 69–79. https://abimperio.net/cgibin/aishow.pl?state=showa&idart=4350&idlang=1&Code=.

Khazanov, Anatoly. 1995. *After the USSR*. Madison: University of Wisconsin Press.

Kirsch, Stuart. 2014. *Mining Capitalism: The Relationship Between Corporations and Their Critics*. Berkeley: University of California Press.

Koch, Natalie, and Tom Perreault. 2019. "Resource nationalism." *Progress in Human Geography* 43(4): 611–631. https://doi.org/10.1177/030913251878149.

Lesutis, Gediminas. 2019. "Spaces of extraction and suffering: Neoliberal enclave and dispossession in Tete, Mozambique." *Geoforum* 102: 116–125. https://doi.org/10.1016/j.geoforum.2019.04.002.

Li, Fabiana. 2015. *Unearthing Conflict: Corporate Mining, Activism, and Expertise in Peru*. Durham: Duke University Press.

Liu, Morgan Y. 2018. "Governance and accumulation around the Caspian: A new analytical approach to petroleum-fueled postsocialist development." *Ab Imperio* 2: 169–198. https://doi.org/10.1353/imp.2018.0032.

Martínez-Alier, Juan. 1997. "Environmental justice (local and global)." *Capitalism Nature Socialism* 8(1): 91–107. https://doi.org/10.1080/10455759709358725.

Martínez-Alier, Juan. 2003. *The Environmentalism of the Poor: A Study of Ecological Conflicts and Valuation*. Cheltenham: Edward Elgar Publishing.

Mestre, Irène. 2017. "Quand les bergers creusent la montagne. Impact des activités minières artisanales sur les systèmes agropastoraux du Kirghizstan. Étude de cas dans la région de Naryn." *Journal of Alpine Research* 105(1): 1–17. http://doi.org/10.4000/rga.3575.

Ministry of Economy. 2014. "Medium and long-term strategy of mining industry development of Kyrgyzstan." Unpublished draft.

NISI. 2013. "Factors of negative attitudes of local communities to mining investors, relationships between miners and local communities." National Institute of Strategic Studies. Bishkek.

Nixon, Rob. 2011. *Slow Violence and the Environmentalism of the Poor*. Cambridge: Harvard University Press.

Penati, Beatrice. 2019. "The environmental legacy of the Soviet regime." In *Kazakhstan and the Soviet Legacy: Between Continuity and Rupture*, edited by Jean-Francois Caron, 51–74. London: Palgrave Macmillan.

Pomfret, Richard. 2018. "Exploiting a natural resource in a poor country: The good, the bad and the ugly sides of the Kyrgyz Republic's gold mine." *IOS Working Papers* 372. Regensburg: Leibniz-Institut für Ost- und Südosteuropaforschung (IOS). http://nbn-resolving.de/urn:nbn:de:101:1-201806190858030680753O.

Rogers, Douglas. 2015. *The Depths of Russia: Oil, Power, and Culture after Socialism*. Ithaca and New York: Cornell University Press.

Tiainen, Heidi, Rauno Sairinen, and Viktor Novikov. 2014. "Mining in the Chatkal valley in Kyrgyzstan—Challenge of social sustainability." *Resources Policy* 39: 80–87. https://doi.org/10.1016/j.resourpol.2013.11.005.

Tsing, Anna Lowenhaupt. 2005. *Friction: An Anthropology of Global Connection*. Princeton: Princeton University Press.

Watts, Michael. 2004. "Resource curse? Governmentality, oil and power in the Niger Delta, Nigeria." *Geopolitics* 9(1): 50–80. https://doi.org/10.1080/14650040412331307832.

Welker, Marina. 2014. *Enacting the Corporation. An American Mining Firm in Post-Authoritarian Indonesia*. Berkeley: University of California Press.

Wooden, Amanda E. 2013. "Another way of saying enough: Environmental concern and popular mobilization in Kyrgyzstan." *Post-Soviet Affairs* 29(4): 314–353. https://doi.org/10.1080/1060586X.2013.797165.

Wooden, Amanda E. 2017. "Images of harm, imagining justice: Gold mining contestation in Kyrgyzstan." In *ExtrACTION: Impacts, Engagements, and Alternative Futures*, edited by Jalbert Kirk, Anna Willow, David Casagrande, and Stephanie Paladino, 169-183. London: Routledge.

Wooden, Amanda E. 2018. "'Much wealth is hidden in her Bosom': Echoes of Soviet development in gold extraction and resistance in Kyrgyzstan." *Ab Imperio* 2: 145–168. http://doi.org/10.1353/imp.2018.0031.
World Bank. 2020. "Kyrgyz Republic: Country economic memorandum—Enhancing the prospects for grown and trade."

2

A Political Ecology of Environmental Law Enforcement

Civil Complaints and Environmental Justice in Post-neoliberal Ecuador

Teresa Bornschlegl

In 2007, Ecuador's newly elected government administration announced the end of the "long neoliberal night" (Correa quoted in "La Hora" 2006). This new government intended to explicitly put an end to the Washington Consensus's program—and the socio-environmental injustices that it brought to the country—through the formation of a post-neoliberal, state-led development model. In 2008, the National Assembly approved a new Constitution aiming to create "a new form of civic coexistence, in diversity and harmony with nature, to achieve *Buen Vivir*" and "a society that respects, in all its dimensions, the dignity of people and collective communities"[1] (Constitution of Ecuador 2008; preamble). Incorporating provisions such as the Rights of Nature and Indigenous Rights, this Constitution came to be counted as one of the most progressive constitutions worldwide (Lalander and Merimaa 2018).

Considerable parts of these progressive provisions resulted from a decade-long social mobilization by Indigenous, social, and environmental movements protesting the environmental degradation and social inequalities caused by the development model based on extractivism (Akchurin 2015; Lewis 2016; Bravo et al. 2017; Radcliffe 2019). Extractivism is a form of resource extraction that causes considerable social and environmental impacts, the costs of which are externalized. It is also export-oriented without adding value, thus generating enclave economies with only scarce benefits for domestic economies (including in terms of job creation) (Gudynas 2013; Hancock et al. 2018).

More than a decade after the approval of this new Constitution, however, the Ecuadorian government administration has still not taken the necessary steps toward a post-extractivist[2] transition, failing to fulfill the hopes of Indigenous, social, and environmental movements that had supported its accession to power. On the contrary, resources were still extracted in an irresponsible way and the dependency on export-oriented resource extraction continued (Larrea 2017). Hence, the initial alliance between these movements and the government administration soon fell apart (Acosta 2016; Arsel et al. 2016). Despite these failures, the progressive constitutional provisions remained, raising a series of questions about their role, efficacy, and outcomes. To what extent did these provisions create leverage for the populations that have historically been excluded in/from the governance of resource extraction (despite being the most negatively affected by it)? Did the progressive constitutional provisions ultimately open pathways toward a more inclusionary resource governance?

This chapter examines one of the multiple institutional mechanisms that are supposed to facilitate such leverage: civil complaints—a standard legal instrument integrated with the enforcement of environmental laws, intending to empower communities adjacent to industrial activities to defend themselves against potential environmental contamination. In Ecuador, any person can submit a complaint (*denuncia*) to the Ministry of Environment if they suspect an environmental violation committed by another person or industry. Subsequently, the Ministry of Environment must investigate the denounced incident. If the environmental violation is corroborated upon inspection, the Ministry of Environment sanctions the entity found in violation, who must also "integrally repair" the environmental and social damage. The concept of "integral reparations" explicitly derives from the progressive constitutional provisions and points to the obligation to restore the rights and dignity of impacted communities. Civil complaints are thus designed to provide communities with some form of environmental justice, raising several questions: first, on the extent to which they result in outcomes that restore environmental and social rights of impacted communities; second, on factors limiting civil complaints from achieving such outcomes; and finally, on whether this legal mechanism can lead to more inclusionary forms of resource governance.

In this chapter, I answer these questions in light of critical state theories that have conceived of the state as a continuously contested terrain in which different projects backed by opposing social forces struggle to gain dominance (Offe and Keane 1984; Jessop 1990). Critical state theories allow the conceptualization of the weakness of environmental law enforcement as the outcome of a constant struggle between extractivist and post-extractivist forces, focusing, in this case, on how weak law enforcement came into being rather than taking it as a given. Methodologically, this chapter draws on data collected during fieldwork in Ecuador from 2013 to 2018, including official reports and a ministerial database related to civil complaints, participant observation as an intern at a provincial office of the Ministry of Environment, and 104 semi-structured interviews. Interviews were conducted with representatives at, for instance, the Ministry of Environment, the provincial governments, the Ombudsman Office, local communities, oil companies, laboratories, environmental consultancies, social and Indigenous organizations, and NGOs in Ecuador. Fieldwork took place at the national level as well as the subnational level focusing on three oil provinces: Sucumbíos, Orellana, and Pastaza.

Scholars who have examined the post-neoliberal turn in Ecuador argue that the governance of extractive industries has mainly remained the same over time, despite the constitutional changes (Castro et al. 2016; North et al. 2016; Villalba-Eguiluz and Etxano 2017; Veda-Almeida 2018). Oil extraction, they argue, continues to be governed by the practices of the oil industry that developed in the early days of Ecuador's history as an oil producer. Accordingly, those who are the most affected by the negative environmental impacts of oil extraction remain excluded from its governance (Lu et al. 2017). Following this argument, one would expect that the Ministry of Environment most likely discards civil complaints, that is, however, not the case. Civil complaints are not disregarded but do lead to an environmental inspection and corresponding enforcement procedures. However, several institutional barriers prevent those enforcement procedures from achieving an outcome that would satisfy communities in the sense of an integral reparation. This chapter argues that, notwithstanding their shortcomings, the progressive constitutional provisions do matter, as they establish legal rules and thus impose constraints on power. As Thompson (1975) put it: "There is a difference between arbitrary power and the rule of law" (266). At the beginning of oil extraction in Ecuador, the industry could operate as "arbitrary power"—without being constrained by any law that would have allowed communities to claim the protection of their environmental and social rights. The constitutional provisions provide such legal basis to communities, thus enabling "the defense of the citizen from power's all-intrusive claims" (Thompson 1975, 266).

CONCEPTUALIZING A POLITICAL ECOLOGY OF ENVIRONMENTAL LAW ENFORCEMENT

Political ecologists and critical geographers hold that the political economy of extractive capitalism relies on the externalization of social and environmental costs to ensure capital accumulation and the delivery of cheap energy into the circuits of the global market (Watts 2005; Huber 2013). Such political economy generates environmental degradation and social inequalities for already marginalized populations, shaping geographies of environmental injustice. Within these debates, the extent to which environmental law enforcement can lead to environmental justice remains a bone of contention (Pulido et al. 2016). On the one hand, scholars regard environmental laws as a political device to green-wash and legitimize capitalism. According to these authors, the idea that "better regulation" can make extraction without environmental degradation possible might elude more political questions of whether extraction is needed in the first place, and if yes, by whom, for whom, how much, and why. (M'Gonigle and Takeda 2013; Stretesky et al. 2013; Lynch et al. 2018). Within these interpretations, environmental law enforcement can hardly benefit local communities and would stall environmental justice claims. On the other hand, scholars argue that environmental laws and enforcement can throw a spanner in the works of capitalism by hindering the externalization of environmental costs (Gudynas 2013; Boyd 2017). Corporations' and governments' frequent attempts to roll back and/or make environmental legislation more flexible evidence the power of such laws in unsettling the operation of extractive capitalism (Gudynas 2015). Environmental laws can delay the issuing of permits or impose a moratorium on resource extraction altogether; these laws can also threaten the industry with sanctions, costly litigation, and facility closure (Broad and Cavanagh 2015; Aldana and Abate 2016; Kauffman and Martin 2016).

Critical state theories, as I show in the analysis that follows in Ecuador, bring nuance to this debate, clarifying the extent to which environmental law enforcement can deliver environmental justice. According to these theories, the state is not a monolithic entity but fragmented into contradictory state projects that often serve directly opposed social forces located outside "the state" (Jessop 1990; Painter 2006). The contradictions inherent in capitalism are also internal to "the state." Accordingly, the state must not only guarantee capital accumulation by facilitating the exploitation of labor and nature but must also discipline such exploitation to first prevent the total depletion of the productive forces that sustain capital accumulation, and second to avoid mass protests (Offe and Keane 1984; O'Connor 1973, 1998). In a post-neoliberal state such as Ecuador, this struggle takes place between extractivist and post-extractivist forces (Larrea 2017; Andreucci and Radhuber 2017; Bornschlegl 2018). On the one hand, the state must facilitate extractivism, which means allowing resources to be extracted in ways that disregard environmental and social standards and guaranteeing the constant expansion of its extractive frontier (Veltmeyer 2016; Ray and Chimienti 2017). On the other hand, the Ecuadorian state must also tame such extractivism to retain some legitimacy as a "green vanguard" by, for instance, following through with some of the constitutional provisions for which Indigenous, social, and environmental movements have been pushing.

Scholars working on Ecuador have argued that those constitutional provisions did not substantially change the environmental governance of oil extraction, demonstrating the continued dominance of extractivist forces (Veltmeyer and Petras 2014; Shade 2015; Castro et al. 2016; Lu et al. 2017). Accordingly, environmental laws would still fail to address environmental injustices. However, a critical state theory perspective points to the importance of not taking such outcomes for granted, but to consider them as a result that arises from constant struggles inside the state. As Jessop notes: "Whether, how and to what extent one can talk in definite terms about the state actually depends on the contingent and provisional outcome of struggles to realize more or less specific 'state projects'" (1990, 9). The question becomes how these struggles inside the state are taking place and why a particular outcome is generated. If the extractivist state project "wins" against the post-extractivist state project, and if environmental law enforcement does not work to the benefit of the people who are affected by environmental degradation, one needs to understand who produced that result, how, and why.

CIVIL COMPLAINTS AND THE LIMITS TO ENVIRONMENTAL LAW ENFORCEMENT

Environmental Governance of Extraction: Past and Present

The analysis of civil complaints, put in perspective with a review of changes in environmental governance since the 2008 Constitution, illuminates the struggles between the post-extractivist and the extractivist state projects in Ecuador. In 1967, a foreign consortium led by Texaco Petroleum—an American subsidiary of Chevron Corporation—was the first company to discover oil in the Ecuadorian Amazon. A robust body of environmental laws was absent at the time, and the government failed to enforce the few available legal requirements addressing environmental protection (Kimerling 1991, 2013; Narváez and Narváez 2012). Texaco's operations within this legal vacuum left such severe ecological damage in the Amazon that areas of oil exploitation came to be commonly referred to as the "Rainforest Chernobyl."[3] In the 1990s, the government introduced the first actual environmental laws and regulations. Their enforcement, however, remained weak: the Ministry of Mines and Energy, rather than the Ministry of Environment, still held the authority to oversee the environmental matters of the hydrocarbon sector, thus creating a conflict of interests that tended to hinder environmental law enforcement. As Ecuador depended on revenues from oil incomes, environmental protection—even if it was recognized as an essential, policy issue—continued to be of secondary importance (Kimerling 2013).

The progressive environmental provisions in the 2008 Constitution attempted to undo these institutionalized power asymmetries, resulting in some significant adjustments in the environmental governance of extraction. The Ministry of Environment gained the authority to oversee the hydrocarbon sector, also receiving a larger budget than before. The responsibility for environmental damages does not expire over time, that is, without any statute of limitation. Furthermore, environmental damages needed to be "integrally repaired" (Constitution of Ecuador 2008, Art. 396 and 397, 178). These changes were essential for the mechanism of civil complaints, in that they increased the capacity of the Ministry of Environment to implement environmental laws. Even more importantly, these changes made it obligatory to restore environmental and social rights in case of an environmental violation; this went beyond the already existing regime of sanctions that encompassed a monetary fine, and, in the most severe cases, required the suspension or revocation of the environmental license. Enforcing integral reparations in cases of corroborated civil complaints can thus be considered part of the post-extractivist state project. However, Ecuador's economy continued to depend on oil revenues and export-oriented resource extraction persisted, if not intensified (Finer et al. 2015; Larrea 2017; Roy et al. 2018), generating a contradiction between post-extractivist and extractivist mandates inside of the Ecuadorian state.

Civil Complaints and Their Limits

In the face of this contradiction, civil complaints do not just collect dust on some state officials´ desks. Once a civil complaint is submitted, inspections follow, samples are taken, technical reports are written, and the Ministry of Environment issues orders to oil companies that violate environmental laws. Civil complaints, nevertheless, do not always lead to outcomes that satisfy affected communities. The issue, I show, is not that the Ministry of Environment does not undertake investigations following complaints, but rather that problems remain unresolved or that remedies involved merely cosmetic changes. In fact, all steps in the process of development, submission, and responses to civil complaints are fraught with "bad practices" (*malas prácticas*) which have persisted since the establishment of the oil industry, and which often lead to inconsequential investigations.

Following the process of submission and response to civil complaints around environmental contamination illustrates this point. The first step, before the submission of a civil complaint about environmental contamination, is the identification of an issue. With this regard, residents and members of communities play an important role through "community monitoring" (*vigilancia ciudadana*). Environmental inspectors from the Ministry of

Environment often call local communities their "eyes in the territory," showing the importance of this mechanism. If local residents suspect or observe an environmental breach, they can then submit a civil complaint.

Once a complaint is submitted, an inspector at the provincial office arranges for an inspection, a process that can take a few days. During the inspection, claimants and affected community members, at least one representative from the involved oil company, and one member of staff from the laboratory that the oil company contracted must be present. Inspections start with a hearing where all parties express their perspectives on the events and the possible responses. After the hearing, the inspector examines the site, and directs the laboratory regarding what samples of soil and/or water to take, how many, and where from based on claimants' knowledge of the terrain. Back at the office, the inspector writes a technical report that they submit to the national headquarters of the Ministry of Environment in Quito. The national headquarters then decides the required actions to follow, the completion of which is subsequently verified through another inspection conducted by the provincial office. If the oil company does not comply with the requirements, the entire procedure is repeated.

While the procedure is clear, the objectives are often unmet, leaving problems unresolved. Multiple barriers make the process unfruitful, starting with its length. Several years can pass between the submission of a complaint and possible responses—oil companies still seem to be able to simply ignore the orders the Ministry of Environment imposes on them. In 2014, for example, community members of Barrio Santa Marianita (Lago Agrio, Sucumbíos) submitted a civil complaint about bad smells, noise, and vibration allegedly caused by the installation of a well pad. Despite several inspections and requirements issued by the Ministry of Environment, the issues were still not resolved in August 2018. One inspection in May 2016 found, among other things, that "there are waste pits of produced/formation water [. . .] of a volume of approx. 8000bls. The area of the waste pits is approx. 6000 m2, and they are without the required cover" (Ministry of Environment 2016, 2). The technical report, therefore, recommended that the National Headquarters should oblige the operating company to "[i]mplement mitigation measures for the waste pit (putting a cover)" (Ministry of Environment 2016, 3). In August 2016, another report determined that the company never fully complied with the previous orders and required the company to mitigate the waste pit problem. In July 2017, three years after the original complaint and after continued pressure, the company indicated that they finally followed the order. "[A]s of June 1 of the current year [2017], the produced water is stored and treated, in a tank yard, implemented to operate with a closed system, to avoid any environmental contamination and the oxygenation of water" (Petroamazonas EP 2017, 1). In August 2018, interviewed community members stated that, in their view, the environmental problems continue and that they had still not received any information—neither on laboratory results nor on the company's responses.

The way laboratories carry out the analysis of water and/or soil samples constitutes another barrier to an efficient process. Laboratory analysis continues to follow the sectorial regulation ("The Ecuadorian Environmental Regulations of Hydrocarbon Operations," RAOHE), not the environmental regulation ("The Unified Body of Secondary Environmental Legislation," TULAS). The requirements of the sectorial regulation are, however, less strict than those of the environmental regulation because they have not been significantly overhauled since 2001.[4] This means that some of the environmental problems denounced in civil complaints based on environmental regulation are not above the permitted limits or are not regulated at all according to the sectorial regulations.

Furthermore, the results of laboratory analyses are not always reliable. Interviewed chemists stated that some laboratories might not have the necessary technical equipment to perform appropriate analysis of the sample. The equipment needed is expensive, and some laboratories might try to keep the prices for their services low, as contracts for oil companies are competitive. Additionally, even if laboratories must be officially accredited, some laboratories might purposefully corrupt results, some interviewees stated, given that it is the oil company that pays for the analysis. For these reasons, most laboratory results indicate that the samples taken during inspections are compliant with the legal requirements.

A third barrier to an efficient process relates to the fact that it remains cheaper for oil companies to simply pay for compensations, clean-ups, and administrative fines than to pay the costs it would take to replace corroding infrastructures. Even if the Ministry of Environment asks a company to expound its plan to mitigate the environmental damage caused by its infrastructure, the outcome is not assured—"because, as you [the company] say, your study of re-engineering will not be executed for the lack of budget allocation" (Ministry of Environment 2017, 1). Interviewees from the corporate realm stated that the divisions responsible for health and environment within oil companies would have the least influence and receive the smallest budget (cf. Bratspies 2011). Some interviewed state officials said mature oil fields would cause a spill each day because of corroding infrastructures. "For example, in one single *bloque* (well pad/extraction site), there are thousands of spills, a real high number of spills. These are not big spills, not big volumes, but this is something daily, quotidian, when it should rather be something that does not often happen, such as an emergency. I think that for the companies, it may just be easier to pay than to change the way they do things" (Government official, interview 2015). That is, a civil complaint about an oil spill tends to lead to the clean-up of the oil spill and some compensation payments. However, the root causes that generate the spills are hardly ever addressed, and thus the next spill is just around the corner.

CONCLUSION

These findings show that the struggle inside the state continues to be won by the extractivist state project; as many scholars have pointed out, "the post-neoliberal state," as aggregate, is neo-extractivist in disguise (Svampa 2019). Several institutional barriers turn inspections and enforcement procedures into some kind of stage performances that remain mostly inconsequential for the extractive sector. Does this mean that civil complaints and environmental law enforcement just help to legitimize what Gudynas (2013) called a "predatory extractivism" (67)? Do they undermine or "trap" the environmental and social demands of impacted communities by diverting them into some empty bureaucratic procedures? Or do they help advance transitions toward post-extractivism by making the externalization of environmental costs more difficult? Does legally asserting the right to integral reparations fortify impacted communities to have more influence in the environmental governance of extraction?

This chapter does not pretend to be able to answer these questions—they are far beyond its scope. It does certainly merit a discussion about whether any research design could be rigorous enough to address such queries without being quickly taken apart and thus delegitimized by anybody siding with extractivism. This chapter only shines a light on one single part among many in terms of how the Ecuadorian Constitutional provisions are translated into realities on the ground. Even though the chapter demonstrated the limits of civil complaints and environmental law enforcement, it nevertheless holds that the progressive constitutional provisions matter. As E.P. Thompson (1975) put it:

> I am insisting only upon the obvious point, which some modern Marxists have overlooked, that there is a difference between arbitrary power and the rule of law. We ought to expose the shams and inequities which may be concealed beneath this law. But the rule of law itself, the imposing of effective inhibition upon power and the defense of the citizen from power's all-intrusive claims, seems to me to be an unqualified human good. To deny or belittle this good is, in this dangerous century, when the resources and pretentions of power continue to enlarge, a desperate error of intellectual abstraction. (266)

The progressive constitutional provisions and the institutional adjustments in the environmental governance they entailed are still better than no environmental laws at all. Due to the constitutional changes, a struggle *is* taking place inside the Ecuadorian state, which would likely *not* take place without those progressive provisions. The constitutional changes provide a legal ground on which communities and state officials can base their claims. Furthermore, struggles for environmental law enforcement do not seem to be mutually exclusive to struggles of social movements. Nor do struggles for environmental law enforcement seem to preclude asking the question of who needs extraction, how much, why, and for what. The only caveat appears to be not to forget one over the other. What is certain is that opening pathways toward more inclusionary resource governance is unlikely to be possible without finding ways to economically diversify away from oil extraction, build alternative economies, and break through established political settlements between extractivist elites and governments (Crabtree and Durand 2017; Bebbington et al. 2018).

NOTES

1. All interviews and documents reflect the author's translation.
2. According to Gudynas (2013), transitions from extractivism toward post-extractivism do "not imply a ban on all extractive industries, but rather a substantial downsizing whereby the only ones left are those that are genuinely necessary, meet social and environmental conditions, and are directly linked to national and regional economic chains. In this way, global export orientation is reduced to a minimum, and the trade in these products concentrates mainly on continental markets" (175).
3. Scholars estimate that Texaco discharged up to 16.000 gallons of crude oil and 3.2 million gallons of toxic wastewater into the environment every day, leaving hundreds of open-air earthen waste pits in the Amazon basin (Kimerling 2013). The Ecuadorian state company that took over Texaco´s operations in the 1990s is said to have mainly continued such practices (Kimerling 2013).
4. For instance, TULAS limits ΣPAHs (C) to 0.1 mg/kg for agricultural soil. In contrast, RAOHE allows up to 2 mg/kg for ΣPAHs (C) for agricultural soil. That is, RAOHE does not seem to be a strict legal base, especially considering that the European classification system of soil contamination determines a level of Σ16 PAHs above 1.0 mg/kg as "heavily contaminated" (Maliszewska-Kordybach 1996; Jiao et al. 2017).

REFERENCES

Acosta, Alberto. 2016. "Post-extractivismo: entre el discurso y la praxis. Algunas reflexiones gruesas para la acción." *Ciencia Política* 11(21): 287–332. https://doi.org/10.15446/cp.v11n21.60297.

Akchurin, Maria. 2015. "Constructing the Rights of Nature: Constitutional Reform, Mobilization, and Environmental Protection in Ecuador." *Law & Social Inquiry* 40(4): 937–968.

Aldana, Raquel, and Randall S. Abate. 2016. "Banning Metal Mining in Guatemala." *Vermont Law Review* 40(597): 1–95. https://ssrn.com/abstract=2817566.

Andreucci, Diego, and Isabella M. Radhuber. 2017. "Limits to 'Counter-Neoliberal' Reform: Mining Expansion and the Marginalisation of Post-Extractivist Forces in Evo Morales's Bolivia." *Geoforum* 84: 280–291. https://doi.org/10.1016/j.geoforum.2015.09.002.

Arsel, Murat, Barbara Hogenboom, and Lorenzo Pellegrini. 2016. "The Extractive Imperative in Latin America." *The Extractive Industries and Society* 34: 880–887. https://doi.org/10.1016/j.exis.2016.10.014.

Bebbington, Anthony, Abdulai Abdul-Gafaru, Denise Humphreys Bebbington, Marja Hinfelaar, and Cynthia Sanborn, eds. 2018. *Governing Extractive Industries. Politics, Histories, Ideas*. Oxford: Oxford University Press.

Bratspies, Rebecca. 2011. "A Regulatory Wake-Up Call: Lessons from BP's Deepwater Horizon Disaster." *Golden Gate University Environmental Law Journal* 5(1):7–60. https://digitalcommons.law.ggu.edu/gguelj/vol5/iss1/3.

Bravo, Elizabeth, Melissa Moreano Venegas, and Ivonne Yánez, eds. 2017. *Ecología política en la mitad del mundo. Luchas ecologistas y reflexiones sobre la naturaleza en el Ecuador*. Quito: Ediciones Abya-Yala.

Broad, Robin, and John Cavanagh. 2015. "Poorer Countries and the Environment: Friends or Foes?" *World Development* 72: 419–31. https://doi.org/10.1016/j.worlddev.2015.03.007.

Bornschlegl, Teresa. 2018. "Petro-Geographies and the Dialectic of the Everyday. Enforcing Environmental Laws in the Hydrocarbon Sector in Post-Neoliberal Ecuador." *Journal of Latin American Geography (JLAG)* 17(3):15–42. https://doi.org/10.1353/lag.2018.0040.

Boyd, David R. 2017. *The Rights of Nature: A Legal Revolution That Could Save the World*. Toronto: ECW Press.

Castro, Fabio de, Barbara Hogenboom, and Michiel Baud, eds. 2016. *Environmental Governance in Latin America*. Basingstoke: Palgrave Macmillan.

Crabtree, John, and Francisco Durand. 2017. *Peru: Elite Power and Political Capture*. London: Zed Books.

Finer, Matt, Bruce Babbitt, Sidney Novoa, Francesco Ferrarese, Salvatore Eugenio Pappalardo, Massimo De Marchi, Maria Saucedo, and Anjali Kumar. 2015. "Future of Oil and Gas Development in the Western Amazon." *Environmental Research Letters* 10(2):1–7. https://doi.org/10.1088/1748-9326/10/2/024003.

Gudynas, Eduardo. 2013. "Transitions to Post-Extractivism: Directions, Options, Areas of Action." In *Beyond Development: Alternative Visions from Latin America*, edited by M. Lang and D. Mokrani, 165–94. Quito: Transnational Institute/Rosa Luxemburg Foundation.

Gudynas, Eduardo. 2015. *Extractivismos. Ecología, economía y política de un modo de entender el desarrollo y la Naturaleza*. Cochabamba: CEDIB.

Hancock, Linda, Natalie Ralph, and Saleem H. Ali. 2018. "Bolivia's Lithium Frontier: Can Public Private Partnerships Deliver a Minerals Boom for Sustainable Development?" *Journal of Cleaner Production* 178: 551–560. https://doi.org/10.1016/j.jclepro.2017.12.264.

Huber, Matt. 2013. "Fueling Capitalism: Oil, the Regulation Approach, and the Ecology of Capital." *Economic Geography* 89: 172–94. https://doi.org/10.1111/ecge.12006.

Jessop, Bob. 1990. *State Theory: Putting the Capital State in Its Place*. Cambridge: Polity Press.

Jiao, Hia, Qi Wang, Nana Zhao, Bo Jin, Xuliang Zhuang, and Zhihui Bai. 2017. "Distributions and Sources of Polycyclic Aromatic Hydrocarbons (PAHs) in Soils Around a Chemical Plant in Shanxi, China." *International Journal of Environmental Research and Public Health* 14(1198): 1–19. https://doi.org/10.3390/ijerph14101198.

Kauffman, Craig, and Pamela L. Martin. 2016. "Can Rights of Nature Make Development More Sustainable? Why Some Ecuadorian Lawsuits Succeed and Others Fail?" *World Development* 92: 130–142. https://doi.org/10.1016/j.worlddev.2016.11.017.

Kimerling, Judith. 1991. *Amazon Crude*. New York: Natural Resource Defense Council.

Kimerling, Judith. 2013. "Oil, Contact, and Conservation in the Amazon: Indigenous Huaorani, Chevron, and Yasuni." *Colorado Journal of International Environmental Law and Policy* 24(1): 43–115. https://ssrn.com/abstract=2332782.

La Hora. 2006. "Termina larga noche neoliberal." *La Hora*, December 9, 2016. https://lahora.com.ec/noticia/509050/noticia

Lalander, Rickard, and Maija Merimaa. 2018. "The Discursive Paradox of Environmental Conflict: Between Ecologism and Economism in Ecuador." *Forum for Development Studies* 45(3): 485–511. https://doi.org/10.1080/08039410.2018.1427622.

Larrea, Carlos, ed. 2017. *¿Está agotado el período petrolero en Ecuador? Alternativas hacia una sociedad más sustentable y equitativa: un estudio multicriterio*. Quito: Universidad Andina Simón Bolívar, Ediciones La Tierra, Pachamama Alliance.

Lewis, Tammy L. 2016. *Ecuador's Environmental Revolutions: Ecoimperialists, Ecodependents, and Ecoresisters*. Cambridge: The MIT Press.

Lu, Flora, Gabriela Valdivia, and Néstor L. Silva. 2017. *Oil, Revolution, and Indigenous Citizenship in Ecuadorian Amazonia*. Basingstoke: Palgrave Macmillan.

Lynch, Michael, Paul Stretesky, and Michael A. Long. 2018. "The Treadmill of Production and the Treadmill of Law: Propositions for Analyzing Law, Ecological Disorganization and Crime." *Capitalism Nature Socialism* 31(1): 107–122. https://doi.org/10.1080/10455752.2018.1545241.

Maliszewska-Kordybach, Barbara. 1996. "Polycyclic Aromatic Hydrocarbons in Agricultural Soils in Poland: Preliminary Proposals for Criteria to Evaluate the Level of Soil Contamination." *Applied Geochemistry* 11: 121–127. https://doi.org/10.1016/0883-2927(95)00076-3.

Ministry of Environment Ecuador. 2016. *Informe Técnico. No. 0265-2016-UCA-DPS-MAE. Quito: Ministerio del Ambiente del Ecuador.*

Ministry of Environment Ecuador 2016. *Oficio No. MAE-DNCA-2016-2302. Quito: Ministerio del Ambiente del Ecuador.*

Ministry of Environment Ecuador. 2017. *Oficio No. MAE-DNCA-2017-2035-O. Quito: Ministerio del Ambiente del Ecuador.*

M'Gonigle, Michael, and Louise Takeda. 2013. "The Liberal Limits of Environmental Law: A Green Legal Critique." *Pace Environmental Law Review* 30(3): 1–69. https://digitalcommons.pace.edu/pelr/vol30/iss3/4.

Narváez, Iván, and Maria José Narváez. 2012. *Derecho ambiental en clave neoconstitucional (enfoque político)*. Quito: FLACSO.

O'Connor, James. 1973. *The Fiscal Crisis of the State*. New York: St. Martin's Press.

O'Connor, James. 1998. *Natural Causes. Essays in Ecological Marxism*. New York: Guilford Press.

Offe, Clauss, and John Keane. 1984. *Contradictions of the Welfare State*. London: Hutchinson.

Petroamazonas EP. 2017. *Oficio No. PAM-SSA-2017-1490-OFI.*

Pulido, Laura, Ellen Kohl, and Nicole-Marie Cotton. 2016. "State Regulation and Environmental Justice: The Need for Strategy Reassessment." *Capitalism Nature Socialism* 27(2): 12–31. https://doi.org/10.1080/10455752.2016.1146782.

Radcliffe, Sarah. 2019. "Pachamama, Subaltern Geographies, and Decolonial Projects in Andean Ecuador." In *Subaltern Geographies*, edited by Tarig Jazeel, and Stephen Legg, 119–142. Athen: University of Georgia Press.

Ray, Rebecca, and Adam Chimienti. 2017. "A Line in the Equatorial Forests: Chinese Investment and the Environmental and Social Impacts of Extractive Industries in Ecuador." In *China and Sustainable Development in Latin America: The Social and Environmental Dimension*, edited by Rebecca Ray, Kevin Gallagher, Andres López, and Cynthia Sanborn, 107–144. New York: Anthem Press.

Roy, Bitty A., Martin Zorrilla, Lorena Endara, Dan C. Thomas, Roo Vandegrift, Jesse M. Rubenstein, Tobias Policha, Blanca Ríos-Touma, and Morley Read. 2018. "New Mining Concessions Could Severely Decrease Biodiversity and Ecosystem Services in Ecuador." *Tropical Conservation Science* 11: 1–20. https://doi.org/10.1177/1940082918780427.

Shade, Lindsay. 2015. "Sustainable Development or Sacrifice Zone? Politics Below the Surface in Post- Neoliberal Ecuador." *Extractive Industries and Society* 2(4): 775–84. https://doi.org/10.1016/j.exis.2015.07.004.

Stretesky, Paul B., Michael A. Long, and Michael J. Lynch. 2013. "Does Environmental Enforcement Slow the Treadmill of Production? The Relationship Between Large Monetary Penalties, Ecological Disorganization and Toxic Releases Within Offending Corporations." *Journal of Crime and Justice* 36(2): 233–247. https://doi.org/10.1080/0735648X.2012.752254.

Svampa, Maristella. 2019. *Neo-Extractivism in Latin America Socio-Environmental Conflicts, the Territorial Turn, and New Political Narratives*. Cambridge: Cambridge University Press.

Thompson, Edward P. 1975. *Whigs and Hunters: The Origin of the Black Act*. New York: Pantheon.

Veltmeyer, Henri. 2016. "Extractive Capital, the State and the Resistance in Latin America." *Sociology and Anthropology* 4(8): 774–84. https://doi.org/10.13189/sa.2016.040812.

Veltmeyer, Henri, and James Petras. 2014. *The New Extractivism: A Post-Neoliberal Development Model or Imperialism of the Twenty-First Century?* Chicago: Zed Books.

Villalba-Eguiluz, C. Unai, and Iker Etxano. 2017. "Buen Vivir vs Development (II): The Limits of (Neo)-Extractivism." *Ecological Economics* 138: 1–11. https://doi.org/10.1016/j.ecolecon.2017.03.010.

Watts, Michael. 2005. "Righteous Oil? Human Rights, the Oil Complex, and Corporate Social Responsibility." *Annual Review of Environment and Resources* 30(1): 373–407.

Part II

OPPRESSORS/OPPRESSED

Gender, Race, and the Extractive/Extracted Body

Rebecca Elmhirst

Recent feminist perspectives that are inspired by postcolonial, decolonial, and anti-colonial thinking have opened up and enriched analysis of life in and around extractive landscapes in ways that center complex subjectivities, violence to bodies, and everyday forms of resistance (Gómez-Barris 2017; Barca 2020). The chapters in this section contribute to and further extend this vibrant and dynamic theme, sharing approaches that challenge and complicate unhelpful binaries around exclusion/inclusion, oppressors/oppressed, and offering generative pathways for deeper analysis and praxis. As this brief introduction outlines, various currents in the loosely configured field of feminist political ecology draw attention to the often disproportionately feminized impacts of large-scale resource extractions. Recent scholarship has marked a political step-change from work that has sought to achieve "feminist goals" by focusing on women: either by recognizing women's historical and contemporary presence in extractive landscapes (Erman 2004; Jenkins 2014; Lahiri-Dutt 2012), or by correcting the erasure of women's experiences through (neo)liberal feminist agendas of empowerment and inclusion through corporate social responsibility initiatives in extractive industries (Hill et al. 2016; Pimpa and Moore 2018).

A more explicitly politicized feminist engagement is being amplified within a constellation of Black, Indigenous, and ecological feminist theory and action around extractive geographies, as the chapters here illustrate. This work draws on and extends an expanded conceptualization of extractivism: a term given to "a complex of self-reinforcing practices, mentalities, and power differentials underwriting and rationalizing socio-ecologically destructive modes of organising life through subjugation, depletion and non-reciprocity" (Chagnon et al. 2022, 4; Willow 2018; Gago and Mezzadra 2017). A focus on extractivist logics across different sectors gives analytical precision by focusing on the relations through which materials and laboring bodies are extracted. An expansive conceptualization of extractivisms opens space for recognizing the ways multiple forms of extraction coexist, are coproduced, amplify harm, and shape registers of resistance. Bringing a feminist lens (broadly conceived) provides the tools for understanding the ways racialized, gendered, and colonial violence underpin the spectacular and banal, catastrophic, and slow violence of extractivism, and how this is experienced and responded to in varied ways. Feminist engagements with extractivism, as exemplified by the chapters in this section, reveal extractive relations that are forged and reproduced within the closer sites, scales, and temporalities of bodies, communities, and the everyday.

In the context of extractivism, feminist political ecology (FPE) has emerged as a convening space for the many ecological feminisms that increasingly draw inspiration through decolonial, postcolonial, and Black feminist thought and what others have termed the environmentalism of the (racialized and gendered) poor, which centers on defending livelihoods, values, place, and territory (Sultana 2021; Harcourt and Nelson 2015; Anguelovski and Martinez Alier 2014; Agostino et al. 2022). A distinct feature of FPE is its focus on understanding and responding to global systems (e.g., extractive logics) and their material consequences (Sundberg 2017), in work that deepens understandings of how power operates within and across the sites and scales that produce and shape extractivist logics and outcomes (Extracting Us Collective). Thus, many strands within FPE share a critical structural and scalar analysis of the roots of environmental degradation and community responses (Rocheleau, Thomas-Slayter, and Wangari 1996) and in various ways, integrate conceptual insights from ecofeminism (Seager 1993), ecofeminist political economy (Mies and Shiva 1993; Katz 2001; Federici 2004), and feminist science and technology studies (Haraway 2001). These strands attend to and trouble binaries around environments and society; nature and culture, human and more-than-human, production and (social) reproduction, and body and territory that endure in Eurocentric colonialist epistemologies and ontologies.

FPE work that explicitly engages postcolonial feminisms (Mohanty 2003) and Black feminist conceptualizations of intersectionality (Cho, Crenshaw, and McCall 2013; Mollett and Faria 2013) presses further to show how multiple intersecting forms of power, including racism, coloniality, patriarchy, and heteronormativity endure and are reproduced and perpetuated within particular landscapes. This has deepened analyses that confront issues such as dispossession (Mollett 2016; Elmhirst et al. 2017; Berman-Arevalo and Ojeda 2021), conservation, access, and exclusion (Sundberg 2004; Elias et al. 2021) and investigation of political-ecological subjectivities (Nightingale 2011), with perspectives attuned to lived experiences in the everyday and in bodies, as sites of emotion, meaning, and affect (González-Hidalgo and Zografos

2020; Sultana 2011; Velicu 2015). Decolonial, postcolonial, and anti-colonial feminisms have, in particular, inspired analysis of how present–past colonial attempts to categorize, control, and extract nature also involve categorizing, controlling, and extracting racialized and gendered bodies (Lugones 2010; Gómez-Barris 2017; Zaragocin 2019; Ojeda et al. 2022).

This evolving and pluralistic constellation of feminisms (characterized here as FPE) that is brought to bear on extractivism is bringing important advances that converge around three distinct but closely related themes, all of which complicate the dualisms that this book and the chapters in this section address. First, feminist perspectives note that much of the literature on extractivism refers to the exploitation of material resources without their reproduction (e.g., Ye et al. 2020), but makes only oblique reference to social reproduction, despite this concept being central to an analysis of the forms of accumulation and dispossession associated with extractive industries. Social reproduction is defined as "the fleshy, messy, and indeterminate stuff of everyday life" (Norton and Katz 2001, 1) and includes the labouring practices that are involved in the survivability of racialized and gendered bodies, communities, cultures and ecologies (Di Chiro 2008; Federici 2004). Contributions from FPE focus on the indifference to life within extractivist logics as these permeate and extract from social reproductive realms and render everyday lives unliveable.

Loss of livelihood, well-being, and belonging are felt as extractive corporate expansion in primary industries (mining, agriculture, and forestry) pollutes ecosystems, disrupts subsistence, displaces communities, and splits families, resulting in an "atmosphere of permanent violence" (Cunha and Casimiro 2021, 80). Almost monotonously repeated stories emerge in relation to mining (Glynn and Maimunah 2021; Lamb, Marschke, and Rigg 2019), agriculture (Berman-Arevalo and Ojeda 2020; Elmhirst et al. 2017; Julia and White 2012). Intersectional analyses reveal how colonial formations of race, Indigeneity, and caste compound and shape the effects of extractivism in the everyday and in quotidian ways, further marginalizing Indigenous and Afro-diasporic communities in the Americas and African contexts (Mollett 2021; Hernández Reyes 2019; Pereira and Tsikata 2021), Indigenous and rural Chinese in Southeast Asia (Glynn and Maimunah 2021; Peluso 2009), and Dalits in South Asia (Roy Chowdhury and Lahiri-Dutt 2021). The realm of social reproduction is contradictory. On the one hand, as Berman-Arevalo and Ojeda (2020) show, social reproduction (understood to include gendered practices of care) sustains extractivism, it is a realm that holds the possibility of transformation (Norton and Katz 2017), and this is seen most clearly in the myriad ways that resistance to extractivism finds its force in the defense of livelihoods and the social reproductive realm (Willow 2018; Barca 2020; Glynn and Maimunah 2021). In the section that follows, Negar Elodie

Behzadi's chapter discusses the ways that intersectional extractive violence reproduces difference within communities, in this case, in Kante, Tajikistan, along gendered, age, and socio-marital lines. While this is associated with the gendered effects of exhaustion of land and resources, and on social-reproductive failures of the (postsocialist) state, it also produces an embodied form of "exhaustion," as young women's labor in the mines threatens their marriageability: a prospect that within local norms and repertoires of relational survival, furthers their precarity.

Behzadi's work draws a link between the realm of social reproduction and the second theme that is explored in FPE analyses of extractivism, that is, the body. The body (and body politics) has long been a key site of feminist engagement and theorizing, from analogies drawn between the female body and the colony in racialized, patriarchal capitalist accumulation (Mies and Shiva 1993) through to questions around sexual and reproductive rights, bodily integrity, health, and violence (Harcourt 2009). Feminist political ecology perspectives on extractivism draw attention to the racialized and gendered body as part of a scalar analysis drawn through engagements with race and environmental justice, intersectionality, and embodiments relating to health, pollution, emotion, and affect, each responding to different forms of power as inscribed on and resisted through the body (Sultana 2011; Mollett and Faria 2013; Doshi 2017; González-Hidalgo and Zografos 2020). As Velicu (2015, 856) writes in relation to the Save Rosia Montana socio-ecological conflict against open cast gold mining in Romania, embodied and emotional FPE questions at the heart of the protest were "'how can we live?', what is development?, 'what is wealth?'", in other words, contested perspectives on "how one is allowed to feel?, what one is allowed to enjoy (doing)?, and how one is supposed to live (spend his or her time)?"

This focus on the scale of the body and embodiment is being developed further within FPE perspectives on extractivism through engagements with Latin American Indigenous epistemologies and ontologies which see the body and territory not as distinct relational entities but as one (Caretta et al. 2020; Zaragocin and Caretta 2020; Mollett 2021). Feminists in the academy have explored the ways that similar ontologies exist between body, land, and territory in other Indigenous places and possibilities for strategic equivalencing (Nirmal and Rocheleau 2019; Leinius 2020). Concepts of body-territory are engaged strategically, for revealing (slow) violence on bodies as following the same logic as extraction from territories, as both are rendered disposable (Gómez-Barris 2017), and also in seeking strategic alliances against extractivism (Leinius 2020; Postar and Behzadi 2022). In the analytics that animate Latin American feminist social movements against extractivism, the body is an anchoring point—the arrival of mining projects and their social and ecological effects "threaten the relationships that weave bodies and territory together into the dynamic web of life" (Leinius 2020, 212). Amber Murrey and Sharlene Mollett's chapter connects with and extends this understanding of body as territory beyond Latin American mining project contexts, to consider the racialized and sexualized logics that render Black women's bodies extractable, in extractive industries in Cameroon, and in extractive tourist sites in Panama.

Both chapters in this section build from and extend a third theme that is associated with feminist political ecology perspectives on extractivism, that is, FPE as praxis (Harcourt and Nelson 2015). As the convening space of FPE is increasingly shaped by those outside the academy (feminist activists, communities) questions persist around ethics and ontological politics. On the one hand, from within frontline communities that fight socio-ecological harms through thinking and being "otherwise," the practices within anti-colonial feminist ecological activisms model ethical relationships with other species and the natural world (Temper et al. 2015; Candraningram 2018) and inspire the practice of FPE. Pluriversal dialogues across and between differently situated researchers and activists involve ongoing critical interrogation of methodological realities and ethical dilemmas in violent extractive contexts (Johnson et al. 2021). The chapters that follow demonstrate unequivocally the point of analyzing from closer scales of the body and from the everyday while understanding that bodies and body-territories are not produced, extracted, or exhausted in the same way. The monotonous sameness of racialized and sexual violence, rooted in present–past colonial exploitation surely enlarges the scope for alliances in challenging extractive logics.

REFERENCES

Agostino, Ana, Rebecca Elmhirst, Marlene Gómez, Panagiota Kotsila, and Wendy Harcourt, eds. 2023. *Contours of Feminist Political Ecology*. London: Ashgate.

Anguelovski, Isabelle, and Joan Martínez Alier. 2014. "The 'environmentalism of the poor' revisited: Territory and place in disconnected local struggles." *Ecological Economics 102*: 167–176.

Barca, Stefania. 2020. *Forces of Reproduction: Notes for a Counter-Hegemonic Anthropocene*. Cambridge: Cambridge University Press.

Berman-Arévalo, Eloísa, and Diana Ojeda. 2020. "Ordinary geographies: Care, violence, and agrarian extractivism in 'post-conflict' Colombia." *Antipode* 52 (6): 1583–1602.

Candraningrum, Dewi. 2018. "The spiritual politics of the Kendeng Mountains versus the global cement industry." In *After the Death of Nature: Carolyn Merchant and the Future of Human-Nature Relations*, edited by Kenneth Worthy, Elizabeth Allison, and Whitney Bauman, 219–234. London: Routledge.

Caretta, Martina Angela, Sofia Zaragocin, Bethani Turley, and Kamila Torres Orellana. 2020. "Women's organizing against extractivism: Towards a decolonial multi-sited analysis." *Human Geography* 13 (1): 49–59.

Chagnon, Christopher W., Francesco Durante, Barry K. Gills, Sophia E. Hagolani-Albov, Saana Hokkanen, Sohvi M. J. Kangasluoma, Heidi Konttinen, Markus Kröger, William LaFleur, Ossi Ollinaho, and Marketta P. S. Vuola. 2022. "From extractivism to global extractivism: The evolution of an organizing concept." *The Journal of Peasant Studies* 49 (4): 760–792.

Cho, Sumi, Kimberlé Crenshaw, and Leslie McCall. 2013. "Toward a field of intersectionality studies: Theory, applications, and praxis." *Signs: Journal of Women in Culture and Society* 38 (4): 785–810.

Cunha, Teresa, and Isabel Casimiro. 2021. "'Cinderellas' of our Mozambique wish to speak: A feminist perspective on extractivism." *Feminist Africa* 2 (1): 71–98.

Di Chiro, Giovanna. 2008. "Living environmentalisms: Coalition politics, social reproduction, and environmental justice." *Environmental Politics* 17 (2): 276–298.

Elias, Marlène, Deepa Joshi, and Ruth Meinzen-Dick. 2021. "Restoration for whom, by whom? A feminist political ecology of restoration." *Ecological Restoration* 39 (1–2): 3–15.

Elmhirst, Rebecca, Mia Siscawati, Bimbika Sijapati Basnett, and Dian Ekowati. 2017. "Gender and generation in engagements with oil palm in East Kalimantan, Indonesia: Insights from feminist political ecology." *The Journal of Peasant Studies* 44 (6): 1135–1157.

Erman, Erwiza. 2004. "Labouring communities: Women's roles in the Ombilin coal mines of Sumatra." In *Labour in Southeast Asia: Local Processes in a Globalised World*, edited by Rebecca Elmhirst and Ratna Saptari, 359–386. London: Routledge.

Extracting Us Curatorial Collective. (2023). "Extracting us: Co-curating creative responses to extractivism through a feminist political ecology praxis." In *Contours of Feminist Political Ecology,* edited by Ana Agostino, Rebecca Elmhirst, Marlene Gómez, Panagiota Kotsila, and Wendy Harcourt. London: Ashgate.

Federici, Silvia. 2004. *Caliban and the Witch*. Brooklyn: Autonomedia.

Gago, Verónica, and Sandro Mezzadra. 2017. "A critique of the extractive operations of capital: Toward an expanded concept of extractivism." *Rethinking Marxism* 29 (4): 574–591.

Glynn, Tracy, and Siti Maimunah. 2021. "Unearthing conscious intent in women's everyday resistance to mining in Indonesia." *Ethnography* 24(1): 23-43. https://doi.org/10.1177/14661381211039372.

Gómez-Barris, Macarena. 2017. *The Extractive Zone: Social Ecologies and Decolonial Perspectives*. Durham: Duke University Press.

González-Hidalgo, Marien, and Christos Zografos. 2020. "Emotions, power, and environmental conflict: Expanding the 'emotional turn' in political ecology." *Progress in Human Geography* 44 (2): 235–255.

Haraway, Donna. 2001. *Simians, Cyborgs, and Women: The Reinvention of Nature*. Abingdon: Routledge.

Harcourt, Wendy. 2009. *Body Politics in Development: Critical Debates in Gender and Development*. London: Bloomsbury Publishing.

Harcourt, Wendy, and Ingrid L. Nelson, eds. 2015. *Practicing Feminist Political Ecology: Moving Beyond the Green Economy*. London: Zed Books.

Hernández Reyes, Castriela Esther. 2019. "Black women's struggles against extractivism, land dispossession, and marginalization in Colombia." *Latin American Perspectives* 46 (2): 217–234.

Hill, Christina, Chris Madden, and Maria Ezpeleta. 2016. *Gender and the Extractive Industries: Putting Gender on the Corporate Agenda*. Melbourne: Oxfam. Accessed October 12, 2022. https://oxfamilibrary.openrepository.com/bitstream/handle/10546/620776/bp-gender-extractives-corporate-agenda-010516-en.pdf?sequence=1&isAllowed=y.

Jenkins, Katy. 2014. "Women, mining and development: An emerging research agenda." *The Extractive Industries and Society* 1 (2): 329–339.

Johnson, Adrienne, Anna Zalik, Sharlene Mollett, Farhana Sultana, Elizabeth Havice, Tracey Osborne, Gabriela Valdivia, Flora Lu, and Emily Billo. 2021. "Extraction, entanglements, and (im) materialities: Reflections on the methods and methodologies of natural resource industries fieldwork." *Environment and Planning E: Nature and Space* 4 (2): 383–428.

Julia, and Ben White. 2012. "Gendered experiences of dispossession: Oil palm expansion in a Dayak Hibun community in West Kalimantan." *The Journal of Peasant Studies* 39 (3–4): 995–1016.

Katz, Cindi. 2001. "Vagabond capitalism and the necessity of social reproduction." *Antipode* 33 (4): 709–728.

Lahiri-Dutt, Kuntala. 2012. "Digging women: Towards a new agenda for feminist critiques of mining." *Gender, Place & Culture* 19 (2): 193–212.

Lamb, Vanessa, Melissa Marschke, and Jonathan Rigg. 2019. "Trading sand, undermining lives: Omitted livelihoods in the global trade in sand." *Annals of the American Association of Geographers* 109 (5): 1511–1528.

Leinius, Johanna. 2021. "Articulating body, territory, and the defence of life: The politics of strategic equivalencing between women in anti-mining movements and the feminist movement in Peru." *Bulletin of Latin American Research* 40 (2): 204–219.

Lugones, María. 2010. "Toward a decolonial feminism." *Hypatia* 25 (4): 742–759.

Mies, Maria, and Vandana Shiva. 1993. *Ecofeminism*. London: Zed Books.

Mollett, Sharlene. 2016. "The power to plunder: Rethinking land grabbing in Latin America." *Antipode* 48 (2): 412–432.

Mollett, Sharlene. 2021. "Hemispheric, relational, and intersectional political ecologies of race: Centring land-body entanglements in the Americas." *Antipode* 53 (3): 810–830.

Mollett, Sharlene, and Caroline Faria. 2013. "Messing with gender in feminist political ecology." *Geoforum* 45: 116–125.

Nightingale, Andrea J. 2011. "Bounding difference: Intersectionality and the material production of gender, caste, class and environment in Nepal." *Geoforum* 42 (2): 153–162.

Nirmal, Padini, and Dianne Rocheleau. 2019. "Decolonizing degrowth in the post-development convergence: Questions, experiences, and proposals from two Indigenous territories." *Environment and Planning E: Nature and Space* 2 (3): 465–492.

Norton, Jack, and Cindi Katz. 2017. "Social reproduction." In *International Encyclopedia of Geography: People, the Earth, Environment and Technology*, edited by Douglas Richardson, Noel Castree, Michael F. Goodchild, Audrey Kobayashi, Weidong Liu, and Richard A. Marston. Chichester, UK: John Wiley & Sons. https://doi.org/10.1002/9781118786352.wbieg1107.

Ojeda, Diana, Padini Nirmal, Dianne Rocheleau, and Jody Emel. 2022. "Feminist ecologies." *Annual Review of Environment and Resources* 47:149–171. https://doi.org/10.1146/annurev-environ-112320-092246.

Pereira, Charmaine, and Dzodzi Tsikata. 2021. "Extractivism, resistance, alternatives." *Feminist Africa* 2 (1): 1–13.

Pimpa, Nattavud, and Timothy Moore. 2018. *Corporate Social Responsibility and the Inclusivity of Women in the Mining Industry: Emerging Research and Opportunities: Emerging Research and Opportunities*. Hershey, PA: IGI Global.

Postar, Stephanie, and Negar Elodie Behzadi. 2022. "'Extractive bodies': A feminist counter-topography of two extractive landscapes." *Geoforum*, No. October 2022. https://doi.org/10.1016/j.geoforum.2022.09.005.

Rocheleau, Dianne E., Barbara P. Thomas-Slayter, and Esther Wangari, eds. 1996. *Feminist Political Ecology: Global Issues and Local Experience*. London: Routledge.

Roy Chowdhury, Arnab Roy, and Kuntala Lahiri-Dutt. 2021. "Extractive capital and multi-scalar environmental politics: Interpreting the exit of Rio Tinto from the diamond fields of Central India." *Third World Quarterly* 42 (8): 1770–1787.

Seager, Joni. 1993. *Earth Follies: Feminism, Politics and the Environment*. London: Routledge.

Shiva, Vandana. 1988. *Staying Alive: Women, Ecology, and Survival in India*. New Delhi: Kali for Women.

Sultana, Farhana. 2011. "Suffering for water, suffering from water: Emotional geographies of resource access, control and conflict." *Geoforum* 42 (2): 163–172.

Sultana, Farhana. 2021. "Political ecology 1: From margins to center." *Progress in Human Geography* 45 (1): 156–165.

Sundberg, Juanita. 2004. "Identities in the making: Conservation, gender and race in the Maya Biosphere Reserve, Guatemala." *Gender, Place & Culture* 11 (1): 43–66.

Temper, Leah, Daniela del Bene, and Joan Martinez-Alier. 2015. "Mapping the frontiers and front lines of global environmental justice: The EJAtlas." *Journal of Political Ecology 22* (1): 255–278.

Sundberg, Juanita. 2017. "Feminist political ecology." In *The International Encyclopedia of Geography: People, the Earth, Environment and Technology*, edited by Douglas Richardson, Noel Castree, Michael F. Goodchild, Audrey Kobayashi, Weidong Liu, and Richard A. Marston. Chichester, UK: John Wiley & Sons. https://doi.org/10.1002/9781118786352.wbieg0804.

Velicu, Irina. 2015. "Demonizing the sensible and the 'revolution of our generation' in Rosia Montana." *Globalizations* 12 (6): 846–858.
Willow, Anna J. 2018. *Understanding ExtrACTIVISM: Culture and Power in Natural Resource Disputes*. London: Routledge.
Ye, Jingzhong, Jan Douwe van der Ploeg, Sergio Schneider, and Teodor Shanin. 2020. "The incursions of extractivism: Moving from dispersed places to global capitalism." *The Journal of Peasant Studies* 47 (1): 155–183.
Zaragocin, Sofia. 2019. "Gendered geographies of elimination: Decolonial feminist geographies in Latin American settler contexts." *Antipode* 51 (1): 373–392.
Zaragocin, Sofia, and Martina Angela Caretta. 2021. "*Cuerpo-territorio*: A decolonial feminist geographical method for the study of embodiment." *Annals of the American Association of Geographers* 111 (5): 1503–1518.

3

A Decolonial Feminist Dialogue-as-Critique

Against the Sexual and Racialized Violence at the Heart of Extractivism

Amber Murrey and Sharlene Mollett

On the Atlantic coast of Panama and the archipelago of Bocas del Toro, a tourism enclave of the same name, foreign land grabbing, and pejorative-carnal imaginaries of local Indigenous and Black women go hand in hand. As ethnographic research describes:

> There is a bar in Bocas, I will call [it] Paradise. While sitting at my table with backpackers from England, we could not avoid hearing a man named Michael [a white American expat who owns and runs a Bed and Breakfast in Bocas], giving some "advice" to male backpackers looking to "hook up" with a "black girl." While we clearly missed the beginning part of the conversation, in a loud voice Michael blurts out "no, no, no, you've got it all wrong. Easy guys, you can get a darky anytime . . . Wari wari girls are wild, you will get more than you want . . . trust me. I know, when I first got here I got with a few wari wari girls and almost died of a heart attack, they clawed me good . . . But after a few months of that . . . .," he says, "as a man, you learn that that's not normal, you have to get sensible, you need a lady." With a full audience, Michael explains that what these newcomers really want is an "Indian girl." He says "they are 'good' girls, ladies, quiet and always happy. If you are going to stay a long time, that's what you need . . . They don't understand much English [everyone laughs], but they aren't looking for trouble either, they just want someone to take care of them." (Mollett 2017, 11)

In a different, yet strangely similar landscape, seemingly banal encounters emerged during ethnographic research along the Chad-Cameroon Oil Pipeline in Nanga-Eboko, Cameroon, another space of extraction. Employed as a cleaner at the temporary worker's camp in Nanga, Silvan, a Cameroonian from Nanga-Eboko explained that the wages he made from the oil consortium were "nothing at all" (Murrey fieldnotes 2012). Yet, shortly after the construction on the pipeline was completed, Silvan bought his first motorcycle and began working as a motorcycle taxi driver. This new money-generating purchase changed his life significantly. His money, he explained, was made indirectly: he worked as a sort of ad hoc middleman, introducing women to some of the migrant workers employed by the oil consortium. The "bosses," as he called them, were a small group of mostly Columbian men employed by the private subcontractor responsible for the wielding of the pipe. These sexual encounters, that Silvan arranged between local women and foreign private subcontractors, resulted in what Silvan referred to as "the generation of *metis*"—or mixed-race—children in the village. Sex punctuated life and death along the pipeline as three young women, former sex workers, died of "mysterious causes" during the pipeline's construction period (Murrey fieldnotes 2012). Such stories bring to life the everyday consequences brought about by the socio-economic context created within and during resource booms. Foreign workers' desires for young women's bodies were substantiated in the pages of nonprofit and advocacy reports (African Forum and Network on Debt and Development 2007, 23).

That sexuality and violence overlap in the spaces of extraction is, of course, not a new phenomenon. Sexual violence and European conquest-cum-extraction shape the histories of places in the Americas and Africa alike (Morgan 2004; McClintock 1995). However, sexuality and sexual violence are often obscured by the pursuits of development and a neoliberal focus on such things as "poverty reduction" and "good governance." The persistent role of sexuality (and the bodily violence and harms attendant to sex work and the imagined sexualities of local women) during ethnographic research ostensibly about land enclosure and an oil pipeline, in Panama and Cameroon, respectively, exposes the entanglements of sexuality, racialized sexual violence, extractivism, and environmental injustice.

Within extractive landscapes, Black and Indigenous women are disproportionately among those subject to multiple kinds of violence, including sexual violence (Simpson 2017; Hausermann, Adomako, and Robles 2020). As Hartman suggests (1997, 57), "the 'givenness' of blackness results from the brutal corporealization of the body and the fixation of its constituent parts as indexes of truth and racial meaning." The same racial tropes that misrecognize black women as only "worthy of servitude" are coupled with settler carnal desires that imagine black women as sexually dangerous, *always* sexually available, and less-than-human (Morgan 2004; Mollett 2021a). Building on decolonial feminist scholarship, this chapter seeks to expand conversations that focus on the toxic consequences of commercial extraction of land

and the environment, to a complicated understanding of extractive practices and their extractivist logics as violent processes that simultaneously harm the land and human bodies (see Mollett 2021a; De Leuuw 2016; Murrey 2015a). Decolonial feminisms have been at the forefront of conceptual and methodological thinking on extractivism, including by unpacking and reframing the ways in which white supremacy and racist-colonial logics create the "conditions for the accumulation and hoarding of resources" (Butler 2015; see also Yusoff 2019); articulating new ways of understanding the complex embodied-landed violence of extraction (for example, *cuerpo-territorio* or, "women's land-body experiences," Castro 2020); and imagining life beyond extractive capitalism through women's embodied and everyday resistances (Simpson 2017; Hernandez-Reyes 2019; Hausermann, Adomako, and Robles 2020; Billo 2020).

Relatedly, we argue, there exists a need to address the cumulative socio-political and ecological effects and logics of extraction beyond subsoil resources, environmental contamination and displaced communities (Mollett 2021a, 2022). We are interested in the resonances between people in Panama and Cameroon and how they lead us to new ways of thinking about black women's experiences of extractivism, in particular, to expand what we take to mean with the term, to include residential tourism as an extension of the logics and practices of extractivism. Here, we draw from emergent work in black geographies articulating the need for cross-fertilizations. Patricia Daley calls for consideration of the "globality of black women's 'lifeways'—especially the interconnections between continental Africa and the African diaspora—and suggests that a more relational approach to the study of black women's lives could inform geographers' understanding of gendered and racial structures of oppression and alternative geographies of resistance and freedom" (2020, 1). We illustrate through a decolonial feminist dialogue-as-critique (see below) between Panama and Cameroon, where both relations of residential tourism and corporate language vis-a-vis an oil extraction are united by colonial, racial, and sexualized logics that would seek to render black life "disposable" (Razack 2016)—a project which is never totalizing nor actualized, as people remain active subjects.

In this chapter, we look at some of the ways that logics and practices of extractivism, in Panama and Cameroon, dispossess and displace through racialized and sexualized asymmetries. While geographers and political ecologists offer strong critiques of how states and foreign interests in extraction reproduce processes of "accumulation by dispossession" at the expense of local and economically precarious populations, we take a slightly different approach. In line with decolonial and postcolonial feminist geographies, we attend to how Black women specifically, in Panama and Cameroon, experience dispossession in embodied ways. To do so, we disclose the ways in which extractive logics frequently initiate and condone forms of racialized sexualized violence. We look at how extractivism settles within internal ecologies (including ecologies of health and disease) in ways that reaffirm colonial logics of race and sexuality. Sexual violence, we argue, is at the center of racialized extractive capitalism.

DECOLONIAL FEMINISMS AND CRITIQUES OF EXTRACTIVISM

Decolonial feminisms expand the analysis of extractive geographies, practices, and logics in fruitful ways. Gomez-Barris (2017), for example, argues that extractivism entails a logic of economic accumulation that relies on the rendering of territories and peoples as *extractible*. For us, extractivism as logic is not limited to economic or accumulative logics but actually informs entire ways of living and being in the world. In this way, we take *extractivism* to denote both the material processes of removal as well as a particular logic (not only of economic accumulation but also of *being*) that renders materials, territories, and bodies *extractible* (on expanding conceptualizations of extractivism, see Zalik 2016; Johnson et al. 2021). This extractivist logic is racialized and sexualized and is colonial in its origins. Leanne Betasamosake Simpson explains,

> Colonialism and capitalism are based on extracting and assimilating . . . the act of extraction removes all of the relationships that give whatever is being extracted meaning. Extraction is taking. Actually, extraction is stealing—it is taking without consent, without thought, care or even knowledge of the impacts that extraction has on the other living things in that environment. There's always been a part of colonialism and conquest. Colonialism has always extracted the Indigenous—extraction of Indigenous knowledge, Indigenous women, Indigenous people. (Interview with Naomi Klein 2013)

A substantial literature within feminist political ecology has delineated the various gendered effects of resource extraction. The relationship between gender power and extraction, for example, has received increased attention within the scholarship on mining and political ecologies of extraction (Zaragocin 2017, 2018; Kelly 2014; Jenkins 2014). The gendered violence of extraction, including women's disproportionate exposure to economic vulnerabilities due to subordinate gender relations within communities and the uneven burdens placed upon women as carers in communities impacted by environmental pollution, toxins, and ecocide, makes up a growing body of scholarship (Coyle 2004; Bagelman and Wiebe 2017; Hernández Reyes 2019; Caretta and Zaragocin 2020). Important work has exposed the permanently shifting relationships between environmental and gendered violence (Caretta et al. 2020; Leinius 2020). This includes work on Latin American anti-mining collectives organizing around an epistemology of "body as territory"; Johanna Leinius (2020, 9) explains,

> The body functions as [the] anchoring point of a broader network of connections and exchange in space and time. It is not individual, because the individual subject does not exist without the inter-subjective ties that bind it to collectively imagined social and territorial geographies . . . The body is but an extension of . . . ties in which community, and thus life itself, is (re-)produced.

Decolonial feminist work advances critiques of the coloniality of gender or the ways in which "coloniality permeates all aspects of social existence," the central role played by gender within the modern/colonial system of oppression and the imperative for decolonial options to offer ways of being and knowing beyond and outside the confines of coloniality (Lugones 2008). Rosalba Icaza and Rolando Vázquez write, "Thinking the coloniality of gender means to think from an embodied experience, it is to think from the ground up, from the body" (2016, 62). In Ghana (Hausermann, Adomako, and Robles 2020) and Tajikistan (Behzadi 2019), women resist by working against gendered norms, actively involving themselves in the mining process and participating in forms of labor deemed masculine within heteropatriarchal societies.

Sexuality, as a form of power that is mutually constitutive of gendered subjectivity has received less attention. When it does, discussions have been largely centered on sexual violence in conflict zones (Cohen and Nordas 2014; Rustad et al. 2016). In the Democratic Republic of the Congo and Colombia, for example, sexual violence is instrumentalized by armed militias to clear, terrorize, and displace people to establish control of mining operations and monopolize valuable natural resources. Outside direct zones of conflict, there has been less attention paid to the intersection of race, gender, and sexual violence in extractive landscapes. We find this curious given that, despite women's agency to resist (see Hernandez-Reyes 2019), extractive landscapes have largely tended to be "portrayed as 'hyper-masculine'" whereby "women in or near mining sites may be more exposed to various forms of sexual abuse or enter into transactional sex of various kinds" (Rustad et al. 2016, 476).

Sex and sexuality are shapers of livelihood security and embodied political ecologies. Buss and Rutherford (2017, 43) argue that sexuality is "an important vector, along which contemporary discourses of self and other, order and criminality; both constituted and inflected with racial and colonial meanings in contemporary forms of global governance." By placing in conversation elite imaginings and lived experiences of Black women in extractive landscape of Panama and Cameroon, we seek to disclose *and contest* the "givenness" of black women in extractive zones. With insights from Panama, we consider how residential tourism shares the embodied political ecologies and historical violence embedded in traditional extractive industries in Cameroon and look at how extractivism in both regions is patterned by racialized and sexualizing logics. We approach our collaboration as a decolonial dialogue, one founded in the critique of the coloniality of being and impetus toward decolonial and post-extractivist futures (Wynters 2003; Gómez-Barris 2017; Castro 2020). We stress the commonalities between sex work, bodily illness, and other racialized and sexualized forms of extractivism, pointing to how imagined sexualities reveal themselves in the misogynist racist tropes about Black women and sex work itself, intensifying the attendant violence of coexistent forms of extraction.

RACE, SEXUALITY, AND THE EXTRACTIVE LOGICS OF DISPOSSESSION

Bocas del Toro, Panama

The Bocas del Toro archipelago is located on the Caribbean coast of Panama. Our focus is on residential tourism development in the archipelago. Known well as "Bocas," the archipelago is a popular tourism destination and growing hub of residential tourism development. From the early 2000s, foreign travelers to Panama began buying land and building homes as both secondary homes or to live permanently (in the case of retirees). This process of residential tourism is seen throughout Panama and is promoted by the Panamanian state through a number of policies and laws that have opened up coastal properties to foreign nationals. While touted as an excellent way to "develop" particularly remote rural regions and offer local people employment, residential tourism is not a benign development venture. Rather, residential tourism in Bocas is made possible by the endurance of overlapping ideologies shaping gender, racial, and carnal power embedded in development policy and practice in extractive landscapes. To be clear, land enclosure is a key part of residential tourism, shifting grounds of access to land that extracts resources from local peoples on behalf of global elites or "residential tourists" a group we refer to in this chapter as "settlers" (Mollett 2017). In Panama, Afro-Panamanians move from having land and embodied autonomy to having less to no land, which deepens their need for full-time employment just to survive. Moreover, they do so in jobs that always require dynamics of servitude that often dehumanize (Mollett 2017, 2022). Like settler state reliance on extraction seen elsewhere (Zaragocin 2019; Simpson 2015), we examine through residential tourism how "the veins of Latin America" remain "open" to foreign extraction (Galeano 1973). Residential tourism is rooted in intersectional settler colonial logics that obscure its own forms of violence through the imaginary of seeing tourism as a "development" strategy and tool of "poverty alleviation."

Natural resource struggles in extractive landscapes are woven into the ongoing "spatial process as colonizers-cum-settlers 'come to stay'" (Mollett 2020, 12; Speed 2017; Gott 2006; Van Noorloos 2017). A tourism priority zone promoted by the Panamanian government, the Bocas archipelago is the homeland of Indigenous Ngabe peoples. Since the late 1800s, Afro-Panamanians, as the descendants of Caribbean economic migrants, have lived on the western Caribbean Coast and islands of Bastimentos and Colon (Bourgois 1989). Once a United Fruit Company (UFC) plantation and outpost, "Bocas" is a residential tourism enclave, populated by affluent settlers from mostly Europe and North America. Settlers own a growing share of land in Bocas facilitated by an array of Panamanian laws and international development funding that prescribes foreign direct investment in land (Guerron-Montero 2014; Spalding 2011). The Panamanian state ensures foreign investors a secure land market and promises domestic residents that these investments will provide employment and economic security that they can enjoy (Guerron-Montero 2014). However, for Afro-Panamanian residents specifically, the promise of jobs as a trade-off from elite land enclosures remains unfulfilled, and the work is largely undignified (Mollett 2017, 2021a). Yet, while many longtime residents desire work in tourism businesses or to start their own, jobs remain scarce (Guerron-Montero 2014; Mollett 2022). These livelihood insecurities intensify with mounting land dispossession, inequality, and a tense and sometimes violent environment shaping Afro-Panamanian land control, access to water, and cultural practices (Guerrero-Montero 2015; Mollett 2017).

Since 1994 and the Panamanian state approval of the Tourism Law (Law 8), state-sanctioned neoliberal land laws fuel foreign investment, both corporate and private in the land (Gaceta Nacional de Panama 1994). In Bocas, in the narratives of both the state and settlers alike, foreign ownership of land is seen as a public good leading to a robust market in land and real estate evidenced by consumption of luxury homes and hotels, ecoleisure and adventure travel, and a plethora of amenities. Such luxuries are found throughout Bocas but concentrated on the Island of Bastimento, and in particular the Red Frog Beach Island Resort and Spa (Mollett 2017). For long-term Ngabe and Afro-Panamanian residents, the impending loss of their historical relations to land, customary forms of land control, and access to nature intensifies with the resort's ongoing expansion and the steady influx of white settlers who, according to Afro-Panamanian residents, come with the assumption that they are happy to see them (Mollett 2022)

On the contrary, land dispossession and enclosure in Bocas increasingly means local people are in need of paid work in order to remain living on the archipelago. For Afro-Panamanian women, their discontent with the diminishing natural resource access is inextricably linked to the sedimented racial and gendered histories that function as a part of the inequitable fabric of everyday contemporary life. For instance, according to many Afro-descendant women, finding a job is difficult, but "finding a job that isn't about cleaning, cooking or looking after children is *very* hard . . . well yes, ok, besides prostitution (laughing)" (Mollett fieldnotes 2012). A lack of jobs outside domestic service makes visible an enduring colonial imaginary that limits black women's subjectivity to tropes that are tied to notions of laboring animals (Morgan 2004). In the historiographies of the Spanish Americas, the disproportionate examples of black women in domestic service are commonly understood as the "afterlife" of colonial slavery (De Santana Pinho 2015; Goldstein 2013; Wade 2013; Hartman 2016). The enduring way Afro-descendant women serve as domestic labor in the homes and businesses of elites "embodies a collective geographic imaginary that assumes certain kinds of labour tasks such as 'cleaning' and 'cooking' are meant to be done by certain kinds of bodies" (Mollett 2021a, 10).

The naturalization of Black and Indigenous women's labor as forms of servitude (simultaneously carnal and domestic) works in tandem with their land dispossession through settler appropriations of land and labor in residential tourism development. Such practices and power relations within residential tourism enclaves resemble how everyday people in traditional extractive zones *jugarse la vida* (wagering life) to reach a life with dignity in the face of precarious conditions engendered by foreign and elite land dispossession and social and environmental processes of dehumanization (i.e., toxins on the body, racial/gender discrimination, and sex and servitude as *domestics*) (Valdivia 2018).

In Bocas, Afro-Panamanian women live in an extractive landscape that demands more than their displacement from land. State logics that drive a foreign-controlled landscape of tourism development and land ownership are also the very same logics that compel their participation in an industry that seeks them to serve (Mollett 2017). The logics of settler tourism development becomes part of the embodied political ecologies of daily survival and livelihood security for Afro-Panamanian women in Bocas. That the images and realities of Black and Indigenous women in Michael's story (from the Introduction) take place in a residential tourism enclave and not in a mining frontier should not exclude residential tourism from our scholarly work on extractivism. Veronica Davidov and Bram Büscher (2013) argue that, far from being polarized practices, natural resource extraction and ecotourism are frequent bedfellows, supported by the same institutions in the same landscapes. They conceptualize this intimate spatial, political, and institutional relationship as the "ecotourism-extraction nexus." In our analysis, we extend this thinking to include the logics of extractivism and the gendered and racialized dimensions thereof. The enduring racial, gendered, and carnal images and ideologies that link the contemporary coast of Panama to the transatlantic slave trade and Indigenous servitude for the purpose of European colonization and settlement in Spanish Americas began with the extraction of people (Indigenous Africans) and lands (Indigenous peoples) for the purpose of wealth pursued through extraction (Galeano 1997). These histories are part of the coast's ontology, where residential tourism development advances in this brutal colonial past present, rendering residential tourism an extractive industry, with its attendant extractivist logics and practices (Mollett 2021a, 2022).

Nanga-Eboko, Cameroon

Nanga-Eboko is an agrarian community of approximately 30,000 people, in the Central Province of the central African country, Cameroon. It lies along the meandering Sanaga River. In 2003, the community was the location of a temporary labor camp for the construction of the Chad-Cameroon Oil Pipeline (see Murrey 2015b). The pipeline right-of-way cleaves through the verdant Atlantic Equatorial coastal forests outside the city limits. The Chad-Cameroon Oil Pipeline is a US$4.3 billion 1,070 km underground steel pipeline, originating in Doba, Chad, and extending 12 km offshore of Kribi, Cameroon, to port facilities on the Atlantic coast. The project was a cooperative effort between the World Bank (which withdrew in 2008, four days after the Chadian government had repaid its project loans), the governments of Chad and Cameroon, and a consortium of private oil companies, including ExxonMobil, Chevron, and Petronas. At the time that the Environmental Assessment (EA) for the project was conducted in 1997, as much as 44 percent of "the Sub-Saharan African disease burden [was a consequence of] infrastructure investments" (Esso Chad EA 1997, 7–48). It has long been recognized within global health and international development that extractivism nurtures the conditions in which sexually transmitted diseases and ecological illness proliferate.

Dehumanization, we argue, is built into the racialized and sexualized logics of extractivism. The World Bank and the oil consortium considered an intensification of the ongoing HIV/AIDS emergency and STDs to be a known consequence of such large-scale infrastructural projects like the Chad–Cameroon pipeline. World Bank reports and other project documents reveal a foundational extractive logic of the abandonment of life and people. A Word Bank Policy Research Report from 1999, for example, offers a window into the dehumanizing econo-centric colonial logics of extractivism: "[S]ufficient net economic returns [should] cover . . . the cost of mitigating negative impacts, including the spread of AIDS" (World Bank 1999, 31). It is worth quoting from this report at length here to demonstrate princely the calculating voiding of life and embodied personhood performed through such equations,

> Other things held constant, the death of people with higher incomes will reduce average income—even though the welfare of those who remain alive has not changed. Conversely, the death of those with lower incomes raises average income, without necessarily improving the lot of any surviving individual and despite the suffering and economic losses of the families of those who died. Further, increased spending on health care and funerals is included in GDP calculations. As a result, per capita GDP may increase, even though overall well-being has not improved and the incomes of survivors have been reduced. (World Bank 1999, 31–33)

After calculating the income generation from funeral and health care from an influx of HIV/AIDS, the authors of the report consider the auxiliary "labor costs" of pursing large-scale infrastructural projects that cause an intensification of HIV/AIDS. Much like in Panama, extractivist logics would relegate women's bodies to forms of subjugated and dangerous labor.

> Because AIDS kills prime-age adults, many of whom are at the peak of their economic productivity, the shock of AIDS to the labor market is one mechanism through which AIDS might affect growth. However, in economies with substantial unemployment, firms should find it easy to replace sick or deceased workers, particularly if they are not key personnel. Other things being equal, the impact of the AIDS epidemic will be small until the economy begins to grow and is constrained by labor supply rather than by insufficient demand ... The departures of lower-skilled workers due to AIDS may have had little effect on firm profits. (World Bank 1999, 34)

Firm profits are prioritized over life, well-being, and community health within the colonial logics of extractivism. The glibly anticipated replacement of deceased workers erases the anticipated violence of bodily illness and naturalizes the capitalist logics of extractivism and accumulation. Moreover, unstated and unacknowledged within the pages of the report is that heightened HIV/AIDS rates are contingent upon heightened levels of sex work in construction zones. The document thus erases women's bodies as critical sites of labor within extractive geographies.

The National Committee for the Fight against AIDS noted in 2004, one year after the pipeline had begun to pump crude, that HIV "prevalence was 19.8 per cent higher along the pipeline corridor than in other regions of Cameroon." Drawing from decolonial and African feminisms, including Lugones (2007, 2010), Amadiume (1987), and Oyewumi (1997), we understand these extractive logics to be echoes of colonial logics, which normalized coloniality and violence as both racialized and sexualized. Mack and Na'puti (2020,

6) write, "The process of colonial subjectification (and dehumanization) is ongoing and constantly renewed, yet erased, obfuscated, and ignored in modernity." Our reading of the World Bank's extractivist logics reveals the acknowledgment and acceptance of gendered and sexual violence at the core of extractivism (and condoned by the state). These colonially inherited extractivist logics would envision women's bodies as *terra nullius.* A decolonial feminist orientation that understand body and territory as intimately interwoven offers a decolonial critique of the colonial gaze that perceives land as open, empty or "wasted" space for occupation and extraction (Said 1994, 253; see also Voyles 2015; Murrey and Jackson 2020, 920; Tilley 2020, 1446). Indigenous feminist resistance in Canada, as Ashley Noel Mack and Tiara R. Na'puti (2020) explain, is articulated through the denial of women's bodies as *terra nullius*—a refusal of the colonial impulse to perceive women's bodies as open territory for accumulation and extraction.

SEXUAL VIOLENCE AND EXTRACTIVISM: AN ENDURING ENTANGLEMENT

In this chapter, we have aimed to contribute to the multidisciplinary body of decolonial feminist work on extractivism. We have emphasized the necessity for a conception of extraction that moves beyond merely the physical movement and removal of subsoil resources to a focus on the logics of extractivism, including how they extend to residential tourism. In doing so, we have critiqued the economic and state-sanctioned logics of racialized sexual violence that are core features of racialized extractive capitalism. By bringing together black women's experiences of extractivism in Cameroon and Panama, we fashion intellectual lineages beyond the methodological nationalism that has long characterized scholarship on extraction.

The multiple power dynamics that are constitutive within residential tourism enclaves cannot be divorced from violent and colonial extractive practices from the past. During the early modern period when European royalty sought gold and mineral extraction in the making of the colonies, the travel accounts of male explorers to Africa and the Americas coproduced European discourses about women that "evoked desire, travellers depicted black women as simultaneously unwomanly and marked by a reproductive value that was both dependent on their sex and evidence of their lack of femininity" (Morgan 2004, 14; see also Mollett 2021b). In Europe, sociosexual deviance seemingly represented "savagery," the same tropes were applied in the colonies as a way to code the boundaries of European-ness. One such image is the trope of the naked African woman with large breasts that both miraculously dangled to the ground in one moment, and in the next, could be whipped over her shoulder to feed her suckling child on her back. Indeed, "the shape of her body marked her deviant sexuality and both shape and sexuality evidenced her savagery" (Morgan 2004, 16). These imagined sexualities remain part of the extractive fabric of daily life in the Americas and across the African continent more generally, and specifically in Bocas.

Similarly, sexuality and sexual violence are at work in the logics underpinning extraction in Cameroon. Our examination of the World Bank Group's archival materials from the Chad–Cameroon Oil Pipeline exposes the institution's preemptive inspection of the risk of heightened HIV/AIDS in communities along the pipeline route. The resulting WB value-assessment is premised upon a classification system of profit, material, value, and resource in which life is disappeared through devaluation of bodily autonomy and dehumanization—not as mere aftereffects or secondary moments but as constitutive of extraction. The territorialization of extraction, or the rendering legible of land into material components and segments for forms of forceful and often violent value-making, extends beyond the subterranean to the corporeal. This extension, we argue, is not accidental but patterned; it is colonial, racialized, sexualized, and gendered. The decolonial feminist concept of the "body-land" helps to unpack the particular experiences of Black and Indigenous women's bodies within geographies of extraction, pollution, and toxicity. These forms of violence are ongoing and make visible a process of racialized governance that weaves through and is co-constitutive of capitalist extractivism violently regulating and controlling the lives, bodies, and cultural practices of Black communities within extractive landscapes in Panama and Cameroon.

Finally, how do we research and write about losses so profound they seem unnameable? This hesitation is at the center of our collaboratively written decolonial feminist dialogue-as-critique. Social historian Cindy Patton (2020, vii) characterized authoring texts on AIDS as *writing under erasure* and Queer theorist Nishant Shahani thoughtfully explores the inadequacy of words and writing "under conditions of black death" (2020, n.p.). Elsewhere, we have critiqued the colonial, racist, and extractive processes and relations entrenched within and indeed constitutive of academic practice. For example, we have sought to demystify the "language of the mouth" of academic work on extraction (Murrey 2017) and the whiteness of geographical fieldwork (Faria and Mollett 2013). Indeed, our collaboration began as a conversation about the ways in which the political ecologies of survival are embodied, for people living in extractive zones. We have sought to contribute to decolonial feminist work that draws inspiration from Lugones' call for "a kind of intersubjective witnessing that might be undertaken as a way to build deep coalitions across those who are oppressed through, and who resist, coloniality" (Mack and Na'puti 2020, 6).

ACKNOWLEDGMENTS

We are grateful to the friends and communities who shared their stories with us. Some of this material is reproduced from Murrey, Amber and Sharlene Mollett. 2023. "Extraction is not a metaphor: Decolonial and black geographies against the gendered and embodied violence of extractive logics" *Transactions of the Institute of British Geographers* https://doi.org/10.1111/tran.12610.

REFERENCES

African Forum and Network on Debt and Development. 2007. "The contribution of oil to debt and under-development in Africa: The case of the Chad-Cameroon oil pipeline project." Harare, Zimbabwe: AFRODAD.

Amadiume, Ifi. 1987. *Male Daughters, Female Husbands: Gender and Sex in an African Society*. London: Zed Books.

Bagelman, Jen, and Sarah Marie Wiebe. 2017. "Intimacies of global toxins: Exposure and resistance in 'Chemical Valley'." *Political Geography* 60: 76–85.

Behzadi, Negar Elodie. 2019. "Women miners' exclusion and Muslim masculinities in Tajikistan: A feminist political ecology of honor and shame." *Geoforum* 100: 44–52.

Bell, Emily, Helena Choi, Cathy Nguyen, Ali Nouri, Rekha Reddy, Rachel Tobey, Tatyana Vinichenko, and Alyssa Wilson. 2004. "Big oil and health: Lessons from the Chad-Cameroon pipeline project." Princeton University: Woodrow Wilson School of Public and International Affairs. Accessed February 1, 2019. http://citeseerx.ist.psu.edu/viewdoc/download?doi=10.1.1.503.1713&rep=rep1&type=pdf

Billo, Emily. 2020. "Patriarchy and progressive politics: Gendered resistance to mining through everyday social relations of state formation in Intag, Ecuador." *Human Geography* 13 (1): 16–26.

Butler, Paula. 2015. *Colonial Extractions: Race and Canadian Mining in Contemporary Africa*. Toronto: University of Toronto Press.

Caretta, Martina Angela, and Sofia Zaragocin. 2020. "Women's resistance against the extractive industry: Embodied and water dimensions." *Human Geography* 13 (1): 3–5.

Caretta, Martina Angela, Sofia Zaragocin, Bethani Turley, and Kamila Torres Orellana. 2020. "Women's organizing against extractivism: Towards a decolonial multi-sited analysis." *Human Geography* 13 (1): 49–59.

Coyle, Fiona. 2004. "'Safe space' as counter-space: Women, environmental illness and 'corporeal chaos'." *The Canadian Geographer / Le Géographe Canadien* 48 (1): 62–75.

Daley, Patricia. 2020. "Lives lived differently: Geography and the study of black women." *Area* 52: 794–800.

Davidov, Veronica, and Bram Büscher. 2013. "Introduction." In *The Ecotourism-Extraction Nexus: Political Economies and Rural Realties of (Un)Comfortable*

Bedfellows, edited by Bram Büscher and Veronica Davidov, 1–14. London: Routledge.

Davies, Thom. 2018. "Toxic space and time: Slow violence, necropolitics, and petrochemical pollution." *Annals of the American Association of Geographers* 108 (6): 1–18.

de Leeuw, Sarah. 2016. "Tender grounds: Intimate visceral violence and British Columbia's colonial geographies." *Political Geography* 52: 14–23.

Esso Exploration and Production Chad Inc. 1997. "Environmental assessment—Chad Export Project, Chad portion. Dames & Moore, commissioned by Esso Chad." Accessed February 1, 2019. http://documents.banquemondiale.org/curated/fr/402921468742900078/text/multi0page.txt

Flint, Adrian, and Vernon Hewitt. 2015. "Colonial tropes and HIV/AIDS in Africa: Sex, disease and race." *Commonwealth and Comparative Studies* 53 (3): 294–314.

Gómez-Barris, Macarena. 2017. *The Extractive Zone: Social Ecologies and Decolonial Perspectives.* Durham: Duke University Press.

Hausermann, Heidi, Janet Adomako, and Maya Robles. 2020. "Fried eggs and all-woman gangs: The geopolitics of Chinese gold mining in Ghana, bodily vulnerability, and resistance." *Human Geography* 13(1): 60–73.

Hernández Reyes, and Castriela Esther. 2019. "Black women's struggles against extractivism, land dispossession, and marginalization in Colombia." *Latin American Perspectives* 46 (2): 217–234.

Icaza, Rosalba, and Rolando Vázquez. 2016. "The coloniality of gender as a radical critique of developmentalism." In *The Palgrave Handbook on Gender and Development: Critical Engagements in Feminist Theory and Practice*, edited by Wendy Harcourt, 62–73. London: Palgrave Macmillan.

Johnson, Adrienne, Anna Zalik, Farhana Sultana, Elizabeth Havice, Tracey Osborne, Gabriela Valdivia, Flora Lu, and Emily Billo. 2020. "Extraction, entanglements, and (im)materialities: Reflections on the methods and methodologies of natural resource industries fieldwork." *Environment and Planning E: Nature and Space* 4 (2): 383–428.

Leinius, Johanna. 2020. "Articulating the body, territory, and the defence of life: The politics of strategic equivalencing between women in anti-mining movements and the feminist movement in Peru." *Bulletin of Latin American Research* 40 (2): 204–219.

Mack, Ashley Noel, and Tiara R. Na'puti. 2020. "'Our bodies are not *terra nullius*': Building a decolonial feminist resistance to gendered violence." *Women's Studies in Communication* 42 (3): 1–21.

Mollett, Sharlene. 2017. "Irreconcilable differences? A postcolonial intersectional reading of gender, development and human rights in Latin America." *Gender, Place & Culture* 24 (1): 1–17.

Mollett, Sharlene. 2020. "Embodied history of land struggles in Central America."In *Historical geographies of, and for, the present,* edited by Levi Van Sant, Elizabeth Hennessy, and Mona Domosh,*Progress in Human Geography* 44 (1): 168–188

Mollett, Sharlene. 2021a. "Tourism as extraction: Unearthing coastal ontologies in the Panamanian Caribbean."In *Extraction, entanglements, and (im)materialities: Reflections on the methods and methodologies of natural resource industries fieldwork*, edited by Johnson, Adrienne, Anna Zalik, Farhana Sultana, Elizabeth Havice, Tracy Osborne, Gabriela Valdiva, Flora Lu, and Emily Billo, *Environment and Planning E* 4: 383–428. https://doi.org/10.1177/2514848620907470.

Mollett, Sharlene. 2021b. "Hemispheric, relational, and intersectional political ecologies of race: Centring land-body entanglements in the Americas." *Antipode* 53 (3): 810–830.

Mollett, Sharlene. 2022. "Swiss human geographies lecture 2019 tourism troubles: Feminist Political ecologies of land and body in Panama." *Geographica Helvetica* 77 (3): 327–340.

Morgan, Jennifer L. 2004. *Laboring Women: Reproduction and Gender in New World Slavery*. Philadelphia: University of Pennsylvania Press.

Murrey, Amber. 2015a. "Narratives of life and violence along the Chad-Cameroon oil pipeline." *Human Geography* 8 (1): 15–39.

Murrey, Amber. 2015b. "Invisible power, visible dispossession: The witchcraft of a subterranean pipeline." *Political Geography* 47: 64–76.

Murrey, Amber. 2017. "A post/decolonial geography beyond 'the language of the mouth'." In *Borders, Borderthinking, Borderland: Developing a Critical Epistemology of Global Politics*, edited by Marc Woons and Sebastian Weier, 79–99. Bristol: E-International Relations Publishing.

Murrey, Amber, and Nicholas A. Jackson. 2020. "A decolonial critique of the racialized 'localwashing' of extraction in Central Africa." *Annals of the Association of American Geographers* 110 (3): 917–940.

Naylor, Lindsay, Michelle Daigle, Sofia Zaragocin, Margaret Marietta Ramírez, and Mary Gilmartin. 2018. "Interventions: Bringing the decolonial to political geography." *Political Geography* 66: 199–209.

Neely, Abigail H. 2015. "Internal ecologies and the limits of local biologies: A political ecology of tuberculosis in a time of AIDS." *Annals of the Association of American Geographers* 105 (4): 791–805.

Oyewumi, Oyeronke. 1997. *The Invention of Woman: Making an African Sense of Western Gender Discourses*. Minneapolis: University of Minnesota Press.

Patton, Cindy. 2020. "Foreword." In *Aids and the Distribution of Crises*, edited by Jih-Fei Cheng, Alexandra Juhasz, and Nishant Shahani, vii-xv. Durham: Duke University Press.

Razack, Sherene H. 2016. "Gendering disposability." *Canadian Journal of Women and the Law* 28 (2): 285–307.

Shahani, Nishant. 2020. "Pre-existing conditions." *Duke University Press Blog*, July 30, 2020. https://dukeupress.wordpress.com/2020/07/30/dispatches-on-aids-and-covid-19-continuing-conversations-from-aids-and-the-distribution-of-crises-dispatch-two/

Tilley, Lisa. 2020. "'The impulse is cartographic': Counter-mapping Indonesia's resource frontiers in the context of coloniality." *Antipode* 52 (5): 1434–1454.

Tsikata, Dzodzi. 2010. "Introduction." In *Land Tenure, Gender and Globalisation: Research and Analysis from Africa, Asia and Latin America*, edited by Tsikata,

Dzodzi and Pamela Golah, 1–34. New Delhi: International Development Research Centre.
Ukeje, Charles. 2005. "From Aba to Ugborodo: Gender identity and alternative discourse of the social protest among women in the Oil Delta of Nigeria." In *Gender Activism and Studies in Africa*, edited by Sidne Arnfred and Babere Kerata Chacha, 67–87. Senegal: CODERIA.
Valdivia, Gabriela. 2018. "'Wagering life' in the Petro-city: Embodied ecologies of oil flow, capitalism, and justice in Esmeraldas, Ecuador." *Annals of the American Association of Geographers* 108 (2): 549–557.
Van Sant, Levi, Elizabeth Hennessy, and Mona Domosh. 2020. "Historical geographies of, and for, the present." *Progress in Human Geography* 44 (1): 168–188.
Voyles, Traci Brynne. 2015.*Wastelanding: Legacies of Uranium Mining in Navajo Country*. Minneapolis: University of Minnesota Press.
World Bank. 1999. "Confronting AIDS: Public priorities in a global epidemic." *A World Bank Policy Research Report.* Oxford: Oxford University Press.
Yusoff, Kathryn. 2018. *A Billion Black Anthropocenes or None*. Minneapolis: University of Minnesota Press.
Zaragocin, Sofia. 2019. "Gendered geographies of elimination: Decolonial feminist geographies in Latin American settler contexts." *Antipode* 51 (1): 373–392.

4

Young Female Miners in Tajikistani Coal Mines

Intersectional Extractive Violence and Ecologies of Exhaustion

Negar Elodie Behzadi

Darya just turned eighteen—"too old to be a miner," she says. She still has hopes and dreams of a future in which she will leave the village, study, and come back to get married. Yet, as for many young female coal miners in Kante, a small village in the heights of the Fann Mountains in post-Soviet Tajikistan, Darya knows that her chances of leaving are as meager as her chances of finding a husband. "Girls like me are the *komorskesh* (coal miners)," she says. When used for a woman, this term feels like an insult in Kante.

In this village of 1,500 inhabitants, which lies on the largest coal reserves in the country, male and female miners of all ages (adults, youth, and children) inhabit the same fragmented postsocialist extractive landscape. Since the fall of the Soviet Union, multiple forms of livelihoods have emerged, including informal/artisanal mining, formal/industrial mining in the local Sino-Tajik mine, and male seasonal labor migration to Russia. All inhabitants, however, do not experience these new forms of labor in the same way.

While villagers often see male migrants and informal miners as strong and proud, they consider women's work in mines as *ayb* (shameful). The nineteen adult women miners in the village work in a hidden area where the coal is harder to extract, but they know they can go unseen. Those occupying the worst position are the eight young female miners, who are reaching the conventional age of marriage (seventeen to eighteen). Among these are Darya (eighteen), Gulpari (sixteen), and Bahar (fourteen). Unlike adult women, they are allowed to mine near men, but their work bears a cost: that of compromising their future prospects. Marriage is crucial for young women in Tajikistan—in a context of economic desolation and widespread unemployment, it constitutes one of the rare viable economic strategies to secure a future. For all of these miners, and for the whole village, everyday life takes place against a constant backdrop of dynamite explosions from the "Chinese mine," and the fear that the black coal-rich ground will disappear from beneath their feet as "the Chinese" blow up the mountain and "take all the coal we've got." "Soon," the villagers say, "there will be none left."

In this chapter, I examine the gendered threats and experiences of exclusion that have appeared in Kante with the emergence of a postsocialist capitalist extractive landscape. First, I consider the threat of coal "disappearing" under the villagers' feet and its impacts in terms of resource access and distribution; and second, I explore the threat of the material and social exclusion of women miners in general, and young female miners in particular. As already revealed by feminist political economists and ecologists, the expansion of capitalist systems of production/reproduction relies on specific gendered exclusions, whereby women are positioned at the bottom of the chain. This includes exclusion from certain places and forms of labor, as well as from access to and the benefits of resources (Rocheleau et al. 1996; McDowell 1999; Nagar et al. 2002). The expansion of this system is also tied to gendered extractive violence, encompassing both sexual and emotional violence through processes that stigmatize, make invisible, and socially exclude certain groups, in particular women (Bashwira et al. 2014; Lahiri-Dutt and Macintyre 2006). While a growing body of feminist political ecologists has attempted to investigate the diverse directions taken by gendered subjectivities with changing relations of resource governance (Elmhirst 2011; Truelove 2011; Sultana 2009), Feminist Political Ecology (FPE) has often been criticized for taking a monolithic approach to gender that erases difference—both in terms of its historical and contemporary construction and its lived experience (Mollett and Faria 2013). In this chapter, I attempt to push the epistemic borders of FPE by asking what a more *holistic* (Verges 2019; Arruzza et al. 2019) political ecology—a *decolonial feminist political ecology* that "messes with gender" (Mollett and Faria 2013) and is attentive to the intersectional construction and experience of difference by/in the colonial extractive capitalist present—can bring to our understanding of gendered exclusion?

To answer this question, I draw on ethnographic research I conducted in Kante over the period 2014–15. At the heart of my response lie the stories of three young female miners, put in perspective with the rise of extractive capitalism in Tajikistan and Kante. In studies of politico-economic change, young people's voices often occupy the margins; yet, as already highlighted by geographers and anthropologists, young people play a crucial role in the (re)working of global capitalism (Dyson 2008; Jeffrey 2010, 2012; Skelton 2013). In the postsocialist context of this study, their standpoint is all the more relevant. Youth in Tajikistan and the rest of the post-Soviet world were born with and grew up alongside the rise, expansion, and formation of capitalist modes of governance (Kirmse 2010).

Drawing on their voices, I argue that a decolonial feminist political ecology reading enables the retheorization of gendered extractive exclusion beyond a binary perspective of exclusion/inclusion, by which women are excluded from access to and the benefits of resources and need to be included (through greater visibility/inclusion). Rather, I develop a systemic analysis of gendered extractive exclusion as a form of intersectional violence embedded in *ecologies of exhaustion*—an inherent feature of extractive capitalism, based on a simultaneous, dialectical, fatigue, and double exhaustion: a productive exhaustion of land/resources and a reproductive exhaustion of women's bodies. I show, however, that within this system, all bodies are not exhausted in the same way: the rise of extractive capitalism fragments landscapes by variously reinforcing existing axes of difference and shaping new ones. Physically and emotionally violent, this fragmentation slowly permeates the intimate geographies of the most vulnerable—in this case, young, underclassed, female, rural, Muslim Tajikistanis.

By adopting this framework, I also show the importance of thinking in terms of colonial presence in our understanding of socio-economic and gendered relations in the borderlands of the former Soviet empire. The Russian tsarist empire's conquest of Central Asia transformed places in ways that were heavily gendered, in particular via the production of the idea of Muslim Central Asian women as being traditional and backward (Tlostanova 2010; Northrop 2004; Zekhni 2022). Debates around whether the subsequent Soviet period which started in the twentieth century could be studied using a postcolonial framework have long animated Central Asian scholars. Those in favor of the use of a postcolonial framework have revealed how the production of gendered and ethnic fragmentation constructed through the Soviet modernization endeavor was partly colonial (Behzadi and Direnberger 2020; Kandiyoti 2007; Yusufjonova-Abman 2018). The specific contemporary form of extractive capitalism that exists in Tajikistan is embedded into this longer history. Yet, coloniality is not only historical. The chapter that follows shows how relations of domination and social hierarchies are still produced and reproduced in the contemporary period, making the use of colonial analogies and decolonial theory particularly pertinent.

The first section of this chapter develops a decolonial feminist political ecology reading. The second then outlines the rise of postsocialist extractive capitalism in Tajikistan and Kante, and the third offers a gendered analysis centered around the stories of three young female miners. The resulting conclusion highlights the (re)production of difference in this context.

"ECOLOGIES OF EXHAUSTION" AND THE SLOW VIOLENCE OF CAPITALIST EXTRACTIVE LANDSCAPES: A DECOLONIAL FEMINIST POLITICAL ECOLOGY READING

Extractive capitalism as a system relies on the simultaneous exploitation of resources and human and nonhuman labor (Moore 2015). According to Marxist theory, this system leads inevitably first to "fatigue"—of working bodies and resources—and second to "exhaustion" (Toscano 2018, 125). While *fatigue* belongs to the domain of the productive, *exhaustion* touches the reproductive domain. *Exhaustion*, in fact, occurs when the boundaries of fatigue have been transgressed, and the damage to reproductive capacities becomes "irreparable," that is, the "reproduction of certain bodily or relational states is no longer possible" (ibid.). Many geographers and other social scientists see the current crisis of capitalism as a crisis of social reproduction, creating a moment in history where the conditions required for the sustenance of life cannot be reproduced (Moore 2015; Meehan and Strauss 2015; Katz 2001). Often inspired by Marxist/feminist theories, such readings highlight the ways in which the exploitation and exhaustion of women's bodies constitute a condition of capitalist expansion (Federici 2004; Moore 2015). This double exhaustion—both productive and reproductive—first of land and resources (minerals, water, and others), and second of women's bodies, is what I refer to, here, as *ecologies of exhaustion*.

When brought into conversation with feminist decolonial and anti-racist scholars' perspectives (Lugones 2010; Verges 2019; Arruzza et al. 2019; McKittrick 2006), the concept of *ecologies of exhaustion* sheds light on the unequal violence of systems of extractive capitalism. Unpacking the category of women and differentiating various forms of feminism, decolonial, and anti-racist feminists highlight that all bodies are not exploited, excluded, and exhausted in the same way (ibid.). Maria Lugones discusses the "coloniality of gender," and its everlasting presence, lying at the intersection of gender, race, class, and other axes of difference. This constitutes one of the "central constructs of the capitalist system of power" (Lugones 2010, 746), under which certain bodies are classed as inferior, made disposable, and denied humanity. This idea of the "coloniality of gender" cuts across all levels of societal organization, shaping economies, ecologies, institutional relations, and everyday lives.

This framework is particularly useful when applied to questions of resource struggles. First, it still allows us to unravel the ways in which resource extraction (e.g., through mining) acts as a central means by which capitalism unfolds and shapes specific gendered struggles around inclusion/exclusion and equality/inequality in access to land, minerals, and labor.

At the same time, decolonial and feminist decolonial perspectives take the discussion further by providing tools to understand the systemic processes through which capitalist and neoliberal extractive development produce further forms of exclusion, both in rhetoric and in practice, including through fracturing the social fabric of life along multiple axes of difference, such as gender, race, class, locality, age, religion, Indigeneity, and ethnicity (Murrey 2016; Anthias 2018; Murrey and Jackson 2020; Behzadi 2018, 2019; Postar and Behzadi 2022; Mollett 2017, 2018). This fracturing and the reproduction of socio-environmental inequalities and differences are a condition of the existence of capitalist and neocolonial forms of resource extraction, as characterized by resource commodification, labor segregation and precarization, informalization, the enclosure of previously commonly owned resources for capital expansion, dispossession, and the exclusion of the most vulnerable.

In this chapter, I approach such fracturing and exclusion as forms of violence, both spectacular and mundane, embedded in colonial history (Mollett 2018), and which I refer to as *intersectional extractive violence*. Expanding on Nixon's theorization of "slow violence," I see the proliferation of conflicts in extractive sites that are "incremental and accretive," "relatively invisible," and temporally dispersed in "situations where the conditions for sustaining life become increasingly and gradually degraded," as forms of violence (Nixon 2011, 2–3). I am similarly attentive to actors who are subject to geographical variations along intersectional lines, and who are the main victims of slow violence. Such violence occurs when an "official landscape" is imposed on a "vernacular landscape," shaping "displacements in place" (Nixon 2011, 19)—that is, displacements that occur without people moving from a place of belonging, but which lead instead to in situ losses of land and resources, to alterations to the social fabric of life, and to the exclusion of certain bodies that are considered to be inferior and disposable (Mollett 2018; Murrey 2015).

AN EMERGENT POSTSOCIALIST EXTRACTIVE CAPITALISM: RESOURCE STRUGGLES AND PRODUCTIVE EXHAUSTION IN TAJIKISTAN AND KANTE

(Trans)National Struggles: The Rise of Extractive Capitalism

In Tajikistan, the fracturing approached theoretically above manifests as a result of the rise of post-socialist extractive capitalism. Here, as elsewhere in the postsocialist world, the end of Soviet rule opened the way for a gradual switch from collective and socialist modes of production/reproduction to hybrid capitalist forms of development and resource governance. The collapse of the Soviet Union led first to a civil war that historians of Tajikistan and other experts have often interpreted as a battle for access to and control over resources, including cotton, drugs, land, and various industries (Nourzhanov and Bleuer 2013; Dudoignon 1998). In 1997, the Tajik government signed a peace treaty monitored by international organizations, and as elsewhere, the neoliberal peace-building agreement precipitated economic reforms and capitalist modernization as instigated by the World Bank and the International Monetary Fund (Heathershaw 2009). Privatization and structural adjustment policies rapidly led to wide-scale unemployment, the retraction of the state from activities relating to social reproduction, the informalization of labor, and the seasonal labor migration of men to Russia.

Post-civil war privatization in Tajikistan took a specific form and happened in "secrecy" (Dudoignon 2004, 133–34), with former commanders taking control of resources, steering international aid toward elites, and creating monopolies (Nakaya 2009). This combination of elite control, authoritarianism, and privatization is what constitutes the particularity of postsocialist extractive capitalism, visible in multiple aspects of Tajikistani economic life, from agriculture (Hofman and Visser 2014) to hydropower development (Ito et al. 2015) and mineral resource extraction (Heathershaw 2011). TALCO Resources is an example of such control in the extraction sector: the state-owned energy and aluminium production company is largely controlled by the president, Emomalhi Rahmon, and has over the past two decades been involved in multiple corruption and fraud scandals (Heathershaw 2011; Nakaya 2009).

New transnational actors and struggles have added to this complex situation, revealing the ways in which Tajikistan is connected to broader global capitalist trends (Heathershaw 2011). In search of new development opportunities, the country geared its efforts toward the exploitation of its rich subsoil and called upon international actors and extractive companies, including Anglo-American as well as Indian, Chinese, and Kazakhstani organizations (Bankwatch 2017). In parallel, the end of Soviet interregional dependency led to the emergence of new struggles over energy. Conflicts surrounding the gigantic Rogun hydropower development in Tajikistan, and its potential impact on neighboring Uzbekistan, caused the shutdown in 2012/3 of the pipeline that was, at the time, providing Tajikistan with Uzbek gas, which in turn expedited a move to coal extraction for national consumption within Tajikistan (Ito et al. 2015). As I explored elsewhere (Behzadi 2019), Chinese companies played a central role in this development, and more generally in resource extraction in Tajikistan, as can be observed by the creation of multiple joint ventures with Tajikistani state-owned companies. This growing role is also embedded in a regional politico-economic strategy driven by the Chinese—the Belt and Road Initiative (BRI), which focuses on investment and infrastructural development including overland highways, pipelines, and railways.

Local Manifestations in Kante

The specific form of extractive capitalism described above—a resource extraction-driven capitalist model embedded in transnational networks and global trends, but still controlled by an authoritarian state—shapes and is shaped by specific localities, as illustrated by the changes experienced in the village of Kante. During the Soviet times, women worked mainly in the local collective tobacco farm while men often commuted to a nearby town to work in the mercury, antimony, and later coal conglomerate. In the past twenty years, however, Kante has experienced a gradual integration into the global market economy. In 1991, Kanteguis started digging coal for subsistence purposes: extended families (*avlod*) opened lateral galleries into the mountains and men worked using pickaxes and donkeys to carry coal bags out. The slow commodification of coal in the country led, however, to the creation of a new extractive landscape where informal and formal mining now coexist.

Since the 2000s, the village has seen the expansion of informal markets: truck drivers come to the village from far and wide, fill their trucks, and sell the coal in other informal markets across the country. Since the closure of the Uzbek-Tajik gas pipeline in 2012, coal demand and prices have risen, leading to a boom in local production and the transition from a subsistence economy to an informal business involving more actors. In parallel, the search for coal at the national level, and the opening up to transnational actors, materialized in Kante in the form of an open-pit Sino-Tajik mine, established in 2013 as a joint venture between TALCO resources and a Chinese company (figure 4.1). According to villagers and the Chinese industrial mine manager, coal production was initially intended to fuel the famous TALCO cement factory in western Tajikistan, previously powered by gas.

Productive Exhaustion

Resource struggles between informal/artisanal miners and the Sino-Tajik mine illustrate the ways in which the rise of this extractive capitalist landscape leads to the destruction of "vernacular landscapes" by "official landscapes" (Nixon 2011, 2–3). The multifocal livelihood setting of Kante resembles many extractive landscapes around the world that emerged with the neoliberalization of resources (Lahiri-Dutt 2018; Behzadi 2019), as characterized by industrial mines encroaching upon informal miners, leading to their exclusion from land and access to resources, and the slow violence of "displacements in place" (Nixon 2011, 19). In Kante, there is widespread evidence of this

Figure 4.1 Open-pit Sino-Tajik Industrial Mining Area. Photo by the author.

violence: the Sino-Tajik mine dynamited an area of the mountains where the informal miners used to work, and today, there is one area at the bottom of the village where informal mining and formal mining overlap. Explosions in the mountains, tumbling stones that crush informal miners' houses and donkeys, constant expropriations, the casualization of labor, and difficult relations with "the Chinese" are all part of the villagers' everyday lives.

The mines are, however, both a blessing and a curse for villagers. "Coal is all we've got," they often say. According to a running village joke, it is better for a woman to marry a donkey from Kante than a man from Shurmashk (a neighboring village that has no access to informal coal mining). Mining, both artisanal and industrial, also provides men with an alternative to work in Russia, where migratory experiences are sometimes harsh, violent, and unsuccessful. "Coal is what allows our men to stay here," say most women in the village. At the same time, the subterranean wealth is a threat to the existence of the village itself with rumors that the village will disappear, eaten up by the mines. This reveals the ways in which the emergence of capitalist modes of production go hand in hand with the threat of the exhaustion of the very means of capitalist production: exhaustion of land and resources by "the Chinese."

The state plays a notably ambivalent role in this system, illustrating one of the main features of this postsocialist form of extractive capitalism: that it relies on elite control and the government's complicity in people's exclusion. Government officials sometimes come to the village, threatening miners and telling them to stop their activities. Often however, the officials only fill their trucks with coal bags and leave—until their next visit. The *hukhamat* (local government) also sends officials to the village to offer families the opportunity to establish new houses in another village down in the valley. "They say there is not enough water here [in Kante], but all they want is our coal [. . .] We know the land is dry there too," one male villager tells me. Men also explain their reluctance to resist the exclusion overtly: resistance would be "useless," they say, aware that the Chinese and the Tajikistani governments are complicit in their dispossession.

INTERSECTIONAL EXTRACTIVE VIOLENCE: WOMEN'S EXCLUSIONS AND THE THREAT OF REPRODUCTIVE EXHAUSTION

Women Miners' Exclusions

The threat of productive exhaustion that the Kantegui live with impacts everyday lives beyond those of men's. The rise of this postsocialist extractive landscape, in its shaping of the gendered and generational local governance of resources, leads to the work of children. Mining from the age of five alongside their parents, children usually lead a donkey back and forth from the galleries and supply bags of coal to the informal markets. However, while villagers normalize children's work in mines, they consider women's mining as shameful. As mentioned at the beginning of this chapter, female miners experience two forms of exclusion: first, a partial exclusion from access to and the benefits of natural resources; and second, a social exclusion. The nineteen women miners who go mining every day hide at the top of the village behind the mountains to work. Villagers refer to their mining area as "another village." Coal in the women's mining area is harder to extract than down in the village, and the informal market where women sell coal is more difficult to access than men's (figure 4.1). To reach the women's market, truck drivers must cross a river that, in the coldest months of the year when coal prices are at their highest, is frozen and unpassable.

Women miners' material exclusion is doubled by social exclusion. They are often excluded from social networks, stigmatized, and laughed at. The reasons for this shaming are multiple. First, though women are excluded from other types of labor (e.g., work in the industrial mines or work found through migration), informal mining remains, for villagers, *kori mard* (the work of men). "Our women do not work, they are *qishlaqi* (country women), say the men and women of the village. "Women in cities work, but here our women are *khonachin* (housewives). Villagers also often refer to the bodies of women miners as too weak to undertake this type of work, which might

Figure 4.2 Child Male Miners Leading Donkeys Out of the Informal Coal Mines to the Informal Market. Photo by the author.

threaten their capacity to "take care of the home" and "make good babies." By working in mines, women also engage with *mardoni begoni* (foreign men)—the truck drivers whom they sell their coal bags to. This contact (both verbal and physical) leads to the women's sexualization.

The causes for these women's exclusion, their experiences, and the gendered dynamics they are embedded within have been ethnographically explored elsewhere (Behzadi 2018, 2019), in accounts that have highlighted the diversity of positions held by women miners. Among women miners in their thirties in Kante, four are women whose husbands left for Russia and have not returned and have stopped sending remittances. Four others worked alongside their husbands as a joint strategy in the women's mining area, while six women miners in their forties ran family businesses with their sons and daughters-in-law. Elements of differentiation in the degree of social exclusion appeared in relation to age and sociomarital position. The most stigmatized

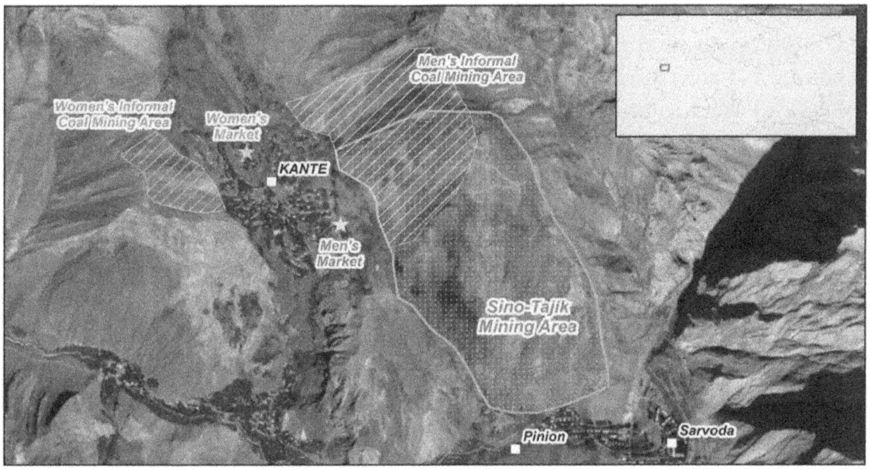

Figure 4.3 Map of the Kantegui Extractive Landscape. Map created by the author.

women were those in their thirties whose husbands had left. They were often called "crazy" and "not reputable" and were considered guilty by the village for their predicament on account of having a bad temper and having failed to attend to their husbands' needs.

Young Female Miners

In this context, young female miners appear as a specific category whose experiences of exclusion both resonate with and differ from those of female adults. Gulpari is a young miner who started work in the mines when she was eight. Now she is fifteen but still goes mining most winter days. She should not have to—her two brothers are thirteen and eight, old enough to mine according to the Kantegui—but Gulpari's family is different from others. Her younger brothers go mining sometimes, but she is the one who goes most often. After school, she returns home, drinks a glass of tea, changes her clothes, ties a scarf around her mouth to protect herself from coal debris, and places her flashlight on her head. She fetches the family donkey from the shed and leads it toward the dark mountain slope. She enters the coal gallery and, emerging with bags full, will go back and forth ten to fifteen times. "It is a tiring job," she says. In the dim gallery, the air is moist and thin. The deeper she goes, the less oxygen there is. "My brothers are both a bit ill," Gulpari explains, with a touch of bitterness. "They cannot work [. . .] one is overweight and cannot breathe, and the other has back pains and cannot carry the coal bags [. . .] so *I* go."

Gulpari mainly mines with her father, Behruz, when he returns from Russia. He left for seasonal migration to Novossibirsk, in Russia, several years ago but comes back every year during the cold months from October to March when coal mining is at its peak and the business most profitable. When he is back, he usually goes mining with his brother in the gallery that belongs to their extended family. They take Gulpari with them—staying for hours at the end of the gallery to extract coal with a pickaxe, while Gulpari goes in and out with the donkey. "We make money faster this way," she says proudly. One of Gulpari's uncles just got married and his wife is pregnant. They are trying to save money to build their own house, so he does not have to return to Russia. He could work for the Chinese instead, but like many men, he does not want to.

Some winters, times are harsher than others and Gulpari's father does not come back from Russia. "He can't afford the ticket and it's too far, so he stays there," she shares. In such meager years, the remittances he sends back home are sporadic and rarely enough for the family to make ends meet. In such times, Gulpari goes mining with her uncle. He pays her 100 somonis (around 20 USD), and she gives 90 somonis to her mother and keeps 10 somonis to buy clothes or books for school.

Gulpari is well aware that she is not the norm. Girls her age usually do not go mining: "17 is the maximum for unmarried girls to work in the mines," she says, and she is not very far from it. "Mines are not a girl's place [. . .] it's tough for girls," she shares with anger. "We don't go out, we have to learn how to do housework [. . .] In life, there is *kori khona* (domestic work) and *kori berun* (public/outside work) [. . . .] Boys only spend time playing outside, we, we work inside." Like all young girls in Kante, Gulpari knows that her time is precious.

Her neighbor, Bahar, fourteen, like other young girls who stopped mining at the right age, is less guarded than Gulpari when I ask her about the fate of young miners. Bahar went mining from the age of nine to eleven until her brother was old enough to take over. Like Gulpari, she worked for her father and uncle. The family stopped altogether, however, when Bahar's father had made enough money to build their new house. Now they just go sometimes to get coal for the home stove. "When girls go mining," says Bahar, "they don't have time to work at home and wash the dishes, do the cooking, sweep the floors [. . .] people don't like that." From around the age of nine, young girls must learn how to undertake housework, recalls Bahar: "When they are young, girls can still work in mines and do both, but from the age of 14, 15, it's not right anymore." "It's the boys' role to go mining and sell the coal, girls' is to do housework." "The women's role is to teach young girls to do so." "Girls who go mining, they don't make them *kelins* (daughters-in-law), they cannot get married." "Coming and going in the mines, you suffer, your legs hurt, that's hard," she says. Young girls get *monda* (tired) and cannot do the housework anymore. Girls who still do it, they are the *kambaral* (the poorest of the poor/the unfortunate), "those who don't have money, or have no one, no big brothers or no brother at all."

Darya, eighteen, whose words begin this chapter, is one of these *kambaral* (the poorest of the poor/the unfortunate). She knows very well what people say about her in the village. She is a *komorkeh*, a miner. People say we're *ganda* (spoiled/bad/exhausted), "that we cannot make good babies." Darya is silent and reserved: while most young women wear the traditional Tajik rumol, she wears her scarf tied in the front of her neck as a sign of devotion. "Like that people know that I'm a good girl." Darya wants to become a nurse one day but knows that it will probably not happen: "Village girls do not have the right to study." "In cities, girls can do things, but *we* cannot." "Our conditions do not allow it anyway." "Girls have to marry, there is no money." "But men don't marry girls like me." "There are lots of men in the mine, and girls get looked at. Men don't want to marry girls who have been looked at." Darya seems resigned— "What else can I do?" she says. "We have to work, otherwise, our father will have to go to Russia." "I don't want him to go," she says. Darya's father, unlike other men in the village, has never been to Russia and does not want to go. Her father, mother, and younger sisters all live together with her in the village, and she carries on working to make sure her father can stay.

Reproductive Exhaustion, Difference, Reworkings

Gulpari, Darya, and Bahar's stories, put in the context of the rising postsocialist capitalist landscape of Kante, highlight the ways in which intersectional extractive violence operates, leading to multiple forms of displacement in place. Slow violence occurs first, through the exclusion of women from access to land and resources, as evidenced by the gendered socio-spatial organization of extractive labor in Kante. But extractive violence also operates through a different form of social fracturing which occurs in tandem and dialectically with the threat of the exhaustion of land and resources: a threat of reproductive exhaustion.

This threat has multiple facets and impacts different women differently. For villagers, work in mines creates fatigue and bodily exhaustion, leading to an alteration of women's reproductive capacity: "women miners are '*monda*' (tired), '*ganda*' (spoiled/exhausted/bad), they cannot make good babies," say the Kantegui. This alteration also touches on their capacity to undertake household domestic labor: "women cannot take care of the house," repeat men, women, and young people in Kante. The term *ganda* goes beyond the idea of fatigue and tiredness. Its multiple meanings include a moral judgment based on the idea of the corruption of the body, and people use it to refer to anything "bad" or that "turned bad," including a spoiled fruit. Applied to women miners, this exhaustion is entwined with their sexualization. For younger women who reach the age of marriage, this sexualization compromises their respectability and ability to find a husband.

Rurality/urbanity is one axis that determines these differentiated experiences of labor: villagers in Kante often mention differences between rural and urban women in terms of their access to and right to work. Darya also mentions differences in mobility and access to education between rural and urban girls. While this idea of rural femininity and how it differs from urbanity appears as a feature of this extractive landscape, I have elsewhere shown that the existence of a gendered and ethnic hierarchy, whereby rural, Muslim Tajikistani women are constructed as traditional and backward, is also a Soviet (post)colonial production (Behzadi and Direnberger 2020).

Intersecting with this notion of rurality/urbanity are two other central axes of difference: age and socio-marital position. The most stigmatized adult women miners in the village are women in their thirties whose husbands have left for Russia and have not returned. Still young enough to be sexualized, their proximity to men is considered unacceptable. Young girls approaching or reaching the age of marriage, such as Gulpari, Bahar, and Darya, are also in a particularly vulnerable position. Work in mines, and its perceived effect upon their bodies, threatens their chances of getting married. Yet marriage, alongside reproductive labor and capacity, has, more than ever in Tajikistan, become a central economic strategy, as well as an element of identity and marker of social status. The retraction of the state from reproductive activities and the absence of employment has led to a re-evaluation of marriage (Cleuziou 2016). Although not new, the gendered division of labor described by young women miners, whereby young women must, from an early age, learn how to grow up to be good daughters-in-law, has been reinforced following the fall of the Soviet Union.

In this context, the young girls whose families accept the threat to their respectability, honor, and future prospects belong to a specific category that has emerged with the rise of the extractive landscape of Kante: the *kambaral* (the poorest of the poor/the unfortunate). Like Darya and Gulpari's families, these are those without brothers capable of performing (sustained) extractive labor, or those where the fathers are either insufficiently successful in their migratory pursuits, or not "strong enough," as villagers say, to become migrants at all, to support the family. The rise of this new category illustrates the ways in which changes in labor relations have led to the shaping of a new axis of difference around elements of class. While successful families (either in business or migration) rise in social status, families with men who do not possess the skills to work in formal employment, or who either refuse migration or have been unsuccessful in its pursuit, occupy an inferior class position in the village.

Darya and Gulpari's stories reveal the emotional distress that results from the social position they occupy—and the bitterness, shame, and anger they feel. At the same time, their stories also highlight their role in reworking the conditions of their social exclusion and the slow violence they are victims of. Through their work in the mines, these young women find spaces to negotiate elements of their identities and challenge their stigmatization. This can be seen in Darya's relationship with the performance of her Muslimness: by adjusting her dress, Darya tries to display signs of modesty that might challenge her social exclusion. These young women also participate in a form of intergenerational solidarity that they experience with pride: their work contributes to helping their fathers avoid the hardships of migration and work with "the Chinese." Through their work, they resist the separation of families that the current conditions of capitalism have imposed upon the Kantegui, the Tajikistani, and many other Central Asians.

CONCLUSION

The stories of Gulpari, Bahar, and Darya, put in perspective with the rise of postsocialist extractive capitalism in Tajikistan, and in Kante in particular, illuminate how slow, gendered, intersectional forms of exclusion and violence emerge—or forms of *intersectional extractive violence*. As the stories of young female miners and other villagers reveal, the rise of the Kantegui extractive landscape leads to the replacement of "vernacular landscapes" by "official landscapes" and creates forms of "displacements in place" (Nixon 2011, 2–3, 19), including through the (re)production of (new) axes of difference, such as gender, age, class, rurality/urbanity, Muslimness, reproductive capacity, socio-marital position, and others. In this system, certain women—those who are young, unmarried, and poor—are more vulnerable than others.

While the stories of Gulpari, Bahar, and Darya are unique, they also occur in an emerging extractive landscape where ongoing changes illustrate the types of exclusions, violence, and struggles experienced by certain categories of actors who are, because of their social position, more vulnerable than others. The system that makes such violence possible presents elements of rupture and continuity with the socialist past. The modernist Soviet development project in the Central Asian South and Tajikistan was based on the extraction of resources and large infrastructure development in the peripheries for industrialization (Kalinovsky 2018), which transformed localities in ways that sometimes produced forms of exclusions. The new ways in which new actors now extract resources is reminiscent of the Soviet past, and it evidences a form of "colonial presence" in the postsocialist present that characterizes extractive capitalist developments.

However, the exclusion of villagers from the benefits of resources, as well as the informalization, precarization, and casualization of labor, and the enclosure of land previously collectively owned for capital expansion, all constitute specificities of this new extractive landscape. In this context, the creation of an underclassed category of women considered as inferior, working in dangerous, illegal, undervalued types of labor, putting their bodies at risk, and threatening their prospects, marriageability, and future, appears as a necessary condition for the development of extractive capitalism. This is what, in this chapter, I referred to as the formation of *ecologies of exhaustion:* a productive exhaustion of land, of resources, and a concomitant reproductive exhaustion of the bodies of young women miners and of their future possibilities. Touching on young women miners' most intimate geographies, this chapter goes beyond binaries of exclusion/inclusion, colonial past/postcolonial present, and in this specific case, postsocialist/capitalist modes of production. It reveals the connectedness between the embodied experiences of extractive violence by marginalized youth, the fragmentation of their everyday lives along multiple axes of difference, and the pushing of new extractive frontiers to areas previously inaccessible to capitalist expansion.

REFERENCES

Anthias, Penelope. 2018. *Limits to Decolonization: Indigeneity, Territory and Hydrocarbon Politics in the Bolivian Chaco*. Ithaca: Cornell University Press.

Arruzza, Cinzia, Tithi Bhattacharaya, and Nancy Fraser. 2019. *Feminism for the 99%: A Manifesto*. London and New York: Verso.

Bankwatch. 2017. "Review of the Coal Sector in the Republic of Tajikistan." Little Earth/Bankwatch. https://bankwatch.org/wp-content/uploads/2017/12/Tajikistan-coal.pdf.

Bashwira, Marie Rose, Jeroen Cuvelier, Dorothea Hilhorst, and Gemma van der Haar. 2014. "Not Only a Man's World: Women's Involvement in Artisanal Mining in Eastern DRC." *Resources Policy* 41 (1): 109–16. https://doi.org/10.1016/j.resourpol.2013.11.002.

Behzadi, Negar Elodie. 2018. "Gendered Coal Struggles at the Margins: The Working Lives of Men, Women and Children in Soviet and Post-Soviet Tajikistan." University of Oxford.

———. 2019. "Women Miners' Exclusion and Muslim Masculinities in Tajikistan: A Feminist Political Ecology of Honor and Shame." *Geoforum* 100 (February): 144–52. https://doi.org/10.1016/j.geoforum.2019.01.001.

Behzadi, Negar Elodie, and Lucia Direnberger. 2020. "Gender and Ethnicity in the Soviet Muslim Peripheries: A Feminist Postcolonial Geography of Women's Work in the Tajik SSR (1950–1991)." *Central Asian Survey* 39 (2): 202–18. https://doi.org/10.1080/02634937.2019.1699777.

Cleuziou, Juliette. 2016. "Mariages, Démariages et Remariages: Rituel, Genere et Parenté Au Tadjikistan Contemporain." Universté Paris-Ouest Nanterre La Défense.

Dudoignon, Stephane A. 1998. "Communal Solidarity and Social Conflicts in Late 20th Century Central Asia: The Case of the Tajik Civil War." *Islamic Areas Studies Working Paper Series* 7: 1–23.

———. 2004. "From Ambivalence to Ambiguity? Some Paradigms of Policy Making in Tajikistan." In *Tajikistan at a Crossroad: The Politics of Decentralization*, edited by L. De Martino, 120–151. Geneva: Cimera.

Dyson, Jane. 2008. "Harvesting Identities: Youth, Work and Gender in the Indian Himalayas." *Annals of the Association of American Geographers* 98 (1): 160–79. https://doi.org/10.1080/00045600701734554.

Elmhirst, Rebecca. 2011. "Introducing New Feminist Political Ecologies." *Geoforum* 42 (2): 129–32. https://doi.org/10.1016/j.geoforum.2011.01.006.

Federici, Silvia. 2004. *Caliban and the Witch: Women, the Body and Primitive Accumulation*. New York: Auto Media.

Heathershaw, John. 2009. *Post-Conflict Tajikistan, the Politics of Peace-Building and the Emergence of Legitimate Order*. London and New York: Routledge.

———. 2011. "Tajikistan Amidst Globalization : State Failure or State Transformation ?" *Central Asian Survey* 30 (1): 147–168. https://doi.org/10.1080/02634937.2011.554070.

Hofman, Irna, and Oane Visser. 2014. "Geographies of Transition: The Political and Geographical Factors of Agrarian Change in Tajikistan." Discussion paper No 151. Halle (Saale): Leibniz Institute of Agricultural Development in Transition Economies (IAMO). http://hdl.handle.net/10419/104004.

Ito, Sonoko, Sameh El Khatib, and Mikiyasu Nakayama. 2015. "Conflict over a Hydropower Plant Project between Tajikistan and Uzbekistan." *International Journal of Water Resources Development* 0627 (April): 1–16. https://doi.org/10.1080/07900627.2015.1076381.

Jeffrey, Craig. 2010. "Geographies of Children and Youth I : Eroding Maps of Life." *Progress in Human Geography* 34 (4): 496–505. https://doi.org/10.1177/0309132509348533.

———. 2012. "Geographies of Children and Youth II : Global Youth Agency." *Progress in Human Geography* 36 (2): 245–53. https://doi.org/10.1177/0309132510393316.

Kalinovsky, A. M. 2018. *Laboratory of Socialist Development: Cold War Politics and Decolonization in Soviet Tajikistan*. New York: Cornell University Press.

Kandiyoti, Deniz. 2007. "The Politics of Gender and the Soviet Paradox: Neither Colonized, Nor Modern." *Central Asian Survey* 26 (4): 601–623. http://doi.org//10.1080/02634930802018521

Katz, Cindi. 2001. "Vagabond Capitalism and the Necessity of Social Reproduction." *Antipode* 33 (4): 709–28. https://doi.org/10.1111/1467-8330.00207.

Kirmse, Stefan B. 2010. "Introduction Bridging the Gap : The Concept of 'Youth' and the Study of Central Asia and the Caucasus." *Central Asian Survey* 29 (4): 381–87. https://doi.org/10.1080/02634937.2010.537136.

Lahiri-Dutt, Kuntala, ed. 2018. *Between the Plough and the Pick: Informal, Artisanal and Small-Scale Mining in the Contemporary World.* Canberra: Australian University Press.

Lahiri-Dutt, Kuntala, and Martha Macintyre. 2006. *Women Miners in Developing Countries: Pit Women and Others*. London and New York: Routledge.

Lugones, Maria. 2010. "Toward a Decolonial Feminism." *Hypatia* 25 (4): 742–59.

McDowell, Linda. 1999. "Work/Workplaces." In *Gender, Identity and Place, Understanding Feminist Geographies*, 123–48. Cambridge: Blackwell.

McKittrick, Katherine. 2006. *Demonic Grounds: Black Women and the Cartographies of Struggle*. Minneapolis: University of Minnesota Press.

Meehan, Katie, and Kendra Strauss, eds. 2015. *Precarious Worlds: Contested Geographies of Social Reproduction*. Athens and London: University of Georgia Press.

Mollett, Sharlene. 2017. "Gender's Critical Edge: Feminist Political Ecology, Postcolonial Intersectionality, and the Coupling of Race and Gender." In *Routledge Handbook of Gender and Environment*, edited by Sherilyn MacGregor, 146–58. London: Routledge.

———. 2018. "The Rio Platano Biosphere Reserve: A Postcolonial Feminist Political Ecology Reading of Violence and Territorial Struggles in Honduras." In *Land Rights, Biodiversity Conservation and Justice: Rethinking Park and People*, edited by Sharlene Mollett and Kepe Thembela. London: Routledge.

Mollett, Sharlene, and Caroline Faria. 2013. "Messing with Gender in Feminist Political Ecology." *Geoforum* 45: 116–25. https://doi.org/10.1016/j.geoforum.2012.10.009.

Moore, Jason W. 2015. *Capitalism in the Web of Life: Ecology and the Accumulation of Capital*. Marston: Verso Books.

Murrey, Amber. 2015. "Invisible Power, Visible Dispossession: The Witchcraft of a Subterranean Pipeline." *Political Geography* 47: 64–76. https://doi.org/10.1016/j.polgeo.2015.04.004.

———. 2016. "Slow Dissent and the Emotional Geographies of Resistance." *Singapore Journal of Tropical Geography* 37: 224–48. https://doi.org/10.1111/sjtg.12147.

Murrey, Amber, and Nicholas A. Jackson. 2020. "A Decolonial Critique of the Racialized 'Localwashing' of Extraction in Central Africa." *Annals of the American Association of Geographers* 110 (3): 917–40. https://doi.org/10.1080/24694452.2019.1638752.

Nagar, Richa, Victoria Lawson, Linda McDowell, and Susan Hanson. 2002. "Locating Globalization: Feminist (Re)Readings of the Subjects and Spaces of Globalization." *Economic Geography* 78 (3): 257–84. https://doi.org/10.1111/j.1944-8287.2002.tb00187.

Nakaya, Sumie. 2009. "Aid and Transition from a War Economy to an Oligarchy in Post-War Tajikistan." *Central Asian Survey* 28 (3): 259–73. https://doi.org/10.1080/02634930903421764.

Nixon, Rob. 2011. *Slow Violence and the Environmentalism of the Poor*. Cambridge, USA: Harvard University Press.

Northrop, Douglas. 2004. *Veiled Empire. Gender and Power in Stalinist Central Asia.* Ithaca: Cornell University Press.

Nourzhanov, Kirill, and Christian Bleuer. 2013. *Tajikistan, A Political and Social History.* Canberra: The Australian National University E Press.

Olimova, Saodat, and Igor Bosc. 2003. *Labour Migration from Tajikistan.* Dushanbe: International Organization for Migration Dushanbe.

Postar, Stephanie, and Negar Elodie Behzadi. 2022 (in press). "'Extractive Bodies': A Feminist Counter-Topography of Two Extractive Landscapes." *Geoforum.* https://doi.org/10.1016/j.geoforum.2022.09.0055

Rocheleau, Dianne E., Barbara Thomas-Slayter, and Esther Wangari, eds. 1996. *Feminist Political Ecology: Global Issues and Local Experience.* London: Routledge.

Skelton, Tracey. 2013. "Young People, Children, Politics and Space: A Decade of Youthful Political Geography Scholarship 2003–13." *Space and Polity* 17 (1): 123–36. https://doi.org/10.1080/13562576.2013.780717.

Sultana, Farhana. 2009. "Fluid Lives : Subjectivities, Gender and Water in Rural." *Gender, Place & Culture* 16 (4): 427–44. https://doi.org/10.1080/09663690903003942.

Tlostanova, Madina. 2010. *Gender Epistemologies and Eurasian Borderlands.* New York: Palgrave Macmillan.

Toscano, Alberto. 2018. "Antiphysis / Antipraxis: Universal Exhaustion and the Tragedy of Materiality." *Mediations* 32 (1): 125–44. www.mediationsjournal.org/articles/antiphysis-antipraxis.

Truelove, Yaffa. 2011. "(Re-)Conceptualizing Water Inequality in Delhi, India through a Feminist Political Ecology Framework." *Geoforum* 42 (2): 143–52. https://doi.org/10.1016/j.geoforum.2011.01.004.

Verges, Francoise. 2019. *Un Féminisme Décolonial.* Paris: La fabrique éditions.

Yusufjonova-Abman, Zamira. 2018. "State Feminism in Soviet Central Asia: Anti-Religious Campaigns and Muslim Women in Tajikistan, 1953-1982." In *The Palgrave Handbook of Women and Gender in the Twentieth-Century Russia and the Soviet Union,* edited by Melanie Ilic, 299–314. London: Palgrave Macmillan.

Zekhni, Malika. 2022. "Invisible Bodies: Civilising Mission, Sexuality, and Prostitution in fin de siècle Russian Turkestan." *Cultural and Social History* 19 (2): 141–159. https://doi.org/10.1080/14780038.2021.2008145.

Part III

HUMAN/NONHUMAN

The More Than Human

Emilie Cameron

The chapters in this section are shaped by, and contribute to, an evolving series of conversations and problematics around the role of the "nonhuman" in extractive processes.

Beginning in the 1980s and 1990s, scholars across the social sciences and humanities began to grapple in new ways with the "more-than-human" and the "posthuman." Although informed by different traditions, this collection of thinkers shared an interest in understanding agency, responsibility, knowledge, and power as not wholly confined to the thoughts and actions of humans. Rather, thinkers like Donna Haraway (1988), Bruno Latour (1987, 1993), and Arturo Escobar (1999) began to conceptualize these processes as refracted across networks of people and things, human and nonhuman. In response to the perceived limitations of humanistic, structuralist, and poststructuralist thought, all of which (albeit in different ways and for different reasons) tended to overstate human agency and silo knowledge about animals, plants, and other nonhuman elements, these new theoretical approaches emphasized the impossibility of separating nature and society and the necessity of understanding the crucial role played by nonhumans in social, political, economic, and environmental processes.

This turn was not simply about taking our relationships with animals seriously or about highlighting the liveliness of nonhuman materialities, although much vibrant work was carried out in this vein. It necessarily involved grappling with some of the foundational dualisms that structure Western thought, including nature versus culture, human versus animal, civilized versus wild, real versus imaginary, and so on. What would it mean, these scholars asked, for our thinking and our being in the world, if we reckoned with the blurriness, the entanglements, and even the absurdity of these foundational dualisms? What if so much of what is taken for granted within masculinist, white,

Eurocentric knowledge systems isn't actually "true"? What would that mean for how we understand our worlds, and how we respond to the challenges and crises of the present? Although quite different in important ways, one can see in, for example, Donna Haraway's work on situated knowledges (1988), Viveiros de Castro's work on perspectivism (2004), and Karen Barad's work on constructivism (1996) a willingness to question what *is*, and how we come to ascertain what is, in new ways. For many, then, the turn toward the nonhuman (and the ontological, which arose alongside this work) was about making room for relations and realities that did not register in previous conceptual frameworks.

Of course, these dualisms do not hold sway across all knowledge systems, and there is a noticeable difference between work aimed at unthinking/rethinking Western dualisms from within Western epistemologies (often involving complex theoretical contortions) and work that theorizes the world from within, for example, Indigenous epistemologies, where to speak of fish or plants as kin is not to speak metaphorically (Todd 2016a; Reo 2019), nor is it to collide with a deeply entrenched denial of human animality in modern Western thought. As Vanessa Watts argues "[Indigenous] cosmologies (and the theories within them) are righteously different and cannot be separated from the stuff of nature" (2013, 32); they are not "about" nature (where "nature" is understood as fundamentally separate from human thought), she argues, but of, entwined with, inseparable from the nonhuman world. Although some earlier work in posthumanism and new materialism drew on Indigenous ontologies in order to think these dualisms differently (e.g., Whatmore 2002; Haraway 1988), fulsome, deep engagement with Indigenous knowledge systems has only begun to inform posthumanist and new materialist literatures to any significant extent. Indeed, Indigenous scholars who have engaged with scholarly debates about the nonhuman have extensively critiqued the superficial, strategic, and abstracted uptake of Indigenous knowledges and the failure to fully reckon with Indigenous epistemologies and their radical difference (Hunt 2014; Todd 2016b; Watts 2013). Today, leading scholarship by Vanessa Watts (2020), Zoe Todd (2018), Kim Tallbear (2015, 2017), Robin Wall Kimmerer (2013), Juno Salazar Parreñas (2018), and others (e.g., Hernandez et al. 2021) not only demonstrates the vitality and centrality of nonhuman kin in human lifeways, but also poses a challenge to theoretical work that, however well intentioned, foundered on the fundamentally Western epistemologies that underpinned them. As Zoe Todd has argued so persuasively, Indigenous scholars have been attending to relations with the nonhuman for decades, and yet they are rarely cited or acknowledged, nor are their profound insights into past and present human/nonhuman relations given more than decorative place in much academic writing (Todd 2016b). This is not accidental or neutral; it reflects the deeply colonial structures of the academy and its valorization of white bodies and knowledges over others.

So the turn to the "nonhuman" is only "new" if decades (or millennia) of knowledge production are disregarded. But this shift in focus within the social sciences and humanities has nevertheless spurred research that challenges some taken-for-granted understandings about how geographies of resource extraction unfold. For example, against tendencies to conceptualize ore bodies as inert targets of development, some scholars have explored the animating and determinative role played by metals and minerals themselves in the development (and after-lives) of various projects, as well as the role played by other nonhuman entities in industrial projects (Sandlos and Keeling 2016; Mitchell 2002; Bridge 2009). Others have queried the role of nonhuman life in the production of value in the extractive sector (e.g., the use of microorganisms in metal extraction, see Labban 2014). Collard and Dempsey (2017) have brought together feminist, Indigenous, and Marxian analyses to think through the different ways in which nonhuman life is valued and sacrificed within capitalist systems. Mario Blaser (2009, 2013) has teased out the ontological basis of conflicts over resource development, calling attention to these as clashes between worlds rather than simply differences in priority or perspective. One can see in these diverse and sometimes divergent works a willingness to think outside of conventional stories about how and why resource extraction unfolds in the way it does, and with what effects.

With their shared interest in the nonhuman, and in probing the gaps between what is and what is recognized, acknowledged, or made true by various institutional and political processes, both chapters in this section are inheritors of this diverse intellectual lineage. As Nuttall demonstrates, the vastness of northwest Greenlandic knowledge about the impacts of proposed mines, knowledge derived from intimate relations with nonhuman kin, largely fails to register within environmental impact assessment processes, but this does not mean it isn't "true" or that it isn't consequential. Indeed, by taking seriously all that fails to register within liberal resource governance regimes, some of the limits of "inclusion" as a goal in resource economies are laid bare. As Nuttall notes, Greenlandic Inuit have had some success in ensuring that their knowledges are included in the formal processes through which extractive projects are assessed. But what does "inclusion" actually mean? On what terms are radically different knowledge systems integrated into assessment processes, and to what ends?

Too often, critics argue, Indigenous knowledges are decontextualized, simplified, and tokenized in environmental assessment processes, and the profound implications of understanding the ecosystemic, sociocultural, and political-economic impacts of a proposed development from within Indigenous knowledge systems are foreclosed in favor of superficial nods to traditional knowledge about particular species (Ellis 2005). Furthermore, critics have also long argued that environmental assessment processes are fundamentally geared toward project approval, and thus the "inclusion" of Indigenous groups and knowledges in the process too often provides a veneer of legitimacy rather than a fulsome engagement with both the risks of a proposed development and the jurisdiction of Indigenous peoples over their traditional territories (Li 2009). Nuttall's chapter engages with these tensions and probes the gap between the rich understandings of the "aquapelago" of northern Greenland held by Kalaallit and the capacity for assessment processes to reckon with these knowledges and their implications. What happens, he asks, when assessment processes claim to engage with peoples and knowledges that far exceed the boundaries placed upon them by the process? On what terms and for what purposes are relations with ice, water, marine mammals, and other kin recognized or occluded by environmental assessment processes in Greenland, and with what effects?

Stephanie Postar's chapter also grapples with the illegibility of consequential human encounters with nonhuman materialities; in this case, claims that unmined uranium is causing a sickness among Tanzanians living near a proposed mine site. But rather than probe the gap between Indigenous and non-Indigenous epistemologies, she demonstrates how impossible it is to define clearly where the human and nonhuman begin and end when it comes to understanding the impacts of mining but also decision-making and relationships with mineral developments. As Postar shows, the "truth" of whether unmined uranium is causing a sickness among people living near a proposed mine site matters less than claims that it does, fears that it does. Unmined uranium, she suggests, circulates and matters, to both mining companies and workers, whether or not its circulation is imaginative, cellular, feared, or known. To conceptualize unmined uranium in this way is to be willing to acknowledge multiple ontologies and to attend to what Povinelli terms the "otherwise" (Povinelli 2011, 10; see also Holbraad, Pedersen, and Viveiros de Castro 2014). Postar is not speaking in metaphors but rather attempting to present an ontology in which the porous boundaries between human and nonhuman are acknowledged. In so doing, some of the knowledges, practices, bodies, and relations that are typically occluded in both resource economies and academic research are given a bit more room to breathe.

REFERENCES

Barad, Karen. 1996. "Meeting the universe halfway: Realism and social constructivism without contradiction." In *Feminism, Science, and the Philosophy of Science*, edited by Lynn Hankinson Nelson and Jack Nelson, 161–194. Dordrecht: Springer.

Blaser, Mario. 2013. "Ontological conflicts and the stories of peoples in spite of Europe: Towards a conversation on political ontology." *Current Anthropology* 54 (5): 547–568.
Blaser, Mario. 2009. "The threat of the Yrmo: The political ontology of a sustainable hunting program." *American Anthropologist* 111 (1): 10–20.
Bridge, Gavin. 2009. "Material worlds: Natural resources, resource geography and the material economy." *Geography Compass* 3 (3): 1217–1244.
Collard, Rosemary-Claire, and Dempsey, Jessica. 2017. "Capitalist natures in five orientations." *Capitalism Nature Socialism* 28 (1): 78–97.
Ellis, Stephen C. 2005. "Meaningful consideration? A review of traditional knowledge in environmental decision making." *Arctic* 58 (1): 66–77.
Escobar, Arturo. 1999. "After nature: Steps to an antiessentialist political ecology." *Current Anthropology* 40 (1): 1–30.
Haraway, Donna. 1988. "Situated knowledges: The science question in feminism and the privilege of partial perspective." *Feminist Studies* 14 (3): 575–599.
Hernández, K.J., June M. Rubis, Noah Theriault, Zoe Todd, Audra Mitchell, Bawaka Country, Laklak Burarrwanga, Ritjilili Ganambarr, Merrkiyawuy Ganambarr-Stubbs, Banbapuy Ganambarr, Djawundil Maymuru, Sandie Suchet-Pearson, Kate Lloyd, and Sarah Wright. 2021. "The creatures collective: Manifestings." *Environment and Planning E: Nature and Space* 4 (3): 838–863.
Holbraad, Martin, Morten Pedersen, and Eduardo Viveiros de Castro. 2014. "The politics of ontology: Anthropological positions." *Theorizing the Contemporary, Fieldsites*, January 13, 2014. www.culanth.org/fieldsights/462-the-politics-of-ontology-anthropological-positions.
Hunt, Sarah. 2014. "Ontologies of Indigeneity: The politics of embodying a concept." *Cultural Geographies* 21 (1): 27–32.
Kimmerer, Robin Wall. 2013. *Braiding Sweetgrass: Indigenous Wisdom, Scientific Knowledge and the Teachings of Plants*. Minneapolis: Milkweed Editions.
Labban, Mazen. 2014. "Deterritorializing extraction: Bioaccumulation and the planetary mine." *Annals of the Association of American Geographers* 104 (3): 560–576.
Latour, Bruno. 1987. *Science in Action: How to Follow Scientists and Engineers Through Society*. Milton Keynes: Open University Press.
Latour, Bruno. 1993. *We Have Never Been Modern*. Cambridge, MA: Harvard University Press.
Li, Fabiana. 2009. "Documenting accountability: Environmental impact assessment in a peruvian mining project." *POLAR: Political and Legal Anthropology Review* 32 (2): 218–236.
Mitchell, Timothy. 2002. *Rule of Experts: Egypt, Techno-Politics, Modernity*. Berkeley: University of California Press.
Parrenas, Juno Salazar. 2018. *Decolonizing Extinction: The Work of Care in Orangutan Rehabilitation*. Durham, NC: Duke University Press.
Povinelli, Elizabeth. 2011. *Economies of Abandonment: Social Belonging and Endurance in Late Liberalism*. Durham: Duke University Press.
Reo, Nicholas J. 2019. "Inawendiwin and relational accountability in anishnaabeg studies: The crux of the biscuit." *Journal of Ethnobiology* 39 (1): 65–75.

Sandlos, John, and Keeling, Arn. 2016. "Toxic legacies, slow violence, and environmental injustice at giant mine, northwest territories." *Northern Review* 42: 7–21.

TallBear, Kim. 2015. "Dossier: Theorizing queer inhumanisms: An Indigenous reflection on working beyond the human/not human." *GLQ: A Journal of Lesbian and Gay Studies* 21 (2–3): 230–235.

TallBear, Kim. 2017. "Beyond the life/not life binary: A feminist-Indigenous reading of cryopreservation, interspecies thinking and the new materialisms." In *Cryopolitics: Frozen Life in a Melting World*, edited by Joanna Radin and Emma Kowal, 179–202. Cambridge: The MIT Press.

Todd, Zoe. 2018. "Refracting the state through human-fish relations: Fishing, Indigenous legal orders and colonialism in North/Western Canada." *Decolonization, Indigeneity, Education, and Society* 7 (1): 60–75.

Todd, Zoe. 2016a. "From a fishy place: Examining Canadian state law applied in the daniels decision from the perspective of Métis legal orders." *TOPIA* 36: 43–57.

Todd, Zoe. 2016b. "An Indigenous feminist's take on the ontological turn: 'Ontology' is just another word for colonialism." *Journal of Historical Sociology* 29 (1): 4–22.

Viveiros de Castro, Eduardo. 2004. "Perspectival anthropology and the method of controlled equivocation." *Tipití: Journal of the Society for the Anthropology of Lowland South America* 2 (1): 3–22.

Watts, Vanessa. 2013. "Indigenous place-thought & agency amongst humans and non-humans (first woman and sky woman go on a European world tour!)." *Decolonization: Indigeneity, Education & Society* 2 (1): 20–34.

Watts, Vanessa. 2020. "Growling ontologies: Indigeneity, becoming-souls, and settler colonial inaccessibility." In *Colonialism and Animality: Anti-Colonial Perspectives in Critical Animal Studies*, edited by Kelly Struthers Montford and Chloë Taylor, 115–128. London: Routledge.

Whatmore, Sarah. 2002. *Hybrid Geographies: Natures, Cultures, Spaces*. London: Sage.

5

Extractive Industry, Impact Assessments, and Exclusion in Northwest Greenland

Mark Nuttall

As the nature, properties, and materialities of polar environments transform under conditions of climate change, many places across the Arctic are being geoassembled as sites of possibility and speculation for extractive industry (Dodds and Nuttall 2019). Greenland is one such place where resource spaces have been marked out for oil exploration and the surveying and assessment of potential mining ventures, both on land and offshore. These activities are often written about as being spurred on by a sense of the increasing accessibility of remote Arctic areas as sea ice melts and glaciers recede. While this view is reproduced with regularity by more popular and media narratives about scrambles for resources in a "new" Arctic, the reality is different. Developing a mining industry and encouraging hydrocarbon exploration have remained stated aims of the various coalition permutations of Greenland's self-rule government over the last decade and significant effort has been put into attracting international companies. Following a snap election in April 2021, a new coalition government was formed with the Inuit Ataqatigiit and Naleraq parties in power. It made a swift move to suspend oil exploration in Greenland's waters and expressed opposition to the extraction of uranium and other radioactive elements, but reaffirmed a commitment to other forms of mining as one pillar on which Greenland's economy should be built.

The presence and influence of extractive industries in Greenland, although not a straightforward nor simple response to the effects of climate change, is materially impacting places and a number of contentious issues have come to dominate national discussion. Among these, the nature of public participation in decision-making processes concerning the development of subsurface resources has been particularly controversial. Many people, for example, do not feel that they are sufficiently informed or consulted about the potential impacts of oil, gas, and mineral extraction projects, even though the Greenland government's updated and revised guidelines for preparing social impact assessments (SIAs) and environmental impact assessments (EIAs) now have a clearer emphasis on a consultation process with affected communities and other stakeholders than their earlier versions outlined (Nuttall 2017).

In this chapter, I discuss why some people feel this way by drawing on anthropological research in northwest Greenland's coastal communities. This is a region where I first carried out research in the late 1980s and continue to do ethnographic work. Since 2014, I have focused on the effects of climate change there, but I have also been examining the interest of extractive industry in the area and people's responses to it. People there often tell me they have been rarely consulted when it comes to discussions of resource development, environmental protection, and wildlife management or the processes that lead to environmental and social impact assessments, and they question whether the revised guidelines will lead to appropriate community engagement and acceptable consultation. For them, meaningful consultation would also involve a conversation with government and industry about the accepted technocentric and scientifically informed ways of viewing the world and allow for an appreciation of the ways people live within surroundings they share and co-constitute with the nonhuman (cf. Kohn 2013). Indeed, nonhuman entities, they say, should also be included in participatory processes and taken into account by social and environmental impact assessments. These local concerns highlight the temporal and spatial limitations of social and environmental impact assessments and point to the need for dialogue on how Indigenous ontologies can contribute to "a more expansive, inclusive approach" (cf. Behn and Bakker 2019, 99).

MAKING THE EXTRACTIVE FRONTIER

Media reports that Greenland is being "opened up" for resource exploration by receding ice may be eye-catching, but extractive industry is not being developed there simply because, or as a result, of climate change. Ideas about the economic potential of subsurface resources are embedded in Greenlandic discourses on future-making and are central to government policy as Greenland pushes forward with economic development strategies for reducing dependency on Denmark, achieving greater autonomy (and possible independence) and enhancing the country's agency in foreign affairs (Nuttall 2017; Sejersen 2019). Mining and oil companies have been interested in Greenland for decades and geological research that assesses resource potential and exploratory activities that have taken mineral samples and drilled for oil predate current concern with climate change. Nonetheless, the trope of a melting

Arctic and increasing accessibility to a new resource frontier has been used as a persuasive device by Greenlandic politicians as they aspire to develop an extractive industry sector (Nuttall 2008).

The interest expressed by mining and oil companies in Greenland has given rise to ideas related to what Gavin Bridge (2004) calls bonanza geographies, places in which investment from the extractive industries sector and activities of exploration and development appear influential in transforming them into resource spaces. The extractive frontier has become part of contemporary Greenland's political life and the subsurface figures in economic calculations for a prosperous future, just as it has for many other countries experiencing resource booms. Even just a few years ago, the near future (which has now arrived as the present) was imagined to be one in which a number of mining projects were up and running in Greenland. In 2012, for example, when I was first doing research in the country's growing capital Nuuk with a focus on the politics, bureaucracies, and narratives surrounding extractive industries, many politicians and business leaders were optimistic that Greenland was on the verge of a minerals and oil bonanza. Mining companies were busy prospecting in several parts of the country, and oil exploration was taking place in west coast waters. There was excitement in the offices and meeting rooms of government departments, geologists, industry consultants, and entrepreneurs that Greenland was regarded by international interests as an emerging resource frontier. Nuuk was being reimagined and redesigned as a hub for scientific, technical, and consultative expertise and a logistics support base for extractive industry.

The bonanza has not yet happened for many reasons, but mainly because of market conditions, a lack of confidence from investors, and technical and environmental challenges in remote locations. Some projects that looked extremely promising, such as the Isua iron ore mine in the Nuuk Fjord, did not, in the end, get started, even though a thirty-year exploitation license had been granted to UK-based London Mining in 2013 to develop and operate it. The project was halted when London Mining went into administration in October 2014. The company no longer had the money to operate its only venture, the Marampa mine in Sierra Leone. A slump in iron ore prices had made things difficult, but the Ebola crisis in West Africa weakened the company further. In 2020, the Covid-19 pandemic also affected logistics for mining operations, limiting activities in Greenland.

Currently, only two mines are operating—a ruby and pink sapphire mine near Qeqertarsuatsiaat in southwest Greenland and the White Mountain anorthosite mine southeast of Sisimiut on the central west coast. Other projects, although proliferating, are making slower progress, even if some of them are likely to begin eventually. Economics, markets, and logistical challenges determine how companies view their prospects, and many consider it only worthwhile to explore and develop ventures in Greenland—which are often in locations that are costly to reach and operate in—if there is a unique mineral deposit (and, of course, having less ice in a fjord does help when shipping ore from an Arctic mine). A dip in global commodity prices and markets over the last few years, as well as the effects of other global processes on plans for resource development in the Arctic have made some companies wary about working in Greenland. While this has challenged politicians to think carefully about relying too heavily on extractive industry projects, the view from Nuuk is that mining should still be one pillar on which Greenland builds its economy.

An exploitation license has been granted to the Australian company Ironbark to develop a lead-zinc mine in Citronen Fjord in Peary Land in the country's far north, and recently to Tanbreez, another Australian company, which is developing a rare earths project in south Greenland. A large ilmenite (an iron titanium oxide) mine near the depopulated and closed settlement of Moriusaq in northwest Greenland has also been approved and exploitation will see it extracted from coastal sand by Dundas Titanium, a company registered in Greenland and 100 percent owned by London-based Bluejay Mining. Alba Mineral Resources has a similar project application under review for ilmenite extraction along the same coast and is also scoping out the prospects for iron ore in the Melville Bay region. It is of interest to note that sand and gravel are being highlighted as providing opportunity for Greenland to become a global exporter of aggregates and help meet increasing global demand (Bendixen et al. 2019). Energy Transition Minerals (another Australian firm, formerly known as Greenland Minerals) is contesting a decision relating to its application for an exploitation license for its uranium and rare earths mine in south Greenland, and several other projects around the country are in various stages of planning and exploration, including at sites north of Qaanaaq, such as Inglefield Land (for copper, gold, cobalt, and nickel), in Melville Bay (for iron ore), the southern part of the Upernavik district (for nickel, copper, and platinum group metals), and a number of others on the east coast.

Extractive industries present challenges to how anthropologists approach and conceptualize the encounters between companies and the people who inhabit the places in which the hunt for oil or probing deep below the subsurface for minerals is taking hold (e.g., Larsen 2015). Greenland's resource spaces are often represented by mining and oil companies as remote frontier zones in areas of wilderness. They are thought of as empty of traces of historic human occupation and contemporary social and cultural presence, yet oil exploration has taken place in offshore waters that are critical for Greenlandic fisheries and for marine mammals, while many mining projects are close to communities, or to the areas people hunt and fish in or travel through. The Isua project is a case in point. The social and environmental impact assessments described the area that would be marked off as the exclusive license zone as a place that was far enough from human habitation that social impacts would be negligible, despite its importance for hunting, fishing, and gathering activities for people from Nuuk and the small community of Kapisillit (Nuttall 2012). These resource spaces are also places described by scientists as ecologically unique, but under threat (Christiansen, Mecklenburg and Karamushko 2014; Laidre et al. 2015; Reeves et al. 2014), or by archaeologists as having significance for cultural heritage (Myrup 2018). Again, such importance is often downplayed in assessments or the potential impacts are considered to be minimal.

PLACES OF HUMAN AND NONHUMAN ENCOUNTERS

Northwest Greenland comprises the two former districts of Upernavik and Qaanaaq (also known as the Avanersuaq area), which are part of Greenland's largest municipal region of Avannaata Kommunia. The administrative headquarters are located at Ilulissat in Disko Bay in central west Greenland and the municipality stretches northwards (and includes vast parts of the inland ice sheet) to Nares Strait, which connects northern Baffin Bay with the Lincoln Sea in the Arctic Ocean. Despite a reorganization of Greenland's municipalities in early 2009, which meant local and regional decision-making was moved away from Upernavik and Qaanaaq to Ilulissat, both areas retain identities as distinct districts, with livelihoods based on hunting and fishing. Indeed, there are specific marine mammal hunting and fisheries management regimes and quota systems that apply to these areas, and so I refer to them by their old district names as people living there continue to do. While people identify as Kalaallit (Greenlanders), and Inuit, the population in Avanersuaq are known as Inughuit. Many people in Upernavik and Avanersuaq often say they feel they also live in regions that are far removed from political concern and daily decision-making in Ilulissat and even further from Nuuk. They often talk about how they think northwest Greenland and people's daily lives and their interests appear peripheral and distant to politicians and decision-makers in these more southerly, increasingly urbanized centers of the country. They imagine that, seen from Nuuk, Upernavik, Qaanaaq, and the other smaller communities of northwest Greenland must be thought of as far off places lying at the outer edge (*nunap isua*) of Greenland.

About 2,800 people live in the Upernavik district—the town of Upernavik has a population of around 1,100 and some 1,700 people inhabit nine smaller villages, ranging in size from about 50 in Naajaat and 450 in Kullorsuaq (Greenland's largest village). In the Avanersuaq area the population is around 800. Qaanaaq's population is about 650, while some 60 people live in Savissivik, around the same number live in Siorapaluk, and some 25 in Qeqertat. There are strong networks of kinship and social relatedness throughout both the Upernavik and Qaanaaq regions, with the communities of Savissivik and Kullorsuaq having especially close family connections and regular movement and the exchange and sharing of hunting and fishing products between them, with people crossing Melville Bay by dog sledge in winter and spring and by open boat during the summer. Upernavik and Qaanaaq are supply centers for the villages in each district and a number of public sector services and private businesses provide some employment.

The economies of the smaller settlements (and many household economies in the towns of Upernavik and Qaanaaq) are based largely on the procurement of living marine resources—marine mammals such as seals, walrus, polar bears, narwhal, beluga, fin, and minke whales, on land animals such as musk-ox and reindeer, and in water Greenland halibut, cod, salmon, Arctic char, Atlantic wolfish, capelin, and other fish species. People, live, work, and move around a complex topography of islands, headlands, fjords, sea, tidewater glaciers, the edges of the inland ice, mountains and fells, lakes, and rivers and, in winter and spring, sea ice. Philip Hayward's (2012) idea of the "aquapelago," an assemblage of marine and terrestrial spaces, seems an apt description. For Hayward, aquapelagic assemblages "are constituted by social units in locations where the aquatic spaces between and encircling islands are fundamentally interconnected with and essential to communities' inhabitation of their locale (and substantially generate their senses of identity and belonging to that place)" (Hayward 2015, 84). The aquapelago as assemblage involves the interactions and entanglements between humans and other actants that "may be animate (living) entities, inanimate ones (such as sand, soil, etc.) or the product of energies (such as individual weather events or larger climatic patterns, such as global warming)" (ibid.). But the northwest Greenland aquapelago is just one assemblage within a dynamic, unfolding world which people living there think and talk about in terms of *pinngortitaq*. This is often just translated from Greenlandic as "environment" or "nature," but it is all that is underneath, above and all around. It refers to more than just the surface of the earth—it encompasses water, ice, soil, rock, sky, and wind; surroundings which include the air, atmosphere, subsoil, mountain interiors, and earth processes; what is above and below and around, and how all of these things intersect, interact, and are entangled. *Pinngortitaq* is always coming in being, forming, and reforming. To live and move within and across *pinngortitaq* is to have a continual encounter with its many moods and habits, and to come to terms with its dynamic, flourishing nature.

Climate change is evident in northwest Greenland in quite often visually dramatic ways. The extent of sea ice cover in winter and spring is diminishing (AMAP 2012; Stroeve et al. 2017), glaciers are melting and retreating, and iceberg calving rates are on the increase (Cowton et al. 2018; van As 2011). Greenland's inland ice is experiencing more surface melt and mass balance loss (van den Broeke et al. 2017). Because the marine ecosystem supports local livelihoods, the changing nature of sea ice—how it forms, takes shape, and persists over winter and through spring—has consequences for hunting and fishing activities and local economies, as well as community life. Greenland halibut are moving further north, for instance, while seals are becoming scarce in some community waters. Climate change, however, has to be understood in a wider context of other drivers of social and economic change, as well as politics and governance that affect daily life in Upernavik and Qaanaaq. For instance, management regulations and quotas for living marine resources combine with environmental and climate change to make for circumstances that restrict the abilities of local people to travel and move around their localities and to hunt and fish certain species (Nuttall 2019). Furthermore, exploratory activity related to extractive industries has brought different kinds of pressures and anxieties, as well as hopes for the future.

MAPPING, SURVEYING, AND IMPACT ASSESSMENTS

Oil exploration took place in northern Baffin Bay in the 1970s and again in the early 1990s, while four short reconnaissance trips by mining companies were made in Upernavik between 1969 and 1981, and mapping work was done by the former Geological Survey of Greenland (GGU) in the late 1970s. The area was once more under the gaze of international oil companies from 2010, and Upernavik town was used as a base for offshore activities, prospecting, scoping projects and site surveys, and environmental assessment work. Extensive seismic surveys then took place in 2012 and 2013. The ships used the town's harbor facilities and a regular turnover of personnel passed through the airport. There was demand for accommodation and a number of private lodgings were established. The oil companies and seismic survey crews also needed logistics support and this provided some local seasonal employment. For a small north Greenland town experiencing the effects of a changing climate on the marine environment, as well as a decline in fish stocks, there was some excitement, mainly on the part of municipal authorities and some local entrepreneurs, that Upernavik would become a permanent base for the oil industry on Greenland's new resource frontier. A plan for the town's development included a new harbor and related oil industry infrastructure and facilities, as well as new housing.

The intense seismic campaigns did not come up with any data that indicated the presence of oil, and no further exploratory activity has been carried out since the surveys of 2013. This had a noticeable impact on a few local businesses, especially those that had pinned some hope not only on an oil boom but also on how people feel about future oil development. While Greenland's government remains committed to developing a mining industry, the decision to suspend oil exploration makes some doubt that the seismic vessels or exploration ships will ever return to northern Baffin Bay. At the same time, many people in the Upernavik and Melville Bay area remain worried that the oil companies will indeed be back some day, although current concerns are also with mining projects near Savissivik and south of Upernavik town. Even if oil exploration may not happen again for some time, in September 2021 the Geological Survey of Denmark and Greenland carried out a survey of some 800 kilometers of seabed off Greenland's west coast near Maniitsoq as a first phase in determining the presence of marine diamond deposits. The survey had been commissioned by De Beers and suggests that Greenland's government may not consider the offshore environment to be off limits to other kinds of investigation by extractive industry.

Based on the experience of the seismic surveys, hunters and fishers talk about the necessity of communities being prepared to deal with industry on an increasingly regular basis. One hunter from the community of Nutaarmiut, which is to the north of Upernavik town, told me in summer 2015 that he anticipated the return of the seismic vessels and the oil company consultants: "The most important thing is to have consultation," he said. "The oil companies have to come and talk with the hunters in the settlements. It is really important to combine knowledge before *anything* starts. It is important to develop the respect and relationships right from the beginning." He spoke, like many other people I know in the area have, of the frustrations they felt when the oil companies had first arrived in the district to talk about their plans. Visits to communities were short, and people said that they had received little information about exploration and the nature of seismic surveys. They felt that the company executives and consultants had no interest in understanding the nature of the environment, and the importance of animals for local livelihoods. In particular, considerable concern has been expressed by hunters about the effects of the surveys on narwhals, which are hunted mainly during the open-water season from August to September, which is the same period when seismic activities can operate in the area.

Following the surveys in both 2012 and 2013, hunters from communities in the Upernavik district, as well as from Savissivik, reported that narwhal behavior was different and some felt that the hunt had been influenced negatively due to the seismic activities in the area (Nuttall 2016). Local observations have continued to indicate that narwhals remain restless and agitated since the seismic survey vessels were operating out at sea, even if it was a few years ago now. Each year, they have been moving closer to the coast, swimming deeper into ice-choked fjords and inlets (which increases the risk of ice entrapment for narwhals when the sea eventually freezes in autumn or early winter).

While environmental changes in the marine ecosystem, such as thinning and declining sea ice, changing water temperatures, and changes in the migration and distribution of fish also likely affect and influence narwhal movement (Greenland halibut are a key prey species for narwhals), hunters say narwhals are not necessarily agitated by such changes alone. They are also aware of how difficult it is to identify a single cause for the changes in narwhal behavior. The general consensus among hunters is that there are many observed changes in narwhal movement, sea ice, weather patterns, and so on, but that there is a multiplicity of factors at work that contribute to this.

Although the seismic vessels have not yet reappeared in these northern waters, the region has seen an increase in activities related to the exploration and surveying of mining potential. During my fieldwork, people have often told me that they feel excluded from decision-making processes surrounding plans for extractive industry projects, and that they have no opportunity to discuss the ways and nature of *pinngortitaq* and animals with consultants. Much discussion about this in northwest Greenland focuses on a desire to see greater emphasis on the inclusion of local knowledge and local observations of change in social and environmental impact assessments.

The government's guidelines on preparing a social impact assessment were first published in 2009, but were revised in 2016. They now begin with this statement:

> Greenland wants to develop the mineral resources area into one of the country's primary and principal business sectors. This is to be done in close collaboration and dialogue with the Greenlandic population. Understanding the interaction between mineral projects and Greenlandic society is therefore essential in order to create sustainable relations between mineral resources companies, municipalities, affected individuals, other stakeholders as well as Greenlandic society in general. (Government of Greenland 2016, 4)

Despite this declaration of the importance of companies and their consultants working in close collaboration with local communities, exploration activities continue to provoke fraught political debate and public concern. Inuit Circumpolar Council-Greenland and local nongovernmental organizations have been especially critical and wonder if a Greenlandic future that includes oil and mineral extraction can avoid significant social, cultural, and environmental impacts and be considered sustainable. It is important to note that the SIA guidelines give no definition or explanation of what "consultation" is and how it should proceed. The guidelines do make reference to traditional and local knowledge:

> When preparing the SIA report, the company must use traditional and local knowledge as far as possible by collecting information through qualitative interviews. It is important to incorporate local knowledge from individuals, commercial hunters and fishermen etc. This knowledge may have been passed on from generation to generation and has not necessarily been described and analysed in publications and public literature. It is recommended that the licensee describes, analyses and uses the traditional knowledge existing in the area in the SIA report. This knowledge may also include municipal planning documents and similar descriptions. (Government of Greenland 2016, 24)

Yet, again the guidelines offer no suggestions or set out a strategy for how local knowledge is to be incorporated, and there is no reference to recognizing Free, Prior, and Informed Consent, which is a prerequisite for engagement with Indigenous communities. In theory, at least, consultation is supposed to offer an inclusive process, enabling Indigenous people to participate in an assessment process when extraction projects with significant socio-environmental impacts are planned. As Flemming and Schilling-Vacaflor (2016) point out, the right to prior consultation and to informed consent represents the basis of a new global model that is shaping state–Indigenous relations, but drawing on research on the hydrocarbon sectors in Bolivia and Peru, they describe the difficulties Indigenous groups have in defending or articulating their concerns, demands, and even their visions. Some of these difficulties arise from asymmetries in power, problems with effective communication, timing, and simplistic assumptions that underlie the consultation approach. For Flemming and Schilling-Vacaflor, this only leads to unfulfilled promises of this new model. In Greenland, Dahl and Hansen (2019, 184) argue that there seems to be a general expectation in the impact assessment guidelines that Indigenous knowledge will be included, but there is a lack of experience in how to utilize it in an appropriate manner once it has been brought into the assessment process.

While companies—or usually the consultants they hire—are expected to carry out baseline studies and environmental and social impact assessments in accordance with Greenland government regulations, they usually proceed to do so without any particular definition of the environment or the social world in mind—there is an implicit assumption that those who carry out such work know what "environment" and "social" mean. The things making up, filling, and constituting an environment or society seem apparent and observable, and the potential impacts and benefits measurable and quantifiable. How environments and societies are sensitive, vulnerable, and likely to be impacted by a mine or oil extraction are assessed, and mitigation and monitoring initiatives are proposed. In short, things appear self-evident to consultants and can be described in terms of "systems," while measures to avoid or mitigate risks and impacts are set out in reports that are submitted to companies and those government departments concerned with approving and regulating extractive industry.

The overall aim of an environmental impact assessment, for example, is to identify, evaluate, and minimize the environmental effects of a project. However, the purpose of an environmental impact assessment carried out in Greenland's resource spaces is also to provide Greenlandic authorities and other stakeholders with information on the environmental impacts of the project including how the mining company intends to promote environmentally sound and sustainable solutions through the identification and implementation of mitigating measures. Greenland's guidelines for environmental impact assessments say that they must contain "a thorough description of the state of the environment before the start-up of mining activities" (Government of Greenland 2015, 7). Essentially baseline studies, they allow certain people—consultants with expertise, an array of scientists—to speak for and describe how the world is and how it is likely or not to be affected by a project.

An environmental impact assessment may describe the ecologies and physical characteristics of potential resources spaces, including some detail of the socio-economics and demographics of the area under assessment, yet water and ice, for example, resist easy classification. In a place like Melville Bay, for example, sea ice, glaciers, icebergs, and oceanic troughs and ridges have involved a considerable investigative labor involving military and scientific personnel from the Cold War to the present, with the aim of sizing up, measuring, and making the region and its nonhuman entities knowable and legible. The environmental and the social co-constitute one another in profound ways, but the slippery, stretchable nature of northern Greenland has often made an accounting of how the shifting, dynamic, and unpredictable elements of the region intermingle and interact a difficult task (Dodds and Nuttall 2019; Nuttall 2019).

Many of the SIAs carried out in Greenland are based on secondary sources. For example, the methodology section of the draft SIA for the Dundas Ilmenite project at Moriusaq states that

> Most of the baseline information presented in this SIA is based on information available from secondary sources. The sources include research reports, relevant studies, official strategies and statistical data from Statistics Greenland. Efforts have been put into presenting the most updated information at the point of writing. At local level, some primary sources have been used to describe the baseline situation. In the baseline it is indicated when information is received from primary sources.[1]

The SIA goes on to explain that data collected from primary sources was done so for the purpose of qualifying the data collected through the secondary sources. In-person interviews and meetings were held during trips to

Qaanaaq in February 2017 and Nuuk and Ilulissat in October 2017. Savissivik, perhaps the community to be affected the most because of the potential impacts on traditional hunting areas, was not visited. The SIA methodology also included telephone and in-person interviews with researchers and key stakeholders between December 2018 and February 2019. Overall, the draft SIA is favorable about the benefits of the project, specifically citing 270 jobs for Greenlandic employees. However, it also highlights that the project will lead to restricted access to the license area, which is used by local hunters, but it concludes that the negative social impacts will be relatively small and easily mitigated. The SIA contains summaries of responses and comments from people in Qaanaaq who attended the open information meeting in February 2017 as well as from semi-structured interviews with eight people in the community (four of the interviewees were hunters; recall that Qaanaaq has a population of around 650—about fifty people are full-time or occupational hunters). There are many in the community who thought the project would bring some economic benefit and employment, but there were many who also expressed concern over its impacts on the environment and on hunting and fishing activities. Yet traditional knowledge or local knowledge is not incorporated in either the draft SIA or the draft EIA for the project—rather, it appears that, for consultants, local knowledge equates a response, concern, or opinion.

Livelihoods and community economies in northwest Greenland are bound up not only with animals but also with cryospheric elements. Absent from assessments, local people say, is a consideration of local knowledge about how such nonhuman entities compose, configure, and animate the world along with humans. A meaningful interest in local knowledge would require consultants to take time to listen to people about how they consider themselves and their livelihoods to be entangled with the nonhuman, and how these livelihoods are also shaped by political, economic, and cultural forces and processes. This is an approach that would need to be attentive to the concern some multispecies scholarship has with the ways in which worlds are made and come into being through the ways the trajectories of many species intersect, an approach that is also informed by Indigenous understandings of the environment and interspecies engagement. Eduardo Kohn has articulated this in his ideas for an anthropology of life that is concerned with an anthropology beyond the human and with the entanglements of a multiplicity of selves. This "requires us to recognize the fact that seeing, representing, and perhaps knowing, even thinking, are not exclusively human affairs" (Kohn 2013, 1). For Kohn, paying attention to "a form of thinking about the world that grows out of a specially situated intimate engagement" (ibid., 73) takes us away from a bounded way of apprehending the world toward a more nuanced understanding of emergence and becomings, an understanding that disrupts our expectations of how the world is, what it looks like, and how it takes shape.

When a draft EIA/SIA Report has been approved as fulfilling the minimum requirements, it is published on the government of Greenland website for an eight-week period of public consultation. During this time, the project proponent is required to hold meetings in the towns and settlements that would be affected by the project. The public consultation period allows people to submit comments in oral form at the meetings or as written submissions, offering another opportunity for traditional and local knowledge to be incorporated (Dahl and Hansen 2019). Following the meetings and the submission of comments, the project proponent revises the draft EIA/SIA and resubmits it to the government of Greenland along with a white paper for review and, if acceptable, approval, which is necessary for submission of an exploitation application license. In theory, the public consultation period allows an opportunity for people to express their opinions, views, and concerns, but they often encounter difficulties and obstacles in doing so. For example, in summer 2020, Greenland's mineral authorities announced the consultation phase for the Dundas Titanium ilmenite project. Hearings were expected to be organized in Ilulissat, Savissivik, and Qaanaaq, but what did take place in the end were information and discussion sessions rather than thorough consultation and opportunities for local residents to dig deep into a range of documents and debate the project with the proponents. The EIA and SIA reports were published on the government's website, along with links to supporting documents on the Dundas Titanium website. However, people in Savissivik and Qaanaaq often encounter difficulty with a reliable Internet connection, so accessing, downloading, and responding to the assessments, various reports and supporting documents is not always an easy task. Furthermore, the Covid-19 pandemic made organizing the consultation meetings difficult. Travel restrictions to and within Greenland meant that representatives from Dundas Titanium, and consultants who were not resident in Greenland could only take part in the consultation meetings online.

CONCLUSION

Greenlandic politics has been closely influenced in recent years by the role and extent of the extractive sector and whether the self-rule government should be working more closely with foreign mining and oil companies and investors to help generate revenue streams that would fund a shift away from economic reliance on Denmark and enable the conditions for greater political autonomy. In northwest Greenland, exploratory activity such as seismic surveys, geological reconnaissance, and geophysical research related to extractive industries has brought different kinds of pressures and anxieties for local people, as well as hopes for the future. Coastal communities are caught up in a geopolitical scripting about frontiers of potential and abundance, and of business opportunities in a "New Arctic." Glacial melt and retreat, disappearing sea ice, narratives about scrambles for territory and oceanic space, and the marking out of subterranean extractive frontiers, inform new oceanic and territorial imaginaries that gesture toward a way of thinking about open polar seas and Arctic lands that may be devoid of ice in the near future. Climate change scenarios unsettle taken-for-granted assumptions about frozen water being a somewhat permanent characteristic of the Arctic and so an environmentalist intervention has brought a different kind of attention to northern Greenland, one that highlights the loss of ice and the possible disappearance of animals such as polar bears and narwhals. Organizations such as the World Wide Fund for Nature (WWF) have defined the region as an exceptional cryospheric space, part of a last area of ice that includes High Arctic Canada and parts of the Arctic Ocean, with globally vital ecosystems and wildlife habitat in need of protection. Marine biodiversity and iconic animals such as polar bears may be front and center in their campaigns, but Indigenous rights and customary hunting and fishing practices are affected in the process.

While Greenland may be viewed as a new resource frontier by many international companies, the region's icy places and its animals also enter into processes and discourses about biodiversity and environmental change that are seeking to shape regional and global approaches to environmental governance and development in the Arctic. Biologists who conduct marine mammal surveys and provide data for management advice, for example, continue to argue that narwhal hunting in Melville Bay is unsustainable and that, combined with the effects of climate change, local practices threaten northern Greenland's sub-populations of narwhal (Heide-Jørgensen et al. 2020). They call for targeted conservation that would manage hunting. Hunters themselves argue that scientists focus on harvesting activities without attempting to understand the agency of narwhals, the ways they think, and how people and narwhals in northwest Greenland interrelate and interact—a familiar case of a clash between Indigenous knowledge, science, and the bureaucracies of wildlife management. By way of response to such calls for the protection of ice and marine mammals, Inuit organizations argue that environmental governance and conservation policy-making should not exclude the local communities in the region and call for the designation of protected marine areas, such as in Pikialasorsuaq (the North Water polynya), that incorporate traditional knowledge and local monitoring as fundamental principles that guide ecosystem management (Inuit Circumpolar Council 2017).

Plans for extractive industry projects, feasibility studies and social and environmental impact assessments, as well as discussions of the environmental and social impacts of seismic surveys and mineralogical mappings, involve discourses about extraction and spatial technologies of power that privilege particular technocentric and economic views of the Greenlandic environment. Places marked out as resource zones become highly contested and social, political, and ideological sites entangled with wider global processes and activities. A broad range of social, discursive, economic, and political practices and procedures are at play through which Greenland's environment and resources are probed, mapped, classified, and audited, and given different meanings, yet which privilege scientific, political, and economic assumptions about them. These do not take into account local community perspectives on human–environment relations and the more than human nature of the surroundings in which resource extraction projects will be carved out. Being attentive to these perspectives and insights would be a major step forward in enacting impact assessments and conservation strategies that acquire local legitimacy as being inclusive and participatory.

ACKNOWLEDGMENTS

This chapter is based on research funded by Project 6400 at the Greenland Climate Research Centre at the Greenland Institute of Natural Resources and also draws on work done as part of the Resource Extraction and Sustainable Arctic Communities (REXSAC) Nordic Centre of Excellence funded by Nordforsk. The chapter sketches out ideas that are elaborated in greater detail and depth in Mark Nuttall (2024) *The Shaping of Greenland's Resource Spaces: Environment, Territory, Geo-security*, London and New York: Routledge.

NOTES

1. The SIA document is available on the website of the Government of Greenland here: https://govmin.gl/2020/06/dundas-titaniums-application-for-exploitation-sia-and-eia-in-public-consultation/.

REFERENCES

AMAP. 2012. *Arctic Climate Issues 2011: Changes in Arctic Snow, Water, Ice and Permafrost*. Oslo: Arctic Monitoring and Assessment Programme.

Behn, Caleb, and Karen Bakker. 2019. "Rendering technical, rendering sacred: The politics of hydroelectric development on British Columbia's Saaghii Naachii/Peace River." *Global Environmental Politics* 19 (3): 98–119.

Bendixen, Mette, Irina Overeem, Minik T. Rosing, Anders Anker Bjørk, Kurt H. Kjær, Aart Kroon, Gavin Zeitz, and Lars Lønsmann Iversen. 2019. "Promises and perils of sand exploitation in Greenland." *Nature Sustainability* 2: 98–104. https://doi.org/10.1038/s41893-018-0218-6.

Bridge, Gavin. 2004. "Mapping the bonanza: Geographies of mining investment in an era of neoliberal reform." *The Professional Geographer* 56 (3): 406–421.

Christiansen, Jørgen S., Catherine W. Mecklenburg, and Oleg V. Karamushko. 2014. "Arctic marine fishes and their fisheries in the light of global change." *Global Change Biology* 20: 352–359. https://doi: 10.1111/gcb.12395.

Cowton, Tom R., Andrew J. Sole, Peter W. Nienow, Donald A. Slater, and Poul Christoffersen. 2018. "Linear response of east Greenland's tidewater glaciers to ocean/atmosphere warming." *Proceedings of the National Academy of Sciences* 115 (31): 7907–7912. https://doi: 10.1073/pnas.1801769115.

Dahl, Parnuna P.E., and Anne Merrild Hansen. 2019. "Does indigenous knowledge occur in and influence impact assessment reports? Exploring consultation remarks in three cases of mining projects in Greenland." *Arctic Review* 10: 165–189.

Dodds, Klaus, and Mark Nuttall. 2019. "Geo-assembling narratives of sustainability in Greenland." In *The Politics of Sustainability in the Arctic: Reconfiguring Identity, Space and Time*, edited by Ulrik Pram Gad and Jeppe Strandsbjerg, 224–241. London and New York: Routledge.

Government of Greenland. 2015. *Guidelines for Preparing an Environmental Impact Assessment (EIA) Report for Mineral Exploitation in Greenland.* Nuuk: Government of Greenland Environmental Agency for Mineral Resource Activities.

Government of Greenland. 2016. *Social Impact Assessment (SIA): Guidelines on the Process and Preparation of the SIA Report for Mineral Projects.* Nuuk: Government of Greenland Ministry of Labour, Industry and Trade.

Hastrup, Kirsten. 2016. "Climate knowledge: Assemblage, anticipation, action." In *Anthropology and Climate Change: From Actions to Transformations*, edited by Susan A. Crate and Mark Nuttall, 36–57. London and New York: Routledge.

Hastrup, Kirsten. 2019. "A community on the brink of extinction? Ecological crises and ruined landscapes in Northwest Greenland." In *Climate, Capitalism and Communities: An Anthropology of Environmental Overheating*, edited by Astrid B. Stensrud and Thomas Hylland Eriksen, 41–56. London: Pluto Press.

Hayward, Philip. 2012. "Aquapelagos and aquapelagic assemblages." *Shima: The International Journal of Research into Island Cultures* 6 (1): 1–10.

Hayward, Philip. 2015. "The Aquapelago and the Estuarine City: Reflections on Manhattan." *Urban Island Studies* 1: 81–95.

Heide-Jørgensen, Mads Peter, Eva Garde, Rikke Guldborg Hansen, Outi M. Tervo, Mikkel-Holger S. Sinding, Lars Witting, Marianne Marcoux, Cortney Watt, Kit M. Kovacs, and Randall R. Reeves. 2020. "Narwhals require targeted conservation." *Letters, Science* 370 (6615): 416.

Inuit Circumpolar Council. 2017. *People of the Ice Bridge: The Future of the Pikialasorsuaq. Report of the Pikialasorsuaq Commission.* Ottawa: Inuit Circumpolar Conference.

Kohn, Eduardo. 2013. *How Forests Think: Toward an Anthropology Beyond the Human.* Berkeley: University of California Press.

Laidre, Kristin L., Harry Stern, Kit M. Kovacs, Lloyd Lowry, Sue E. Moore, Eric V. Regehr, Steven H. Ferguson, Øystein Wiig, Peter Boveng, Robyn P. Angliss, Erik W. Born, Dennis Litovka, Lori Quakenbush, Christian Lydersen, Dag Vongraven, and Fernando Ugarte. 2015. "Arctic marine mammal population status, sea ice habitat loss, and conservation recommendations for the 21st century." *Conservation Biology* 29 (3): 724–737.

Larsen, Peter Bille. 2015. *Post-Frontier Resource Governance: Indigenous Rights, Extraction and Conservation in the Peruvian Amazon*. London: Palgrave Macmillan.

Myrup, Mikkel. 2018. *Moriusaq Archaeological Survey Report 2018*. Nuuk: Nunatta Katersugaasivia Allagaateqarfialu/Greenland National Museum and Archives, 18pp.

Nuttall, Mark. 2008. "Climate change and the warming politics of autonomy in Greenland." *Indigenous Affairs* 1-2/08: 44–51.

Nuttall, Mark. 2012. "The Isukasia iron ore mine controversy: Extractive industries and public consultation in Greenland." *Nordia Geographical Publications* 41 (5): 23–34.

Nuttall, Mark. 2016. "Narwhal hunters, seismic surveys and the Middle Ice: Monitoring environmental change in Greenland's Melville Bay." In *Anthropology and Climate Change: From Actions to Transformations*, edited by Susan A. Crate and Mark Nuttall, 354–372. London and New York: Routledge.

Nuttall, Mark. 2017. *Climate, Society and Subsurface Politics in Greenland: Under the Great Ice*. London and New York: Routledge.

Nuttall, Mark. 2019. "Icy, watery liquescent: Sensing and feeling climate change on Northwest Greenland's coast." *Journal of Northern Studies* 14 (2): 71–91.

Reeves, Randall R., Peter J. Ewins, Selina Agbayani, Mads Peter Heide-Jørgensen, Kit M. Kovacs, Christina Lydersen, Robert Sydam, Wendy Elliot, Gert Polet, Yvette van Dijk, and Rosanne Blijleven. 2014. "Distribution of endemic cetaceans in relation to hydrocarbon development and commercial shipping in a warming Arctic." *Marine Policy* 44: 375–389. https://doi.org/10.1016/j.marpol.2013.10.005.

Sejersen, Frank. 2019. "Brokers of hope: Extractive industries and the dynamics of future-making in post-colonial Greenland." *Polar Record* 56: E22. https://doi.org/10.1017/S0032247419000457.

Stroeve, Julienne C., John R. Mioduszewski, Asa Rennermalm, Linette N. Boisvert, Marco Tedesco, and David Robinson. 2017. "Investigating the local-scale influence of sea ice on Greenland surface melt." *The Cryosphere* 11: 2363–2381.

Van As, Dirk. 2011. "Warming, glacier melt and surface energy budget from weather station observations in the Melville Bay region of northwest Greenland." *Journal of Glaciology* 57 (202): 202–220.

Van den Broeke, Michael R., Jason Box, Xavier Fettweis, Edward Hanna, Brice Noël, Marco Tedesco, Dirk van As, Willem Jan van de Berg, and Leo van Kampenhout. 2017. "Greenland ice sheet surface mass loss: Recent developments in observation and modelling." *Current Climate Change Reports* 3 (4): 345–356. https://doi.org/10.1007/s40641-017-0084-8.

6

Intimate Encounters with Uranium at an Anticipated Uranium Mine in Tanzania

Stephanie Postar

In 2018, thirty-two Tanzanian men wrote a letter to their district government office, complaining of injuries they claim to have sustained while working at the Mkuju River Project, Tanzania's first uranium mine. District officials in the East African country acted on behalf of their constituents and referred the complaints—of male infertility and skin rashes—to the mining company. The mining company responded to the district, denying the allegations.

The Mkuju River Project has been quiet, "mothballed" in an extended state of care and maintenance. Although the mine officially opened in 2015, commercial extraction has yet to begin.[1] The mine is located wholly within the UNESCO World Heritage Site Selous Game Reserve, considered sub-Saharan Africa's oldest and largest protected area, covering much of the Rufiji River watershed.[2] In 2012, the Tanzanian government successfully lobbied the World Heritage Committee to alter the border of the UNESCO site to allow for the Mkuju River Project to move ahead, representing the first time the World Heritage Committee permitted such an annexation of a natural heritage area to allow for extraction. Expanding extractive industries appeared as a pillar of Tanzania's strategy to become a middle-income country by 2025. However, in response to low prices of uranium since the 2011 triple disaster in Japan, in 2016 the local mining subsidiary announced plans to delay mining at Mkuju River until uranium prices rise. In 2017, the parent mining company confirmed that it applied to suspend the project for at least three additional years, again due to the low price of uranium (Basov 2017). Nonetheless, the mining company appears committed to the project. The company anticipates bringing US $1 billion in foreign direct investment and 1,600 workers at peak times of construction and decommissioning over the projected twelve- to fifteen-year life of the mine.

In the context of the uncertain future of the Mkuju River Project, young and old hear of a sickness called *homa* (Kiswahili for fever) caused by polluted water linked to the mine, and the allegations about the contamination of the bodies of local men who work at the mining site. Regardless of the health concerns, locals remain outwardly eager for work at the mine. Economic inclusion in the project is a key concern for citizens living in the area, whose livelihoods generally revolve around small-scale agriculture. Based on ethnographic research from 2014 to 2016 in the villages closest to the uranium mine, this chapter examines how uranium has power before it is extracted and what this means for inclusion and exclusion in a postponed mining project. By tracing the material form of unmined uranium in the fear of a sickness linked to the mine, this chapter argues that the rumors of this sickness can be read as a quiet resistance to the terms of industrial extraction.

To trace how unmined uranium impacts people and ecosystems, this chapter examines the allegations of workplace harm and rumors of contaminated water. Tracing is method and material: "objects that remain, marks of passage, residual amounts, vanishing inscriptions" (Geissler and Lachenal 2016, 15). The chapter first introduces young men who still hope for jobs at the mine despite hearing of the risks, then considers efforts by the mining company to contest the rumors of water contamination, and finally examines how gender and labor link to fears of the future impact of industrial extraction. In the spreading and contestation of the rumors of how unmined uranium is already harming health in the villages nearest the mine, the body materializes the unseen but still powerful mineral as well as the unknown socio-economic impacts of industrial extraction on a rural area. Before focusing on the rumors, the materiality of resources, the social and physical body, and pollution contextualize uranium mining and its environmental health impacts.

POLLUTION, THE BODY, AND POWERFUL MATERIALS

This chapter situates itself in terms of the physical, cosmological, and metaphysical approaches to the materiality of mineral resources. By focusing on uranium and its contaminations, I follow the tradition of questioning a binary nature/society framing by placing mineral resources at the center of critical inquiry (Bakker and Bridge 2006; Richardson and Weszkalnys 2014). Moving beyond materiality focused on historical, human-centered trajectories (marketing, ownership, and access to resources, among others) this "resource materialities" approach recognizes resources as distributed over time and space, always in formation, and relational, including "how resources are experienced and embodied by people who work with, transform, or (deliberately or accidentally) ingest them" (Richardson and Weszkalnys 2014,

16). This chapter builds upon the work of social scientists and Indigenous scholars examining how natural resources are more than their material uses for humans and their impacts on environmental health (Nash 1979; Mauss 2016[1925]; Tallbear 2017). As with other academic engagements with "more than human" entities (Whatmore 2002), this chapter contributes to the study of materiality as it broadens our understanding of the generative capacities of nonhumans lacking intentionality (Chen 2012), while still recognizing the unevenness of power and colonial presences (Willey 2016; Abrahamsson et al. 2015; Sundberg 2014).

In the rumors of a sickness linked to a mining site, we can recognize fears of pollution from industrial waste and contamination. Waste (identified as disordered and dirty) is increasingly central to concerns about materiality and the body (physical and social). Traces of waste embedded in or leaching from ruined landscapes like extractive zones reveal normative expectations of the ordering of the physical world (Edensor 2005). Heavy metals with the capacity to harm, mediate, and destabilize categories (e.g., life, race, class, ability, sexuality), challenging understandings of pollutants as inanimate (Chen 2012). Waste and contamination from mining activities have sparked protest movements around the world; however, around the Mkuju River Project, local people are vocally and overwhelmingly supportive of the mine. Despite outward support for the mine, the allegation of mining pollution suggests a level of anxiety about the project. Doubt and uncertainty are the norm in the study of toxic exposures (Auyero and Swistun 2008). This ambivalence is also characteristic of the "resource affect" of anticipated extractive booms (Weszkalnys 2014).

The study of toxicity may be defined by uncertainties, except in terms of the social and physical interlacing of (industrial) waste, (more than human) environments, and bodies. Indigenous activists and scholars increasingly link extractive violence to the land (in terms of landscape change and pollution from industrial waste) with violence to female bodies, highlighting contemporary gendered experiences of extraction (Deer and Nagle 2017; Konsmo and Pacheco 2016). The body is historically situated and infused with social meaning while remaining a site of social and biological (re)production. Bodies are not bounded or clearly defined, particularly in light of new medical technologies (Sharp 2000) and degraded environments (Lock 2018). The prism of personal experience of illness in postcolonial environments attends to the mingling of pollutants among and within bodies (Mavhunga 2018).

Where porous bodies interact with toxics at sites of extraction, the allegations of the men who claim they were injured before commercial uranium mining began to fit into the context of the nuclear world. Uranium, though inanimate, is, for many, connected to fears of radiation-induced mutation (Masco 2004), apocalyptic nuclear weapons attacks (Gusterson 1999), and power plant meltdowns (Petryna 2002). The singularity of the uranium associated with nuclear disaster often contrasts with what Gabrielle Hecht (2009) calls the "banality" of uranium in mines, where uranium could be treated as any other commodity. Still, much remains to understand about the spectrum between the extraordinary and banal natures of nuclearity (Pitkanen and Farish 2017; Malin and Alexis-Martin 2020).

The symptoms of contamination recounted at the beginning of the chapter bear notable similarity to some of the symptoms of radiation exposure and contamination (Agency for Toxic Substances and Disease Registry 2013). Scientific debate continues regarding the impacts of radiation, particularly the health impacts of long-term, low-dose radiation, like that which may be experienced in villages near a mine and yellowcake processing facility (Goldstein and Stawkowski 2015). The lack of consensus in radiation science and the lack of systematic health screening in this case do not make the possibility of harm less real to those at risk and do not slow the long-term processes that ultimately lead to contamination.

The "slow violence" (Nixon 2011) of incrementally accumulating pollution from industrial activity and the unequal burden of this violence (Bullard 2000) shape the connections between industrial activity, environment, and marginality. Recent scholarship highlights forms of politics emerging in communities subjected to slow violence that are neither event-based nor highly visible. Rather, emotions can be the basis for long-term resistance to extractive violence in a kind of "slow dissent" (Murrey 2016). These kinds of everyday refusals to accept life in a damaged world and (seemingly small) efforts to intervene to generate alternative futures, relations, and kinships in contaminated environments (Murphy 2017) inform the way this chapter makes sense of rumors (and fear) of a sickness thought to originate at the delayed uranium mining project. Therefore, this chapter traces the materialization of (unmined) uranium in the fear of the *homa* in the present and in unknown futures. In the next section, young men recount the rumors of a sickness they associate with the mining site and address how they relate to growing anxiety about the risks of working at the mine.

Homa

In the villages closest to the Mkuju River Project, young men eagerly anticipate jobs at the mining concession, even though they are familiar with the rumors of a sickness somehow tied to the mine. Hamidu, a twenty-two-year-old farmer, explains that despite cultivating six acres, he struggles to get ahead.[3] Elephants and baboons destroyed three acres of his pigeon pea crop in 2015 and 2016, even though his farms are not all in the same area.

Hamidu is married and has one small child. His parents fled Mozambique's civil war, and he was born in Tanzania in this refugee camp-turned village bordering the Selous Game Reserve. Hamidu says the soil is increasingly depleted, but there is enough water and forest to meet their current needs. He tries to push ahead, he explains, but always ends up getting shoved behind. Asked about the Mkuju River Project, he says, "We would like to go [to work at the site]. We are losing out on going [to the mine]. We hear there is a *homa* [Kiswahili for fever, see below] that comes and reduces the strength of the body." He explains that he heard about the *homa* from other people in the village and that the *homa* "reduces male strength." Despite the threat of this illness, he confirms he still would accept an opportunity to work at the mine.

In the same village, Bakari, a twenty-four-year-old, discusses how he cultivates a three-acre farm as his main livelihood. In 2016, he explains, elephants ate nearly one and a half acres of his crop. He filled out a form to request compensation for the damages from the government that year but, he explains with frustration, he never received payment. Asked about the possibility of the mine, he explains that the working-age people in his village have not received employment opportunities. He would want to work at the mine, even though he heard the rumors of the *homa*. He says that some people in the village are afraid of the illness, but not him.

A village leader, Rajabu, listened with me to these young men. We later spoke about the *homa* and he explained that it is an illness people say is impacting men who worked at the Mkuju River Project, and also that the *homa* is tied to water. People think the polluted water comes into the villages, reducing overall strength in men's bodies, by which he means that it reduces male fertility.[4] Another village leader confirmed that there were young men in the village who were afraid to go work at the mine because they fear they will be sickened at the site. *Homa* is generally translated as "fever" but has wider connotations in this area and throughout the Kiswahili-speaking world. *Homa* can be an illness itself or may be used to describe an array of illnesses, not only an above-average body temperature (Winch et al. 1996).

A CORPORATE RESPONSE TO RUMORS OF WATER POLLUTION

The mining company did not allow the rumors of *homa* to spread in the villages without attempting to intervene to calm anxieties about the safety of drinking water. The day before the young men mentioned the rumors of the *homa*, Rajabu visited the mining concession. He and a few others from the villages closest to the mine toured the mining concession, focusing on water. The mining company picked them up in the villages and drove them through the Selous Game Reserve and into the mining site. At the site, they took in the various facilities, including a 900L water tank meant for drinking water. Rajabu recalled that the water came from a nearby river and that the water was measured and cleaned, using a chemical agent. He explained that the group of visitors was instructed to tell people back in the villages about their visit to the mining site, their experience drinking clean water there, and seeing the pristine environment at the site. He added that the mining company would come to test the water quality in the rivers running through the villages.

Another local leader, a young woman who joined the group brought to the mining site, confirmed the mining company tested the groundwater at the mining site, and that they said they would start measuring water quality in the villages. When the mining company makes these measurements, she recalled, they will bring representatives from each village, not only village leaders but also regular villagers. The purpose of the visit, she twice said, was so that regular villagers could be taken to the mining site so that in the future, people cannot complain about their water quality. During her time at the mining site, she said she drank clean water, and she did not see any problem at the mining site but noted that people in the area are still worried about their water.

Both village leaders understood that their trip to the mine site was part of the mining company's response to the rumors of the *homa*. Testing water in rivers running through the villages seems to reflect the company's awareness of the power contaminated water may have on people in the area (and their willingness to tolerate their extractive neighbor), as well as the unevenness with which information about mining activities and their impacts are spreading through the villages. The mining company seemed to explain that they would undertake periodic testing of the water at the site and in the villages even before mining started.[5] Taking villagers to the site of the mine would seem to be an effort to dispel rumors directly related to fears that people may be harboring about going to work at the site in the future. However, the company's decision that they would test water quality in the villages indicates that they are aware that people are or may be skeptical of the safety of their domestic water sources. Establishing baseline data about water quality could conceivably show that uranium impacts water sources before it is mined. However, limited studies of the hydrological context have been completed and made public through the government-mandated environmental and social impact assessment. Therefore, much remains unknown about the relationship between the villages, their above and below-ground water resources, and possible pollution from the mining project.

THE IMPACTS OF UNMINED URANIUM, WATER, AND LABOR

Despite efforts by the mining company to change local perceptions of the project and its impact on water sources, the thirty-two workers who made a written complaint about workplace safety grievances indicate the persistent fears of the *homa*. Nonetheless, local men like Hamidu and Bakari, mentioned earlier, who voice anxiety about the possibility of being excluded from mining jobs are also worried about missed economic opportunities. It is also possible that these young people are foreseeing that their (gendered) position as men without work in a mining boomtown may have social repercussions. These seemingly contrasting sentiments—of fear of *homa* as a hazard of mine work and dread of exclusion from mine work—highlight how uncertainty about labor associated with the mining project impacts how people speak about the *homa*.

For men who are able to find work at the mine and appear to be included in the project, it remains possible that they see the mine as an entity that extracts something from them. The transactional nature of wage work at the anticipated mine demands the input of labor from (generally male) bodies in exchange for a cash wage and the prestige of working for the commercial endeavor. Furthermore, abstracting from the rumors of the *homa*, men seem to fear that their inclusion in the project means there is an additional transaction in the process of the labor-capital exchange not only in the extraction of their labor but also in the extraction of their vitality (in the form of their reproductive capacity). Perhaps through the *homa,* as an ambiguous threat to their masculinity, these men see that their employment at the mine comes at an unknown, feared cost to their capacity to maintain their roles as full adult men.

The focus on men—the male worker, the male body, and the excluded male villager—only presents part of the picture. Village leaders, speaking on behalf of their constituents in a meeting, asked the mining company representative if women would have the opportunity to work at the mine. The mining company stated that women could apply for jobs, but it was broadly understood that men would obtain more jobs from the project than women. Still, some young women hoped for jobs at the mine, and anticipated work including cleaning, washing, and cooking.

In addition to future work at the mine, women's roles in the villages may also be impacted by the mining project. Collecting water in the villages closest to the mine is overwhelmingly a woman's task. Women and children collect water from rivers, open wells, and closed wells. Sections of streams are known to be used for washing clothing, bodies, and other items, while other sections are reserved for collecting drinking water in one to five-gallon plastic buckets and jerry cans. Wells are monitored by each of the villages' committees on water. Each of the villages in the ward closest to the mine, home to between 4,000 and 7,000 inhabitants, have fewer than ten closed, hand-pump operated wells, all in various states of (dis)repair.

Women and some men in the three villages nearest to the mine agree that, while there is enough water for rainfed irrigation of crops, there is not sufficient clean water for drinking in the villages. In response to these concerns, in the village closest to the mine the mining company provided funding for a well, as part of its corporate social responsibility portfolio. Also intervening in local water distribution, in 2019, the Tanzanian government began a project of bringing well water into two-thirds of the homes in the village closest to the mine.

The strongly voiced need for more water is foundational in the context of rumors of water contaminated by the mining company's activities. Thinking through the flows of tainted water, the water can be expected to first come into contact with women and girls in these villages, as they are responsible for domestic water. It would be the women and girls who bring the contaminated water into the home for drinking by the entire family, and it would be women and girls who cook with the water and introduce it into food for family meals. This could mean women in the area who might expect to be excluded from wage work at the mine may still find themselves adversely included in the project through their contact with water, with unknown impacts on health, gender roles, and sociality.

Through contact with water, local people already feel connected to the mining project, and this entanglement does not appear purely to their benefit. Even before commercial mining begins, risks of uranium mining seem to materialize in non-extractive work at the mining concession as well as in drinking water in areas surrounding the mine. Limited efforts to communicate with villagers about uranium mining and radiation, to mitigate concerns about the *homa*, and to publicize grievance mechanisms come inconsistently from the mining company, as well as Tanzanian government agencies and civil society. As with the water itself, mining company outreach efforts amount to a kind of strategic but not necessarily favorable inclusion of neighboring communities. Water and outreach efforts appear to materialize as well as contest the uncertainty related to uranium not yet extracted. The banal, ubiquitous necessity that is water becomes a powerful, potentially insidious force because of its contact with unmined uranium. Although uranium remains hidden before extraction, at the mining site, and in domestic spaces in nearby villages, contaminated water emerges with capacities to shape society and individual lives amid fears of *homa*.

SOCIAL ILLNESS AND TRACES OF RESISTANCE

People in the villages closest to Tanzania's first uranium mine expect this major extraction initiative to impact them by creating jobs but also fear that they will experience health problems due to mining pollution. Most people vocally support the mine and are eager for jobs, amid dependence on small-scale agriculture and limited alternative economic opportunities, however, hopes for the future of the mine are tempered. Unpacking the concerns about the *homa*, a sickness said to be linked to the delayed mine, raises questions about how unmined uranium takes material, sensible forms. These seemingly contradictory attitudes speak to the ambivalence of local people toward their various shades of inclusions and exclusions from the extractive project (Postar and Behzadi 2022). Rather than trying to identify an ultimate "truth" regarding the validity of the *homa* as evidence of toxic contamination, or evaluating the embodied experiences of the men who launched the formal complaint about skin rashes and infertility following employment at the mining site, this chapter acknowledged that the felt risk of *homa* and actual fact are blurred in a way that demonstrates how unmined uranium impacts everyday life. Fear materializes unmined, unseen uranium in the bodies of people living in the villages closest to the mine. In the fear of the *homa* and the spreading rumors of how the mining site is harming health, lie traces of resistance to the violence of extraction before industrial extraction begins.

The connection between *homa* and (in)fertility, targeting youth, full of vitality and strength, who would be capable of physical labor at the mining site, foregrounds a perceived threat to masculinity. African masculinities emerge as "hegemonic and subordinate" (Ratele 2014, 118), reflecting differentiation by exclusive structures including race and class. Workers at the mine claim skin diseases and illustrate how they may be marked externally as polluted by their work at the mine, giving an outward indication of possible internal transformations. Invisible, unknowable, and perhaps slowly accumulating toxic agents (thought or able to perforate the body) raise questions about what contaminated bodies and spaces look like, and what happens when those bodies and spaces cease to be productive and, themselves, turn to waste. The relationship between bodies, toxicity, waste, and value, from a superfluous worker (Ferguson 2015) to the human-as-waste (Yates 2011), could inform the ways in which human bodies are economically (and sometimes physically) displaced from work or the wage-earning world and also reveal the ways this might bleed into a social displacement.

Attending to the ways in which people report their bodies are internally impacted by nonhuman forces, we can briefly acknowledge the particular anxieties about biological reproduction. (In)fertility is political, and male infertility is often deeply connected to masculinities (Inhorn and Wentzell 2011). In addition to the letter of complaint from mineworkers, the body, permeable and unbounded, gives the rumors of *homa* and risks of unmined uranium a material form. Beyond the suggestion of the symbolic body (Douglas 1966), physically bearing the signs of societal sensitivity by way of physical violations of the body, Scheper–Hughes (1993) provides a theoretical scaffolding for which to deconstruct a "social" illness linking nonhumans to social reproduction. Where sickness is a threat to the social, moral, and reproductive order, illness is political. *Homa* could be seen as speaking to people's feelings of precariousness as well as the social and economic contradictions percolating in the villages surrounding the delayed mine. The *homa* might also be a commentary by local people on their bodies and the way that future employment, inclusion, or proximity to the mine may sap their life force or vitality, or the way an exclusion from alternatives to extraction-related job opportunities is also felt to be narrowing their capacity for social reproduction. Where protest against the mine may risk discrimination in future employment, or the ire of neighbors who support the project, the spread of rumors of *homa* can serve as a "safe way" to express fear or discontent with the development of the mine. For men who say that they have heard of the *homa* but still hope for jobs with the project, sharing the rumor can be a permissible expression of (otherwise disavowed) feelings of apprehension about the future by casting them as rumors of a sickness causing infertility.

Without needing to confirm or deny the scientific validity of claims of mine-related pollution, unmined uranium takes a material form in the experience of and rumors of water pollution and the *homa*. The uranium, though still underground, impacts people living near the mining project in part by turning water into a suspicious commodity. The introduction of industrial activities in this rural area and the ensuing fear of illness (regardless of its basis in science) reveals anxieties about new actors, social reproduction, ecosystems, and bodies. The ambiguous role of the uranium miner who exchanges a cash wage for his labor and manhood raises questions about the unknown powers and legacies of the radioactivity associated with uranium, and the unknown futures of the mineworkers and their families. The focus of concerns around the contamination of water suggests a questioning of how people will meet their most basic needs in the future. The fear of *homa* might be the basis of future resistance to industrial extraction because it traces the intimate interactions between people and natural resources and concerns about the kinds of new, unforeseen challenges that villagers may encounter as industrial extraction inserts itself into their rural periphery.

CONCLUSION

From the whispers of young people of a rumored illness linked to a uranium mine in Tanzania, this chapter traces how uranium has power even before it is extracted. This uranium is unseen, hidden inside a wilderness area and unextracted, at a mine site delayed indefinitely. Working-age people express both dread at being excluded from the economic opportunities of mine work and fear of adverse inclusions in the health hazards associated with extractivism. Uranium's faculties for chemical and social pollution, leaching from the earth and also radiating from the socio-economic changes on a mining boom town to-be, here are felt in the body: in a sickness called *homa.* The *homa,* which is thought to cause male infertility and skin lesions, appears as a threat to achieving standards of masculinity, having children and being able to economically support them. Even before mining, uranium influences the thoughts and affairs of humans. The unmined, yet still powerful uranium considered here invites us to examine how porous ecosystems and (more than) human bodies encounter global commodities, reorganizing genetic material and social worlds. By centering attention on uranium amid living and nonliving beings, this chapter suggests in the fears of an illness linked to the mining site also lie traces of resistance to the current terms of industrial extraction.

NOTES

1. Russian news reports a pilot project at the site may begin in late 2023 ("Rosatom May Start" 2023).
2. In July 2019, Tanzanian President John Magufuli split the Selous Game Reserve: the northern half would become Nyerere National Park, permitting only photographic safaris, while the southern part would remain the Selous Game Reserve and continue to permit big game hunting.
3. All names appearing here are pseudonyms.
4. Young Tanzanian men discussing threats to fertility with a white, female researcher raises difficult intersectional questions that are unlikely to be resolved here.
5. By August 2019, the mining company had yet to test the water in the villages.

REFERENCES

Abrahamsson, Sebastian, Filippo Bertoni, Annemarie Mol, and Rebeca Ibáñez Martín. 2015. "Living with Omega-3: New Materialism and Enduring Concerns." *Environment and Planning D: Society and Space* 33 (1): 4–19. https://doi.org/10.1068/d14086p.

Agency for Toxic Substances and Disease Registry. 2013. "Toxicological Profile for Uranium." *ATSDR's Toxicological Profiles*. https://doi.org/10.1201/9781420061888_ch157.

Auyero, Javier, and Debora Swistun. 2008. "The Social Production of Toxic Uncertainty." *American Sociological Review* 73: 357–79.

Bakker, Karen, and Gavin Bridge. 2006. "Material Worlds? Resource Geographies and the 'Matter of Nature.'" *Progress in Human Geography* 30 (1): 5–27. https://doi.org/10.1191/0309132506ph588oa.

Basov, Vladimir. 2017. "Russian State Corporation Suspends $1.2 Billion Uranium Project in Tanzania." MINING.com. 2017. http://www.mining.com/russian-state-corporation-suspends-1-2-billion-uranium-project-tanzania/.

Bullard, Robert D. 2000. *Dumping in Dixie: Race, Class, and Environmental Quality*. 3rd ed. Boulder: Westview Press.

Chen, Mel Y. 2012. *Animacies: Biopolitics, Racial Mattering, and Queer Affect*. Durham and London: Duke University Press.

Deer, Sarah, and Mary Kathryn Nagle. 2017. "The Rapidly Increasing Extraction of Oil, and Native Women, in North Dakota." *The Federal Lawyer* 64 (3): 35–37.

Douglas, Mary. 1966. *Purity and Danger: An Analysis of Concepts of Pollution and Taboo*. London: Routledge & Kegan Paul.

Edensor, Tim. 2005. "Waste Matter - The Debris of Industrial Ruins and the Disordering of the Material World." *Journal of Material Culture* 10 (3): 311–32. https://doi.org/10.1177/1359183505057346.

Geissler, P. Wenzel, and Guiaume Lachenal. 2016. "Brief Instructions for Archaeologists of African Futures." In *Traces of the Future: An Archaeology of Medical Science in Africa*, edited by P. Wenzel Geissler, Guillaume Lachenal, John Manton, Noémi Tousignant, Evgenia Arbugaeva, and Mariele Neudecker, 15–30. Bristol: Intellect.

Goldstein, Donna M., and Magdalena E. Stawkowski. 2015. "James V. Neel and Yuri E. Dubrova: Cold War Debates and the Genetic Effects of Low-Dose Radiation." *Journal of the History of Biology* 48: 67–98. https://doi.org/10.1007/s10739-014-9385-0.

Gusterson, Hugh. 1999. "Nuclear Weapons and the Other in the Western Imagination." *Cultural Anthropology* 14 (1): 111–43. https://doi.org/08867356, 15481360.

Hecht, Gabrielle. 2009. "Africa and the Nuclear World: Labor, Occupational Health, and the Transnational Production of Uranium." *Comparative Studies in Society and History* 51 (4): 896–926. https://doi.org/10.1017/S001041750999017X.

Inhorn, Marcia C., and Emily A. Wentzell. 2011. "Embodying Emergent Masculinities: Men Engaging with Reproductive and Sexual Health Technologies in the Middle East and Mexico." *American Ethnologist* 38 (4): 801–15. https://doi.org/10.1111/j.1548-1425.2011.01338.x.

Konsmo, Erin Marie, and A.M.K. Pacheco. 2016. "Violence on the Land, Violence on Our Bodies: Building an Indigenous Response to Environmental Violence." Women's Earth Alliance and Native Youth Sexual Health Network. http://landbodydefense.org/uploads/files/VLVBReportToolkit2016.pdf.

Lock, Margaret. 2018. "Mutable Environments and Permeable Human Bodies." *Journal of the Royal Anthropological Institute* 24 (3): 449–74. https://doi.org/10.1111/1467-9655.12855.

Malin, Stephanie A., and Becky Alexis-Martin. 2020. "Assessing the State of Uranium Research: Environmental Justice, Health, and Extraction." *Extractive Industries and Society* 7 (2): 512–16. https://doi.org/10.1016/j.exis.2020.02.006.

Masco, Joseph. 2004. "Mutant Ecologies: Radioactive Life in Post–Cold War New Mexico." *Cultural Anthropology* 19 (4): 517–50. https://doi.org/10.1525/can.2004.19.4.517.

Mauss, Marcel, and Jane I. Guyer, transl. 2016[1925]. *The Gift*. Chicago: HAU Books.

Mavhunga, Clapperton Chakanetsa. 2018. *The Mobile Workshop: The Tsetse Fly and African Knowledge Production*. Cambridge: MIT Press.

Murphy, Michelle. 2017. "Alterlife and Decolonial Chemical Relations." *Cultural Anthropology* 32 (4): 494–503. https://doi.org/10.14506/ca32.4.02.

Murrey, Amber. 2016. "Slow Dissent and the Emotional Geographies of Resistance." *Singapore Journal of Tropical Geography* 37 (2): 224–48. https://doi.org/10.1111/sjtg.12147.

Nash, June C. 1979. *We Eat the Mines and the Mines Eat Us: Dependency and Exploitation in Bolivian Tin Mines*. New York: Columbia University Press.

Nixon, Rob. 2011. *Slow Violence and the Environmentalism of the Poor*. Cambridge: Harvard University Press.

Petryna, Adriana. 2002. *Life Exposed: Biological Citizens After Chernobyl*. Princeton: Princeton University Press.

Pitkanen, Laura, and Matthew Farish. 2017. "Nuclear Landscapes." *Progress in Human Geography* 42 (6): 862–880. https://doi.org/10.1177/0309132517725808.

Postar, Stephanie, and Negar Elodie Behzadi. 2022. "'Extractive Bodies': A Feminist Counter-Topography of Two Extractive Landscapes." *Geoforum*, October 2022. https://doi.org/10.1016/j.geoforum.2022.09.005.

Ratele, Kopano. 2014. "Hegemonic African Masculinities and Men's Heterosexual Lives: Some Uses for Homophobia." *African Studies Review* 57 (2): 115–30. https://doi.org/10.1017/asr.2014.50.

Richardson, Tanya, and Gisa Weszkalnys. 2014. "Resource Materialities." *Anthropological Quarterly* 87 (1): 5–30.

"Rosatom May Start Uranium Mining in Tanzania in Coming Months — Ambassador." *TASS Russian News Agency*, 23 July 2023. https://tass.com/russia/1651987.

Scheper-Hughes, Nancy. 1993. *Death Without Weeping: The Violence of Everyday Life in Brazil*. Berkeley: University of California Press.

Sharp, Lesley A. 2000. "The Commodification of the Body and Its Parts." *Annual Review of Anthropology* 29: 287–328.

Sundberg, Juanita. 2014. "Decolonizing Posthumanist Geographies." *Cultural Geographies* 21 (1): 33–47. https://doi.org/10.1177/1474474013486067.

Tallbear, Kim. 2017. "Beyond the Life / Not-Life Binary: A Feminist-Indigenous Reading of Cryopreservation, Interspecies Thinking, and the New Materialisms." In *Cryopolitics: Life in a Melting World*, edited by Joanna Radin and Emma Kowal, 179–202. Cambridge: MIT Press.

Weszkalnys, Gisa. 2014. "Anticipating Oil: The Temporal Politics of a Disaster Yet to Come." *The Sociological Review* 62 (June): 211–35. https://doi.org/10.1111/1467-954X.12130.

Whatmore, Sarah J. 2002. *Hybrid Geographies: Natures, Cultures, Spaces*. Thousand Oaks, CA and London: SAGE.

Willey, Angela. 2016. "A World of Materialisms: Postcolonial Feminist Science Studies and the New Natural." *Science Technology and Human Values* 41 (6): 991–1014. https://doi.org/10.1177/0162243916658707.

Winch, P. J., A. M. Makemba, S. R. Kamazima, M. Lurie, G. K. Lwihula, Z. Premji, J. N. Minjas, and C. J. Shiff. 1996. "Local Terminology for Febrile Illnesses in Bagamoyo District, Tanzania and Its Impact on the Design of a Community-Based Malaria Control Programme." *Social Science and Medicine* 42 (7): 1057–1067.

Yates, Michelle. 2011. "The Human-As-Waste, the Labor Theory of Value and Disposability in Contemporary Capitalism." *Antipode* 43 (5): 1679–95. https://doi.org/10.1111/j.1467-8330.2011.00900.x.

Part IV

STATIC MATERIALS/ DYNAMIC MATERIALS

Resource Materialities, Temporalities, and Affect

Gisa Weszkalnys

Although mostly localized and "grounded," the process of mineral resource extraction is hardly fixed or static. Its pace and intensities are shaped, among other things, by notions of geological potential as well as practical, moral, and aesthetic value; by the cyclicality of commodities markets, state policy, and fiscal regimes; by the availability of equipment, infrastructures, and expertise; by epistemic and risk management techniques; by expectant imaginations of profit and progress; and by the physical limits of laboring bodies. Critical resource scholarship in anthropology, geography, and related social sciences has sought to integrate these material, temporal, and affective dynamics that together constitute extraction (e.g., Bakker and Bridge 2006; Barry 2014; Ferry and Limbert 2008; Richardson and Weszkalnys 2014). This work aims to deepen our understanding of the uneven and unequal processes that connect the "above-ground" and "below-ground" in extractive encounters. It departs from an essentialist category of resources embedded, for example, in an earlier Marxian historical materialism or in resource management approaches, and from the deterministic assumptions of resource curse theories. Rather than assuming resources to be pregiven in nature, ready to be discovered and put to use, it reveals resource making as both spatially and temporally dispersed, resulting in historicized and deeply contested "resource materialities" (Richardson and Weszkalnys 2014, 8).

The generation of distributed resource materialities has been central to the refiguration of nature–society relations in the Anthropocene. Building on a range of intersecting ideologies of exploitation, exclusion, and dispossession, it has buttressed the violence of imperial and nation-building endeavors, modernist government as well as racial capitalism (Davis and Todd 2017). As activists and academics have demonstrated, this refiguration is ongoing (e.g., Anthias 2018; Estes 2019; Gardner 2012; Li 2015; Peluso and Watts 2001; Sawyer 2004; Voyles 2015). Resource making continues to be underpinned by unequal, racialized power relations that translate, all too frequently, into forms of epistemic injustice and conflict, sometimes prior to their extraction (Behrends 2008). Struggles over extractive projects have proliferated in recent decades, while public awareness of their entwined ecological and sociocultural cost has sharpened—even as international financial organizations continue to push such projects as a source of "development." Importantly, these struggles concern not merely the allocation and division of material resource substances. They are struggles for survival, aimed at maintaining people's right to determine individual and collective forms of existence.

Contemporary capitalist resource exploitation at once conceals the violence necessary to derive value from nature and is partisan in its definition of what that value should be. It invokes a calculative logic, confidence in scientific and technological knowhow, and the pursuit of limitless growth. Its infrastructures fix developmentalist visions in place and reproduce the modern idea of nature as subservient to human desires and standardized technical solutions (Carse 2017). Dominant understandings of what make a resource comprise both the abstraction of substances from existing ecologies *and* an ontological abstraction (Richardson and Weszkalnys 2014). The two cannot be held apart. Crushing, grinding, sorting, separating, and other forms of biotechnological and chemical modification contribute to substances' biophysical transformation. This is underwritten by processes of analyzing, naming, and standardizing required to generate fungible commodities while infusing them with new utility and value. Extracted substances are dissociated from a range of attributes and connotations, thus fracturing previous cultural and sensorial appraisals of quality and worth. As resources, they are reimagined to hold generative potentiality that requires human activation and channeling and is operationalized, for example, in notions of prospectivity, de-risking, and assetization (cf. Gilbert 2020; Wood 2019).

By starting from a notion of resource materialities *in flux*, critical resource scholarship reveals the indeterminate nature of resource development and challenges the binaries of included/excluded, powerful/marginalized, among others, that have commonly framed critical analyses (Kneas 2018). This research details the diverse and often-overlooked practices of those laboring and living on the sidelines of extraction (e.g., Anthias 2018; Jaramillo 2020; Powell and Curley 2008; Smith 2011) and reveals the multiple junctures at which extractive worldings are formulated, negotiated, and eclipsed (Sundberg 2014; Todd 2016). This is borne out by Willow's chapter (this volume) on the controversial Site C Dam hydropower development on Peace River in British Columbia. Driven by an alliance of corporate and state power, the British Columbia government has touted the project as a prospective source of economic growth through "green" energy production, generating a diversity of both support and opposition. Willow highlights the longevity of this vision of resource-based progress, which has been maintained and recycled over decades. Once seen, such a vision has become hard to unsee. It lingers as strategic retentions that bolster purported national and corporate histories (Kama 2016, 2020). At the same time, it serves as a frame through which fresh extractive possibilities are given political and economic force; their benefits are portrayed as indisputable as they are inevitable. Nature continues to be reframed and put to work in the name of economic growth, despite the project's likely destructive effects on local environments and Indigenous communities. Site C Dam is implicated in the violent construction of national histories and the reproduction of an exclusionary national community (e.g., Coronil 1997; Ferry and Limbert 2008; Shever 2012), reinforcing settler resource ontologies.

Questions about the ontology of resources and their affective power, therefore, need to be centered in enquiries about extractive capitalism in ways mindful of the specificities of diverse customary principles, environmental understandings, and Place-Thought (Watts 2013) as well as the dissonances *and* overlaps between them (see also High 2017; Powell and Curley 2008; Todd 2016). Communities who ostensibly "sell out" may do so with the hope of securing livelihoods and of reversing histories of exploitation and marginalization (Anthias 2018; Gardner 2012). Corporations give substance to investable assets invoking market projections of high prices and future earnings, geophysical samples, and other technoscientific measurements. Among their multiple audiences, including investors as well as local communities, these generate affective states vis-à-vis prospective resource futures (see also Gilbert 2020; Kneas 2016; Weszkalnys 2015, 2016). West and Crowhurst discuss the revival of zinc mining in the Parina and Riso Valleys of the northern Italian Alps on the back of investment by Energia Minerals, an Australian mining company. Energia Minerals' interest in the Valleys was driven by the mineral commodities boom that swept across the globe in the first decade of this millennium. The project's local appeal derived not least from its affective force. Mining continued to have significant cultural salience in the region, despite decades of deindustrialization and decline. Residents projected their own sense of the past and hopes for the future onto the agreements with the company and its promise not only to revitalize extraction and generate jobs but also to invest in the preservation of industrial heritage.

Mines, such as those of the Parina and Riso Valleys, thus carry affect, and not just as assets in an investment portfolio (Weszkalnys 2016). Consider June Nash's description of the noises and smells experienced by Bolivian tin miners: "The pulsating thump of the . . . compressor, that breaks up the rock like a heartbeat . . . the odors of hydrogen sulphide are expelled like the gases from an over-taxed digestive system. The reddish-brown . . . liquid [that] drains from the interior like pus-filled blood from suppurating wounds" (1979, 170–1). Industrial labor and extractive technologies co-constitute lively resource environments that can be awe-inspiring, sacred, threatening, and beckoning respect, in ways that are both deeply cultural and visceral. Miners' own practices of sensing, adapting to, and skillfully navigating such agentive, hostile resource environments, and of caring for each other, help "turn the danger into socially productive matter" (Jaramillo 2020, 57; Rolston 2013). While circumscribed by modes of capitalist exploitation, they point to the different ethical and social possibilities present in the materialities of the mine (see also Curley 2019). From this perspective, northern Italian miners' calls for a musealization of their mining history may be seen as an attempt to give durability to, rather than to elide, such deeply embodied, cultural and affective experiences. It is an active manipulation of ambiguities embodied in the material ruins and ecological shadows of extractive projects, which persist even as mines close and investment has moved elsewhere.

Resources register and index time in unique ways and across a number of temporal scales. The pace and pattern of extractive operations shape the rhythms of life of extractive workers, reaching into the most intimate relations of family and community life (Dahlgren 2019; Shever 2012). The modernist promise has framed resource development in the Anthropocene as a social contract. However, this hides a deeply speculative experiment. It is an experiment constituted within the heterogenous temporalities of the state, technoscientific expertise, and corporate finance, and local ways of life. Here, corporate logics are densely linked with the power of the state, which legitimates their colonizing drive. Political and legal arrangements permit capitalist extraction to press on under a licit veneer, sometimes regardless of its failures (Appel 2019). Contracts, licenses, and strategic development plans can become "gestures of resource potentiality without compelling a predetermined outcome" (Weszkalnys 2015, 612); they suggest linearity and inevitability where delay and failure may be just as common (see also Kama 2020; Kneas 2018; Powell 2018).

As the studies in this section show, attempts to align collective hopes with the "ethics of probability" (Appadurai 2013, 295) of extractive companies and global commodities markets often result in bitter disappointment. In the Parina and Riso Valleys, residents' hopes for a revival of the region's mining legacy and its economic growth were quickly dashed. Key gestures that would have helped reimagine this potential, such as a feasibility study, never came to pass. By contrast, in the case of Site C Dam on the Peace River, the legal process effectively suppressed concerns over the financial risk and environmental destruction. The construction of the dam has been enabled by a change in the law, which entails a potential infringement of First Nations' culture, treaty rights, and demands for democratic inclusion. The visions for "improvement" based on extraction promoted by state, corporate, and local actors were thus enfolded in complex historical configurations of race, nation, territory, and development (Anthias 2018), modes of oppression and resistance, and disparate senses of lives worth living and places worth preserving.

The visionary activism highlighted by Willow demands attentiveness to such complexity. It juxtaposes future ecological and social disasters with the cosmological and physical nourishment people have drawn from the Peace River. Embedded within this activism lie cultural ontologies, epistemologies, and affects that permit a querying of the inevitability of extractivist worlding (see also Estes 2019; Li 2015; Powell and Curley 2008). Transgenerational experiences of past, present, and future harm reveal the developmentalist vision to be deeply flawed. People anticipate not collective growth but, rather, a stymying of the generative capacities of natural environments and substances. Instead of labeling this an anti-extractivism; however, visionary activism is better understood as a matter of care (Puig de la Bellacasa 2011); it results from a commitment to, and curiosity about, the resource environments with decidedly ethicopolitical intent. As they counter threats of expropriation and displacement, Indigenous residents are worldmakers imagining a different relationality with both future generations and more-than-human kin.

In this context, time itself can become a political resource. Extractive projects are characterized by extensive pauses and delays in their planning and realization (Weszkalnys 2015). They also increasingly include drawn-out phases of community consultation on proposed mines to achieve "Free, Prior, and Informed Consent" and community "buy-in." These anticipatory techniques are partly aimed at regulating people's affective responses to promising resource futures. They dovetail with corporate concerns with sustainability, "expectations management," corporate social responsibility, and Environmental, Social, and Governance (ESG) scores (e.g., Leins 2020; Okenwa 2020; Welker 2009; Weszkalnys 2016). However, they also provide opportunities for information sharing, inquiry, protest, blockage, and sowing doubt regarding the feasibility of projects among investors. Kirsch (2014) speaks of a new "politics of time" that emerges in the hiatus that precedes extraction and may prevent it from being realized. When narratives of a progressive temporal unfolding of extractive projects are questioned, a space opens up "for the emergence of other histories and projects that development, humanism, and progress marginalized" (Hetherington 2019, 7).

The chapters in this section speak of the interweaving of material, temporal, and affective dimensions at the heart of contemporary modes of resource extraction. In doing so, they point to the multiple and intersecting levels at which exclusion as well as inclusion can occur. Simultaneously, they splinter any simplistic conceptualization of these processes in binary terms. Understanding these dispersed and contested resource materialities will continue to be key to critical research praxis as extractive geographies are now again experiencing a reconfiguration. The apparent retreat of coal mining and hydrocarbon extraction in Europe and the United States, for example, goes hand in hand with a notable advancement of new extractive hotspots for minerals that help power a digitized and decarbonized global economy. Talk about rehabilitation, reclamation, and transitioning seeks both to point to future modes of value creation and to divert attention from pressing liabilities in the present. There is a risk that mining's apparent end in certain locales may produce forms of epistemic injustice and violence similar to those that accompanied its beginnings. This unbecoming refiguration of resource potentialities calls for research that moves beyond the familiar troubles associated with the cyclicality of boom and bust in order to consider more expansive temporalities, material tensions, and affective possibilities generated in contexts of dereliction, decommissioning, aftermath, and reassembly. This requires an imaginative engagement with the material and temporal ethics of what exists and endures, and with what could be otherwise.

REFERENCES

Anthias, Penelope. 2018. *Limits to Decolonization: Indigeneity, Territory, and Hydrocarbon Politics in the Bolivian Chaco*. Ithaca and London: Cornell University Press.

Appadurai, Arjun. 2013. *The Future as Cultural Fact: Essays on the Global Condition*. London and New York: Verso.

Appel, Hannah. 2019. *The Licit Life of Capitalism: US Oil in Equatorial Guinea*. Durham and London: Duke University Press.

Bakker, Karen, and Gavin Bridge. 2006. "Material worlds? Resource geographies and the 'matter of nature'." *Progress in Human Geography* 30 (5): 5–27.

Barry, Andrew. 2014. *Material Politics: Disputes Along the Pipeline*. Malden and Oxford: Wiley Blackwell.

Behrends, Andrea. 2008. "Fighting for oil when there is no oil yet: The Darfur-Chad border." *Focaal* 52: 39–56.

Carse, Ashley. 2017. "Keyword: Infrastructure – How a humble French engineering term shaped the modern world." In *Infrastructures and Social Complexity: A Routledge Companion*, edited by Penelope Harvey, Casper Jensen, and Atsuro Morita, 27–39. London and New York: Routledge.

Coronil, Fernando. 1997. *The Magical State: Nature, Money, and Modernity in Venezuela*. Chicago: Chicago University Press.

Curley, Andrew. 2019. "T'áá hwó ají t'éego and the moral economy of Navajo coal workers." *Annals of the American Association of Geographers* 109 (1): 71–86.

Dahlgren, Kari. 2019. "Digging deeper: Precarious futures in two Australian coal mining towns." PhD diss., Anthropology, London School of Economics and Political Science (LSE).

Davis, Heather, and Zoe Todd. 2017. "On the importance of a date, or, decolonizing the anthropocene." *ACME: An International Journal for Critical Geographies* 16 (4): 761–780.

Ferry, Elizabeth Emma, and Mandana E. Limbert. 2008. "Introduction." In *Timely Assets: The Politics of Resources and Their Temporalities*, edited by Elizabeth Emma Ferry and Mandana E. Limbert, 3–24. Santa Fe: School for Advanced Research Press.

Gardner, Katy. 2012. *Discordant Development: Global Capitalism and the Struggle for Connection in Bangladesh.* London: Pluto Press.

Gilbert, Paul Robert. 2020. "Speculating on sovereignty: 'Money mining' and corporate foreign policy at the extractive industry frontier." *Economy and Society* 49 (1): 16–44.

Hetherington, Kregg. 2019. "Introduction. Keywords of the anthropocene." In *Infrastructure, Environment, and Life in the Anthropocene*, edited by Kregg Hetherington, 1–16. Durham and London: Duke University Press.

High, Mette. 2017. *Fear and Fortune: Spirit Worlds and Emerging Economies in the Mongolian Gold Rush.* Ithaca: Cornell University Press.

Jaramillo, Pablo. 2020. "Mining leftovers: Making futures on the margins of capitalism." *Cultural Anthropology* 35 (1): 48–73.

Kama, Kärg. 2016. "Contending geo-logics: Energy security, resource ontologies, and the politics of expert knowledge in Estonia." *Geopolitics* 21 (4): 831–856.

Kama, Kärg. 2020. "Resource-making controversies: Knowledge, anticipatory politics and economization of unconventional fossil fuels." *Progress in Human Geography* 44 (2): 333–56.

Kirsch, Stuart. 2014. *Mining Capitalism: The Relationship Between Corporations and Their Critics.* Berkeley, CA: University of California Press.

Kneas, David. 2018. "Emergence and aftermath: The (un)becoming of resources and identities in northwestern Ecuador." *American Anthropologist* 120 (4): 752–764.

Leins, Stefan. 2020. "'Responsible investment': ESG and the post-crisis ethical order." *Economy and Society* 49 (1): 71–91.

Li, Fabiana. 2015. *Unearthing Conflict: Corporate Mining, Activism, and Expertise in Peru.* Durham and London: Duke University Press.

Nash, June. [1979] 1993. *We Eat the Mines and the Mines Eat Us: Dependency and Exploitation in Bolivian Tin Mines.* New York and Chichester: Columbia University Press.

Okenwa, Doris. 2020. "Contentious benefits and subversive oil politics in Kenya." In *Land, Investment & Politics: Reconfiguring Eastern Africa's Pastoral Drylands*, edited by Doris Okenwa and Jeremy Lind, 55–65. London: James Curry.

Peluso, Nancy, and Michael Watts, eds. 2001. *Violent Environments*. Ithaca and London: Cornell University Press.

Powell, Dana E. 2018. *Landscapes of Power: Politics of Energy in the Navajo Nation.* Durham and London: Duke University Press.

Powell, Dana E., and Andrew Curley. 2008. "K'e, Hozhó, and non-governmental politicals on the Navajo nation: Ontologies of difference manifest in environmental activism." *Anthropological Quarterly* 81 (30): 109–138.

Puig de la Bellacasa, Maria. 2011. "Matters of care in technoscience: Assembling neglected things." *Social Studies of Science* 41 (1): 85–106.

Richardson, Tanya. 2016. "The politics of multiplication in a failed Soviet irrigation project, or, how Sasyk has been kept from the sea." *Ethnos* 81 (1): 125–151.

Richardson, Tanya, and Gisa Weszkalnys. 2014. "Resource materialities." *Anthropological Quarterly* 87 (1): 5–30.

Rolston, Jessica Smith. 2013. "The politics of pits and the materiality of mine labor: Making natural resources in the American West." *American Anthropologist* 115 (4): 582–594.

Shever, Elana. 2012. *Resources for Reform: Oil and Neoliberalism in Argentina.* Stanford, CA: Stanford University Press.

Smith, James H. 2011. "Tantalus in the digital age: Coltan ore, temporal dispossession and 'movement' in the Eastern Democratic Republic of Congo." *American Ethnologist* 38 (1): 17–35.

Sundberg, Juanita. 2014. "Decolonizing posthumanist geographies." *Cultural Geographies* 21 (1): 33.

Todd, Zoe. 2016. "An Indigenous feminist's take on the ontological turn: 'Ontology' is just another word for colonialism." *Journal of Historical Sociology* 29 (1): 4–22.

Watts, Vanessa. 2013. "Indigenous place-thought & agency amongst humans and non-humans (first woman and sky woman go on a European world tour!)" *Decolonization: Indigeneity, Education & Society* 2 (1): 20–34.

Welker, Marina A. 2009. "'Corporate security begins in the community': Mining, the corporate social responsibility industry, and environmental advocacy in Indonesia." *Cultural Anthropology* 24 (1): 142–179.

Weszkalnys, Gisa. 2014. "Anticipating oil: The temporal politics of a disaster yet to come." *The Sociological Review* 62 (S1): 211–235.

Weszkalnys, Gisa. 2015. "Geology, potentiality, speculation: On the indeterminacy of 'first oil'." *Cultural Anthropology* 30 (4): 611–639.

Weszkalnys, Gisa. 2016. "A doubtful hope: Resource affect in a future oil economy." *Journal of the Royal Anthropological Institute* 22 (S1): 127–146.

Wood, Caura. 2019. "Orphaned wells, oil assets, and debt: The competing ethics of value creation and care within petrocapitalist projects of return." *Journal of the Royal Anthropological Institute* 25 (S1): 67–90.

7

Contested Futures in British Columbia's Hydro-Extractive Corridor

Progress, Pragmatism, and Visionary Activism

Anna J. Willow

We put in where the Halfway River pours into the much larger Peace. The water was frigid, but the summer sun warmed the air. My boat was one of 200 to make the ninety-minute trip east to our designated landing site at Bear Flats. Both a celebration of the free-flowing Peace River and a protest against the Site C Dam, 2017's Paddle for the Peace was a chance to reconnect with research contacts, experience the river up close, and support a cause I believed in. Perhaps I'd fallen victim to stale metaphors and 1980s lyrics ("time is a river, rolling into nowhere," one singer tried to convince us), but floating downstream in a borrowed canoe seemed like the perfect opportunity to contemplate the past, present, and future of a waterway that could soon be permanently transformed.[1]

When I think back to that July day, it is neither the stirring speeches by local chiefs, politicians, and activists nor the satisfying meals served by Prophet River and West Moberly First Nations that stand out. Instead, my clearest memory is the bittersweet sensation of knowing that the wonders I was beholding—towering south banks, lush riparian foliage, active beaver colonies—would be submerged beneath rising reservoir waters if the new dam's gates close as scheduled in 2024. My odd, anticipatory grief was tempered by the recognition that northeastern British Columbia residents on all sides of the ongoing debate over generating additional electricity from the movement of water traveling from the Rocky Mountains to the Arctic Ocean have a much greater stake in the future of the valley than a visiting anthropologist ever could.

This chapter is about relationships with the future and their implications. Once we accept that "our lives are constructed around knowledges of the future that are as full (and flawed) as our knowledges of the past" (Rosenberg and Harding 2005, 4), it follows that decisions we make today are inevitably influenced by predictions about what tomorrow will bring as well as by past experiences that condition what kind of future we desire and expect. This is as true when we choose a spouse as it is when we invest in—or take to the streets to protest—uranium mines and oil pipelines. We find ourselves permanently suspended between past and future, stuck in the precise moment when the fixed forces of "what has been" meet the mutable prospects of "still to come." Although what we make of all three temporal planes is contoured by cultural assumptions, values, and justifications, the present is our one and only chance to act. It is in the here and now that we fulfill destinies and chart courses.

In the pages that follow, I demonstrate that diverse responses to industrial resource extraction are based not only on past experiences and received cultural values but also on visions of the future and expectations about who will produce it. Drawing on ethnographic fieldwork conducted in 2016 and 2017, I consider three different reactions to British Columbia's proposed Site C Dam to suggest that individuals' relationships with the future—and, more precisely, their interpretations of themselves as temporal actors—play important roles in debates surrounding resource extraction and, ultimately, in the culturally constituted (but materially manifested) process of creating the socionatural worlds we leave behind. While some regional residents' visions of the future are underlain by a belief in environmental transformation as preordained progress or by an acknowledgment of the need for pragmatic adaptation to unavoidable change, others refuse to accept the dam as inevitable. Highlighting this latter option, I identify visionary activism as a distinctive way of being in time in which concerned citizens undertake present actions to influence future realities, thus decreasing the conceptual distance between present and future and transcending taken-for-granted temporal binaries. Like other anthropologists of the future, I claim no special ability to predict what will transpire in the moments and millennia to come (Wallman 1992). Rather, I aim to understand how and why northeastern British Columbia citizens imagine different futures, the definitive contributions they expect to make, and the consequent actions their visions inspire.

TIME AND THE FUTURE

Time—the ubiquitous yet idiosyncratic context in which all human lives unfold—has captured the attention of several generations of anthropologists (Gell 1992). Within the anthropology of time, however, considerably more attention has been directed backward than forward. The enduring anthropological propensity for origins, tradition, and cultural reproduction has resulted in an "anthropological neglect of the future" (Munn 1992, 115) in which tomorrows are overshadowed by yesterdays. For many years, anthropology and the future could thus be analogized as ships passing in the night, with conventional topics of anthropological interest (e.g., shamanism, subsistence horticulture) assumed to have no future worth considering (Collins 2007). In recent years, numerous anthropologists have recognized our discipline's disregard for the future as a serious problem. Notably, Arjun Appadurai calls on us "to place futurity, rather than pastness, at the heart of our thinking about culture" (2013, 194).

Over the past decade, the anthropology of the future has emerged as a vibrant area of study. While the future has sometimes been bypassed as empty, uncharted, and ultimately inscrutable, this has never been the case (Rosenberg and Harding 2005). Like the nostalgic past, the hopeful future is a vehicle for critiquing the subjective shortcomings of the current moment; both past and future "mirror what we praise and reverse what we condemn in the present" (Lowenthal 1992, 31). Expressed through a diverse range of future-oriented emotions—from hope, excitement, and optimism to forbidding, fear, and dread—and varying considerably between (sub)cultural groups, all societies embrace their own version of what "the good life" looks like (Appadurai 2013, 187). People everywhere, it seems, eagerly attend to some futures while refusing to entertain others. As societal power structures determine whose visions are made to seem inevitable and whose are made to seem impossible (Westman 2013), futures become profoundly contested.

Examining divergences in how futures are imagined can help make sense of natural resource disputes, especially when visions of the future prove to be mutually incompatible. Because such images guide how resources are exploited, managed, and/or preserved in the present, they are extremely consequential. People everywhere aspire to futures that they create first in their mind's eye and only later in the world at large. It is, at least in part, the ongoing process of envisioning possible futures that paves the way for these futures' emergence. Given the structural power currently held by those who seek to ensure the inevitability of an extractive future and the impossibility of any alternative that stands in its way, envisioning the future has become a deeply political act.

HYDRO-EXTRACTIVE ETHNOGRAPHY

Hydroelectricity has rarely been contemplated by extraction anthropologists, yet it shares much in common with the logging, mining, and oil and gas industries. Hydroelectric dams are promoted as nonpolluting, renewable energy sources, but they cause irreparable environmental damage and are responsible for the displacement and impoverishment of tens of millions of people, a majority of them Indigenous or ethnic minorities (Scudder 2005). While water itself is not necessarily appropriated during the hydroelectric generation process, the energy of water's movement is extracted and exported, thus exemplifying an extractivist pattern of resource procurement based on removing as much material as possible for as much profit as possible. What I've come to call *hydro-extractivism* parallels nonrenewable resource removal's inexorable imbalance of costs and benefits; local destruction is undertaken at the direction of and for the benefit of distant decision-makers. Every megawatt of electrical power generated by hydroelectric dams is infused with the political "power of dam proponents and beneficiaries over local people who are powerless to stop the destruction of rivers that nourish their bodies, minds, and souls" (Willow 2018, 69).

In the fall of 2015, BC Hydro (British Columbia's provincially owned electric utility) commenced construction on the Peace River's third large hydroelectric dam. This was despite the fact that the Site C Dam had been previously proposed and rejected over concerns that the scheme was too financially risky and its environmental impacts too great. At a time when many developed nations are decommissioning dams and restoring riverine ecosystems, British Columbia changed its own laws to eliminate independent review requirements and allow Site C to proceed (Cox 2018). It was also despite the fact that very serious questions concerning the dam's effects on First Nations culture and treaty rights remain unresolved, with legal challenges slowly moving through the Canadian court system. Citing the dam's infringement on Indigenous rights, the United Nations Committee on the Elimination of Racial Discrimination called in December 2018 for the project to be halted until such time as the impacted First Nations grant Free, Prior, and Informed Consent (Canadian Press 2019). This request has not been heeded. For all these reasons, the Site C Dam is extremely controversial.

For two summers, I conducted ethnographic research to illuminate northeastern British Columbia residents' experiences of and attitudes toward the region's major extractive industries. I stepped into a context of enduring external control and intensifying cumulative effects (Willow 2017).[2] With the ultimate objective of comprehending why some people accept (or even embrace) externally imposed extractive development while seemingly similar others vehemently oppose it, I deliberately sought out individuals with diverse perspectives. Although I aimed to understand multiple outlooks, I never claimed neutrality regarding large-scale hydroelectric development, nor did I hide my pro-environment, pro-treaty rights stance. In addition to focused participant-observation research, I conducted forty-eight interviews with northeastern British Columbia residents.[3] In most of these conversations, Site C dominated the discussion. In nearly every interview, I had the opportunity to ask how individuals envisioned the future of their region. The following discussion draws on interviewees' responses to this prompt and presents one small segment of a much larger research project. I also draw on press releases concerning the project, media coverage of construction highlights and anti-dam protests, and reports from environmental and treaty rights groups.

CONTESTED FUTURES

Important decisions leave us keenly aware of just how tightly the future is bound to the present. The ordinarily taken-for-granted flow of time is disrupted by the realization that we are forever at a crossroads, with no option but to choose (even if through inaction) what happens next. This is true not only for individuals but also collectively, as communities chart courses for land and life. When groups that share a landbase disagree about how the land and its resources should be utilized and are unable to coexist and unwilling to compromise, impassioned struggles erupt over which tomorrows are deemed ideal, acceptable, or dystopian. People often go to great lengths to turn their own visions into reality—and prevent the realization of others' visions. As the Site C story shows, expectations about the future are foremost among the factors that determine whether hydro-extractivism is viewed as progress, as a necessary evil, or as an intolerable affront.

PROGRESS

Dam proponents view extracting electricity from flowing water through the temporal lens of progress. The dam, for them, embodies steady movement through time toward an ultimate goal; hydro-extractivism is a chance to contribute to continuing economic growth and keep the dream of a high-producing, high-consuming industrial civilization alive. Proponents evoke the proud certainty of inevitability in explaining their support for the Site C project and envision the Peace River's third massive hydroelectric dam as a logical and necessary extension of modernity's teleological transformation of the environment to advance the human condition.

Site C was first imagined in the 1950s, when W.A.C. Bennett (British Columbia premier from 1952 to 1972) set out to develop the province's vast northeast (Loo 2007). By 1958, fourteen potential dam sites had been identified along the Peace River and in 1961, BC Hydro was created to lead the hydroelectric charge. The first massive dam (at site "3A") was completed in 1968 and ultimately given Bennett's name. A second, smaller dam was completed at Peace Canyon (site "1") in 1980. From the beginning, BC Hydro acted to implement what it had already decided, with the W.A.C. Bennett Dam envisioned even during its construction as the first in a series and the second dam at Peace Canyon justified to the public as "an integral part of the much larger Peace development" (Sawatsky 1974, 7).

Pursuing a dam at the site "C" location was recommended in the 1972 BC Energy Report and formal planning commenced in 1980. Motivated by fresh memories of the devastation caused by the W.A.C. Bennett Dam—memories of inundated settlements, flooded gravesites, drowned wildlife, and curtailed connections among communities and to the land (Loo 2007)—contingents of First Nations and Euro-Canadian residents objected strongly to the project. The Peace Valley Environment Association soon organized to publicly oppose it (Cox 2018; Pollon and Matheson 1989). After many months of hearings, the British Columbia Utilities Commission (an independent agency with the mandate of reviewing all major energy projects and regulating the province's public utility and petroleum industries) concluded in 1983 that there was no immediate need for the dam, nor was there evidence that a dam at Site C was the province's only or best source of energy (Dusyk 2011). The rebuff put the project on hold for over two decades.

Still, the possibility that a dam might someday stand at Site C loomed, residents say, like a "dark cloud over the valley" (interview June 18, 2016). Renewed interest in the project emerged in 2007, when the British Columbia government began discussing climate policy solutions and promoting non-hydrocarbon energy sources. In its 2010 Clean Energy Plan, "clean energy" was interpreted to mean not the exploration of solar, wind, or other renewable energy alternatives, but rather the inevitable perpetuation of the province's established trajectory of large-scale hydroelectric generation (Dusyk 2016). Yet none of the factors that had halted the dam in its earlier iteration had changed: The project is still incredibly costly (estimated at around $9 billion CAD, it will not be paid off until at least 2094 (Levin 2016)). The environmental and social concerns—and the project's infringement on First Nations culture and rights—are identical to those expressed in past decades.

Nevertheless, a new generation of political leaders appeared determined to build a dam at Site C. In January 2016, Christy Clark (British Columbia premier from 2011 to 2017) used her speech at Bill Bennett's (British Columbia premier from 1975 to 1986, not to be confused with W.A.C. Bennett) memorial service as a platform to promote the Site C Dam. "Premier Bennett, you got it started and I will get it finished," she declared, "I will get it past the point of no return" (News 1130 Staff and the Canadian Press 2016). With proponents arguing that a third Peace River dam is essential to keep the wheels of socio-economic progress turning and build the kind of future that any rational individual would want (one made possible by competent and capable utilization of British Columbia's rich natural resources), the conjoined concepts of progress and inevitability served to push the project forward. BC Hydro made sure that constituents could easily picture the dam's presence; even before construction began, it was frequently depicted as an already existing entity.[4] And while much of the material produced by BC Hydro during the planning and assessment process considered what *would* happen if the dam were built, more recent discussions treat it as a forthcoming reality. Later iterations of BC Hydro's project webpage (https://www.sitecproject.com) consistently describe the form the dam *will* take and the effects it *will* have, almost never allowing for the possibility left open by *may* or *might*.

Individuals who support Site C almost always cite its provision of jobs and revenue to justify their stance. One man pointed to the thousands of formerly desperate workers now employed at the construction site. Given "the stress that people are under economically," he told me, "they're very thankful for these jobs" (interview June 11, 2016). It is true that the project has put numerous employees to work—over two thousand were on staff in early 2019.[5] Yet opponents argue that almost all of these jobs are necessarily temporary given the dam's eight-year construction timeline. As a woman from an area First Nation explained, people who support Site C because of jobs are shortsighted and "misguided." While the jobs will last a little while, she said, "the dam is going to be there for a long, long time. Once it's built and done, they're stuck without jobs. So, what are they gonna eat? They've gotten rid of their future!" (interview June 21, 2017).

As the debate wore on, proponents' rationale shifted from providing the energy needed to fuel the liquified natural gas (LNG) industry (now lagging in the province), to exporting electricity to Alberta's unconventional hydrocarbon epicenter (seen by many as antithetical to any claim of "clean" energy), to meeting a projected increase in energy demands in southern British Columbia (which will only materialize in several decades, if then), causing skeptics to speculate that the dam wasn't really needed at all. Critical of the provincial government's determination to forge ahead without considering alternatives or long-term consequences, Clark's "point of no return" speech became a rallying cry for opponents.[6] Roland Willson, West Moberly First Nations' outspoken chief, railed against Clark's pronouncement that the dam was a *fait accompli*. As he declared in a press release issued in October 2016, "despite Christy Clark's plans to push this project past the point of no return, it is obvious that this project is anything but. We intend to stop the infringement to our Treaty Rights. And we will stop Site C once and for all" (Prophet River First Nation/West Moberly First Nations 2016). We will return to this perspective below.

PRAGMATISM

Northeastern British Columbia residents who take a more passive stance regarding Site C also subscribe to the doctrine of inevitability. Yet their imaginings of the dammed future tend toward anxiety and anticipatory grief rather than pride and purpose. They envision a future with a dam they do not desire. Along with project proponents, members of this largest but least vocal group accept existing political and socio-economic structures, not because they endorse and enjoy the status quo but because they see no realistic alternative. Far from a full-blown fatalism, however, their relationship with time is marked by the proactive process of pragmatic adaptation to a landscape-altering and life-changing project they feel powerless to stop.

The Site C Dam was initially approved by the province in December 2014. As noted above, preliminary construction activities commenced the following autumn. Still, debates over the future of the Peace River continued throughout the region as well as in British Columbia's more populous south. In May 2017, British Columbia's voters elected new provincial leaders who campaigned on promises to subject the Site C project to independent assessment. After an expedited review ordered by the new government—and several more months of uncertainty—the government made the "difficult decision" to proceed with the dam, citing the economic impracticality of cancelling what was already underway (McElroy 2017). With the project approved not once but twice and construction activities normalized in and around Fort St. John, many northeastern British Columbians resigned themselves to the reality of the impending dam.

I was initially taken aback to hear interviewees lament the ecological and cultural damage they expected to follow the dam's completion and then, almost in the same breath, concede that their communities needed the jobs and other economic benefits that flowed from Site C. These individuals framed their acquiescence not as capitulation but as level-headed compromise; it wasn't that they didn't care about the impending ecological and cultural losses, but simply that they could see no practicable way to prevent them. They were prepared to adapt to circumstances deemed outside of their control and determined to make the most of an imperfect situation. One First Nations woman, for instance, spoke of the need to make realistic decisions with everyone's best interest in mind. The knowledge that some area First Nations had already made agreements with BC Hydro and will consequently receive construction contracts and compensation for impacts led her to think carefully about how scarce financial resources should be utilized. Why spend money on lawyers, she rationalized, "when there's not much of a chance that [the fight against the dam is] going to be successful. And all of that funding could be going to health and education, things that are really needed in the

Figure 7.1 The 12th Annual Paddle for the Peace (July 8, 2017). Photo by the author.

community" (interview June 28, 2017). Belief in the dam's inevitability—rather than any desire to see it built—underlies this pragmatic response.

While a wide range of factors (including the degree of personal impact, the value of economic versus noneconomic dimensions of well-being, worldview, trust in the ability of regulators, views of imposed industrial development as injustice, and connections to the land) influence individuals' responses to the Site C Dam, so do their beliefs concerning what constitutes a future worth aspiring to. Many northeastern British Columbia residents view the dam—like environmentally destructive industry more generally—as an unavoidable means to the end of socio-economic sufficiency. Convinced that economic development and full participation in modern industrial society are true necessities—and in some cases understandably desperate to avoid the suffering that poverty brings—they see no other way to put food on tables and roofs over heads. Those who pragmatically accept Site C believe (all puns aside) that they are damned if they endorse the dam, but even more damned if they don't.

VISIONARY ACTIVISM

In stark contrast with those who—whether eagerly or reluctantly—accept Site C, anti-dam activists counter declarations of inevitability with vibrant visions of alternative futures for the Peace Valley. The significance they accord their here-and-now actions is elevated by an activist relationship with time in which the present moment is recognized as the crucial connection between a valorized past and an optimistic but uncertain future. Profoundly aware of their legacy, dam opponents take action today to ensure the best possible tomorrow for the land and the human (and nonhuman) communities it sustains. At least as important as their ability to articulate plans for a free-flowing river is the shared belief that their actions can influence outcomes; while residents who see the dam as inevitable concentrate on how they can make the most of uncontrollable circumstances, those who take a stand against the project are much more likely to advocate ways of life that actively challenge what "progress" has handed them.

Anti-dam activists do not share a singular cultural identity, nor do they have a common relationship to the environments they inhabit. They are instead united by a sense of belonging to a shared landscape that is now under threat—what Grossman calls "place membership" (2005, 25). Recognizing that a free-flowing river sustains diverse ways of existing in the Peace River Valley, they contend that it is not *one* future but rather *multiple* futures that stand to be erased. The value of the valley's agricultural lands—nearly 16,000 farmable acres lie within the designated flood zone—was mentioned repeatedly by interviewees; with their rich alluvial soils and mild microclimate, this irreplaceable asset may someday prove vital to northern food security (Holm 2018). The physical, cultural, and spiritual sustenance supported by land-based subsistence—and the fact that Treaty 8 of 1899 promised Indigenous signatories that they would be "as free to hunt and fish after the treaty as they would be if they never entered into it" (Fumoleau 2004, 87–88)—were often noted by First Nations citizens and allies. And the gains that could come from realizing the recreation and tourism potential of the valley were matched by descriptions of habitat (for wildlife) and peaceful enjoyment (for humans). According to one interviewee, the Peace River region should be an ideal location in the coming decades, given that "climate change is here. Winters are less severe, storms are more severe, and forest fires are getting way out of hand. With good management, this will be *the* area to live in the future" (personal communication July 20, 2016). All of these possibilities would be mutually precluded by the Site C Dam and its 83-kilometer reservoir.

One clear expression of anti-dam activists' vision took the form of a joint presentation developed by Treaty 8 Tribal Association, Peace Valley Landowners Association, Peace Valley Environment Association, and the Yellowstone to Yukon Conservation Initiative for the Site C Joint Review Panel hearing held on January 20, 2014.[7] Entitled "Alternative Vision for the Peace River Valley," the presentation both celebrated what collaborators value about the river's past and present and considered how these values could be extended. Liz Logan (chief of Treaty 8 Tribal Association) and Ken Boon (president of the Peace Valley Landowner Association) highlighted not only the beauty of the valley but also its importance for wildlife conservation, archaeology and heritage, agriculture, ecotourism, regional identity, and First Nations subsistence and cultural activities. The "alternative vision" was not one vision but multiple diverse yet compatible ones.

BC Hydro proudly proclaims that the Site C Dam will be a "source of clean, reliable and cost-effective electricity for more than 100 years."[8] For dam detractors, however, 100 years seems a dreadfully short temporal horizon, especially when juxtaposed against the river's timeless flow. Both Euro-Canadian and First Nations interviewees who oppose the dam often trace their stance to concerns for future generations. Sometimes this means, as it does for one Euro-Canadian woman whose farmstead lies within the flood zone, celebrating emplaced family roots and "passing the place on to our next generations; our kids and our grandkids . . . You just look after the land so you can pass it on. You make it a better place for the next generation" (interview June 24, 2017). Other times it means hoping that the future one planned for can eventually come to pass, as it does for a young First Nations man who imagines someday fishing the river with children and grandchildren who do not yet exist (interview July 20, 2016).

For some First Nations citizens, concerns for future generations are intertwined with stories from the past. Several interviewees described British Columbia's ongoing industrial invasion as the fulfillment of prophecies foreseen long ago and transmitted as oral teachings. Spiritual leaders told that a time would come when

> You will see many lights all over the land. It will look like matches, or lights like fire. You'll see a black snake in the ground—those are all the pipelines. Then you will see that the animals will start to leave and not come back. That's what they said! They warned [us] of this coming up, and I really believe that's what we're going to see. We're going to see our inability to have moose meat, to have fish, to have clean air, to have clean water. We're going to be living in a wasteland that is not going to support our children. It's not going to support anyone's children. (Interview August 1, 2017)

These visions of industrial devastation are not grounds for despair, my interviewee calmly reassured me, but rather a warning and a directive to prevent further damage. We are shapers of tomorrow, in this view, standing at the switch between two (or more) tracks.

Even after the Site C Dam was approved and preliminary construction commenced, its opponents continued to believe they could positively influence the future of the Peace. Evocative of the acutely asymmetrical power struggle surrounding the Site C project, dread of worst-case scenarios, nostalgia for foreclosed futures, and ideas of the future as an entity that can be "lost" or "stolen" are common drivers of anti-dam sentiment. Angered by the divide-and-conquer strategies employed by BC Hydro to subvert coordinated opposition, one First Nations woman asserted that the provincial utility is "creating a future that [is] non-existent for our future children and great grandchildren" (interview August 1, 2017). What is stolen, of course, is not time itself, but the ability to aspire to the version of tomorrow that is most desirable today.

Every moment, every action undertaken by ourselves or by others expedites some possibilities and negates others. Whether we linger to mourn or let alternate options slip quietly away, some futures are no longer possible. We are haunted by the ghosts of past futures, Rosenberg and Harding tell us, so that "more and more, our sense of the future is conditioned by a knowledge of, and even a nostalgia for, futures that we have already lost" (2005, 3). Yet even in the face of anticipatory grief and dread, visionary activists find the courage to envision futures they know may never transpire. Taking strength not only from their elders but also from their children, they will continue to defend their visions of tomorrow—even if the future does end up dammed. In northeastern British Columbia, the power of now—the sense that the present moment may be our only chance to create the kind of future we want to see—is a prominent component of anti-dam activists' experience of being in time.

CONCLUSION: OTHER WORLDS ARE POSSIBLE

While some readers may be intrigued by the hydro-extractive context and tripartite analysis presented here, this chapter's broader goal is to encourage anthropological attention to how diverse envisionings of and relationships to the future guide equally diverse contemporary responses to industrial resource extraction projects. Given that ideas about the future shape actions undertaken in the present, such research sheds light not only on ongoing debates surrounding the Site C Dam but also on countless additional intersections of extraction, environment, and economy.

Together with other Anthropocene citizens, we tie present to future in two interrelated ways. On the one hand, today's decisions are made with disparate envisionings of the future in mind. On the other, these same decisions—intentionally or unwittingly—set in motion the tangible work of creating the socionatural realities we (or our descendants) will soon face. While we cannot accurately predict what the future will bring, we can be certain—or *make* certain—that some prospects will never be realized. From the multitude of possible futures, we select some to reject and others to nurture. "The possible passes into the real through limitation," Gilles Deleuze tells us, via "the culling of other possibilities" (1991, 187). Along the Peace River, anti-dam activists point out that "flooding the valley for a single use—electrical generation—will destroy all the current uses that the valley serves" (Dusyk

2011, 878). Multiple futures, they argue, would be sacrificed to satisfy a singular goal. While both proponents and opponents of Site C take inspiration from comprehensive imaginings of what they believe the river's future should entail, the incompatibility of their scenarios ensures that debates about the dam are more profoundly contests over who has the power to convert vision into reality. Natural resource disputes, in this view, are about "who has the power to tell the story of the future and then to enact it" (Westman 2013, 112). In northeastern British Columbia and far beyond, alternative futures that challenge the politically dominant extractivist status quo are deliberately—and all too often effectively—excluded from the realm of possibility.

What roles, then, might anthropologists who share their research participants' concerns for the future of a river (or the future of the world) play? More broadly, as Samuel Gerald Collins asks, "Can anthropology change the world by writing about the future" (2007, 122)? Collins' answer is affirmative. Other worlds, even those that currently seem unrealistic (a world without Western imperialism or capitalism, for example) are not just possible, he argues, they are "virtualities waiting to be actualized" (Collins 2007, 122). In this sense, looking forward to tomorrow's engaged anthropologies is also looking back on the things that make our discipline unique. Convincing audiences that we can live differently has long been on our list of essential duties. Reminding readers that countless human lives have been—and will necessarily be—premised on socionatural relationships that do not entail relentless exploitation. Assuring anxious students that it doesn't have to be like this. Wherever existing power structures conspire to make some prospects appear "inevitable" and others "impossible," we can endeavor to expand the range of foreseeable futures. We can inspire more people to cling to enticing visions of what tomorrow could bring and less to succumb to the doctrine of inevitability. We can promote less "progress" and more possibilities. An anthropology that amplifies, perhaps even innovates, alternative futures takes a bold but necessary first step. It dares people to dream.

NOTES

1. These lyrics come from Steve Winwood's hit song "The Finer Things" (Winwood and Jennings 1986).
2. In addition to hydroelectric generation, forestry, oil and natural gas, and coal are major industries in the region.
3. A total of 60 percent of my interviewees had a First Nations affiliation, 50 percent were female, and ages spanned five decades.
4. An artistic rendering of the completed dam graced the cover of project's environmental impact summary statement. Released in 2013, this report is available at https://www.sitecproject.com/sites/default/files/site_c_eis_executive_summary_amended_%28july_2013%29.pdf (accessed January 30, 2019).

5. For updated employment statistics, see https://www.sitecproject.com/site-c-employment-statistics.

6. Amnesty International picked up on the quote in a scathing indictment of the consequences of the dam for First Nations people in the region; see https://www.amnesty.org/download/Documents/AMR2042812016ENGLISH.PDF.

7. A hearing transcript is available at http://www.ceaa.gc.ca/050/documents / p63919/97984E.pdf (accessed February 4, 2019). The Peace Valley Landowner Association represents landowners affected by the proposed Site C Dam. Treaty 8 Tribal Association is a support and advisory organization that represents multiple First Nations communities in the area. The Peace Valley Environment Association is a local environmental group. Yellowstone to Yukon Conservation Initiative (Y2Y) is an environmental nonprofit focused on protecting the wildlife and ecosystems of the northern Rocky Mountains.

8. See https://www.sitecproject.com/faq (accessed February 6, 2019).

REFERENCES

Appadurai, Arjun. 2013. *The Future as Cultural Fact: Essays on the Global Condition.* London: Verso.

Canadian Press. 2019. "UN Panel Warns Canada That Site C Dam Project May Violate International Agreements." *Global News*, January 14, 2019. https://globalnews.ca/news/4846627/site-c-indigenous-consent-united-nations/ (accessed January 18, 2019).

Collins, Samuel Gerald. 2007. *All Tomorrow's Cultures: Anthropological Engagements with the Future.* New York: Berghahn Books.

Cox, Sarah. 2018. *Breaching the Peace: The Site C Dam and a Valley's Stand Against Big Hydro.* Seattle: University of Washington Press.

Deleuze, Gilles. 1991. *Bergsonism.* New York: Zone Books.

Dusyk, Nichole. 2011. "Downstream Effects of a Hybrid Forum: The Case of the Site C Hydroelectric Dam in British Columbia, Canada." *Annals of the Association of American Geographers* 101, no. 4: 873–881.

Dusyk, Nichole. 2016. "Clean Energy Discourse in British Columbia, 1980-2014." *BC Studies: The British Columbian Quarterly* 189: 77–101.

Fumoleau, Rene. 2004. *As Long As This Land Shall Stand: A History of Treaty 8 and Treaty 11, 1870-1939.* Calgary, AB: University of Calgary Press.

Gell, Alfred. 1992. *The Anthropology of Time: Cultural Constructions of Temporal Maps and Images.* Oxford: Berg.

Grossman, Zoltan. 2005. "Unlikely Alliances: Treaty Conflicts and Environmental Cooperation Between Native American and Rural White Communities." *American Indian Culture and Research Journal* 29, no. 4: 21–43.

Holm, Wendy. 2018. "Rape of the Commons: The Impact of Site C on Farmland and Food Security." In *Damming the Peace: The Hidden Costs of the Site C Dam*, edited by Wendy Holm, 95–117. Toronto: James Lorimer and Company.

Levin, Dan. 2016. "Canada's $7 Billion Dam Tests the Limits of State Power." *New York Times*, December 10, 2016. https://www.nytimes.com/2016/12/10/world/canada/canadas-7-billion-dam-tests-the-limits-of-state-power.html?_r=0 (accessed January 3, 2017).
Loo, Tina. 2007. "Disturbing the Peace: Environmental Change and the Scales of Justice on a Northern River." *Environmental History* 12, no. 4: 895–919.
Lowenthal, David. 1992. "The Death of the Future." In *Contemporary Futures: Perspectives from Social Anthropology*, edited by Sandra Wallman, 23–35. London, UK: Routledge.
McElroy, Justin. 2017. "B.C. Government to Go Ahead with Site C Hydroelectric Dam Project." *CBC News*, December 11, 2017. https://www.cbc.ca/news/canada/british-columbia/site-c-dam-decision-1.4435939 (accessed December 12, 2017).
Munn, Nancy D. 1992. "The Cultural Anthropology of Time: A Critical Essay." *Annual Review of Anthropology* 21: 93–123.
News 1130 Staff and the Canadian Press. 2016. "Former B.C. Premier Bill Bennett's Accomplishments Celebrated at Memorial." *CityNews*, January 31, 2016. https://www.citynews1130.com/2016/01/31/former-b-c-premier-bill-bennetts-accomplishments-celebrated-at-memorial/ (accessed December 4, 2018).
Pollon, Earl K., and Shirlee S. Matheson. 1989. *This Was Our Valley.* Calgary, AB: Detselig Enterprises.
Prophet River First Nation/West Moberly First Nations. 2016. "AFN National Chief Perry Bellegarde Visits the Site C Dam to See the Destruction." Press Release, October 27, 2016. https://globenewswire.com/news-release/2016/10/27/974747/0/en/AFN-National-Chief-Perry-Bellegarde-visits-the-Site-C-Dam-to-see-the-Destruction.html (accessed December 11, 2018).
Rosenberg, Daniel, and Susan Harding. 2005. "Introduction: Histories of the Future." In *Histories of the Future*, edited by Susan Harding and Daniel Rosenberg, 1–18. Durham, NC: Duke University Press.
Sawatsky, John. 1974. "Hydro Airs Plan to 'Milk' Peace." *Vancouver Sun*, February 9, 1974, pp. 1 and 7.
Scudder, Thayer. 2005. *The Future of Large Dams: Dealing with Social, Environmental, Institutional and Political Costs.* London: Earthscan.
Wallman, Sandra. 1992. "Introduction: Contemporary Futures." In *Contemporary Futures: Perspectives from Social Anthropology*, edited by Sandra Wallman, 1–22. London: Routledge.
Westman, Clinton. 2013. "Social Impact Assessment and the Anthropology of the Future in Canada's Tar Sands." *Human Organization* 72, no. 2: 111–120.
Willow, Anna J. 2017. "Cultural Cumulative Effects: Communicating Energy Extraction's True Costs." *Anthropology Today* 33, no. 6: 21–26.
Willow, Anna J. 2018. *Understanding ExtrACTIVISM: Culture and Power in Natural Resource Disputes.* London: Routledge.
Winwood, Steve, and Will Jennings. 1986. "The Finer Things." *Back in the High Life.* London: Island Records.

8

Zinc's Time in the Sun?

Tracing the Reopening of the Riso and Parina Valleys' Mines from Precarious Optimism to Affective Indeterminacy

Robin West and Isabel Crowhurst

On our visit in 2017, an office in the village of Oltre il Colle in the northern Italian Orobic Alps and a makeshift sign on the road to the hamlet of Zorzone signalled the presence of the Australian mining company, *Energia Minerals*. The sign pointed to a dirt road and the graffiti-covered shell of the *laveria*

Figure 8.1 The Ruins of the *laveria* in Oltre il Colle. Photo by Robin West.

(figure 8.1), a former flotation plant now reclaimed by nature. The ruin is a reminder of the once-thriving mining economy that shaped and continues to structure core aspects of the sociocultural life of the Parina and Riso Valleys. The surrounding slopes bear witness to a mining history now only immediately visible in the abandoned pylons that once carried excavated ore. Further, along the graveled track, sat *Energia Minerals*' compound framing one of a number of entrances to the extensive galleries and their mineral deposits. Workers wearing high-vis vests, portable buildings, diggers, and clusters of equipment indicated fresh activity and the possible revival of extractive operations, with a sign announcing the present holder of the area's mining research permit as *Energia Minerals*: one of *Altamin*'s prior incarnations.[1] In 2013, *Energia Minerals* acquired exploration rights to the existing 230 kilometers of underground galleries and one mining concession setting up a presence in the valley in 2015.

This chapter addresses a peripheral feature of global extractivism and of the exclusionary practices discussed in this volume: the revitalization of European mining via renewed extraction activities in long-closed mines. The Valleys' mines were closed in the early 1980s as Italian resource policy shifted priorities. We locate the Parina and Riso Valleys case as lying on one of the "multiple frontiers of extraction" that represents the extractive industries' intensification and expansion into new geographical and conceptual domains (Mezzadra and Neilson 2017). Extractivist economies and social exclusion-related research in the Global North have typically focused on marginalized communities, threats to traditional livelihoods, property rights, and contested governance (Rodon and Therrien 2015; Tiainen et al. 2015). Our approach draws inspiration from such studies and neo-extractivist literature (e.g., Veltmeyer and Petras 2014) by reflecting on mining communities' perspectives, economic optimism, and potential conflict in the cultural sphere.

The sense of "precarious optimism" suggested by the chapter's title hints at the cautious hopes that the local communities projected onto the reopening of the mines. We suggest on the basis of our preliminary findings that faith in anticipated benefits brought by the return of mining activity is overoptimistic despite *Altamin*'s presence instilling hope in the resurgence of the mineral market and the financial prospects of the mines' reopening. These anticipated benefits become visible on both material and symbolic levels in terms of the Valleys' economic revival and the sustainability of a collective mining identity. However, the emergent "optimism" is no longer grounded in the feeling of trust suggested by early interactions between the company and community, undermined as the reality of disparities of power became recognized. We conclude by arguing that as a consequence of these shifting relationships, the local community is caught in the grips of "affective indeterminacy" and contemplating the reality of cruel optimism (Berlant 2010, 2011) as hopes of a revitalized collective subjectivity and material security remain in limbo.

Drawing on observations, informal conversations, and documentary analysis,[2] this chapter traces developments aimed at reopening extractive activities in the Valleys in the late 2010s. Not only did the prospected revival of the mines bear economic implications for the valley, but it also dissects at a phenomenological level what Raymond Williams (1961, 1977) identified as the "structure of feeling."

For Williams, the structure of feeling serves as a reference point for the "felt sense of the quality of life" at "a particular place and time" wherein the arrangement of material activities intersect with ways of thinking and living, generating a specific cultural space and its affective dimensions of experience (Williams 1961, 63). The "structure" articulates the feel of presence and relational practice constituted in the past yet enabling a sense of continuity in terms of a "structure of identification" and investment in "common social culture" (Grossberg 2010, 24–28; Highmore 2016, 147[3]). This notion of "feeling" as form-giving and determined by the residual experience of a labor process invites attention to the power of affect as it is generated, sustained, and in this case confronted by the arrival of, for better or worse, an external actor.

Our focus here is therefore on the role that the structure of feeling plays—as the embodiment of collective subjectivity—within local communities as the plan to reopen the mines unfolds. We also identify what we refer to as "affective indeterminacy," a useful concept in making sense of the complex intersection of hope, anticipation, fear, uncertainty, and disillusionment that forms the fabric of such a structure.

THE RISE, FALL, AND LONG-AWAITED RESURGENCE OF MINING IN THE VALLEYS

Active since before the Roman period, the exploitation of the zinc and lead reserves of the Parina and Riso Valleys' mines was significantly expanded in the nineteenth century when the prices of minerals rose under increased demand across Europe. In the early twentieth century, Belgian and British companies bought the mines from local entrepreneurs and expanded infrastructures further. Under the autarkic policies of the fascist government in Italy, mining concessions were acquired by an Italian public–private company and, at the close of World War II, substantial Allied reconstruction investment channeled through the Marshall Plan allowed the Valleys' mines and related activities to flourish. In the late 1970s, following a succession of management companies, the mines became part of the state-owned *Italian*

National Energy (*ENI*). At that time a new national resources strategy focusing solely on oil and natural gas meant that all mineral extraction enterprises across the country were terminated. Despite infrastructural modernization in the 1970s and the discovery of new deposits of zinc, lead, and silver, activities in the Parina and Riso Valleys' mines ceased between 1981 and 1982 (Alta Zinc 2019b; Furia 2013; Locatelli 2012).

The economic impact of the mines' closure was not immediately felt as the then-booming economy, bolstered by thriving textile and construction industries and related job opportunities, maintained high levels of regional employment. From the late 1980s onward, economic contraction, locally, nationally, and globally, resulted in job losses followed by de-population as people relocated to larger cities in search of better employment prospects (informal conversations with locals). Today, the two main municipalities of the Valleys, Oltre il Colle and Gorno, count just below 1,000 and 1,500 inhabitants, respectively, having reached over 1,800 and 2,000 inhabitants in the 1960s (TuttItalia 2021a, 2021b).

Global Mineral Markets, Local Economies, and Affective Attachments

As this historical snapshot suggests, the fate of mining in the valleys has been subject to economic variance and the changing priorities of exogenous institutions over an extended period of time. While locals worked in and around the mines with mining central to livelihoods and ways of life, over the decades a succession of national and international companies owned, took control of, and made decisions over extractive activities in the valleys in response to ever-shifting global geopolitical conjunctures. In this respect, the arrival of *Energia Minerals* in the late 2000s following several decades of inactivity represents a reiteration of centuries-long operations decided by people and organizations external to the local community, as has generally been the case with extractive activities globally (Harvey and Press 1990).

To place this in a broader context, a rise in mineral commodity prices at the turn of the millennium, following extended periods of weak prices in the 1980s–1990s, triggered a boom in the fortunes of the global mining industry (Humphreys 2015). This resulted in the latter's structural transformation with increased investor interest amid heightened competition in the global mineral market as the growing economies of countries such as China, India, Russia, and Brazil vied for an increased share of raw materials (Tiess 2010). In some instances, this led to the "dusting off" of long-sidelined projects and the buying-up of rival companies resulting eventually in what appeared to be a relatively stabilized market (Humphreys 2015). The subsequent economic optimism would survive through a series of inevitable market fluctuations following the initial boom period. *Energia Minerals*' interest in the Valley's historic zinc deposits is representative of how today's industry surfs the peaks and troughs of the market seeking investment opportunities by breathing "new life" into neglected European mining environments. As *Energia*'s managing director put it in a press release at the time, "it's going to be zinc's time in the sun" (Sanderson 2015).

The industry is also newly sensitized to the public's poor perception of mining's exploitative and environmentally dubious practices inviting new perspectives on its responsibility toward communities (Broad 2014; Bice 2016). This seeming concern for substantial relationships underwrites a core thread of our argument in the case of *Altamin* and the Valleys' population. Communities' acceptance of the mining (and other) industries' presence—commonly referred to as "social license to operate" (SLO)—is increasingly important, being centered on expectations of derived economic value and a sensitivity to local identity and values. An SLO articulates in part a community's "feeling" of possessing a vested interest ranging from increased employment and ancillary services (Zhang et al. 2015; Litmanen et al. 2016; Heffron et al. 2021) to recognition of cultural characteristics (Waterton and Smith 2010). Any sense of trust suggested by the SLO is therefore largely conditional, not merely on the realization of economic benefits, but on the company's capability for "quality" dialogue and will to comprehend local culture and history (Moffat and Zhang 2014; Litmanen et al. 2016; Mercer-Mapstone et al. 2017). Below we not only point to the structure of feeling as implicated in the Valleys' initial level of acceptance but also indicate that the interactions that define the relationship between company and community suggest the relative hollowness of SLOs and their intangible, opportunistic, nature (Bice and Moffat 2014; Owen 2016[4]).

Through our documentary research and conversations with locals, it seems clear that the early stages of consultations with locals and a demonstration of interest in the region's mining legacy amounted to a substantial outreach aimed at procuring the project's legitimacy within the community. Subsequently, the company's operation was framed in an appeal to the structure of norms, values, and indeed deeper beliefs that shape aspects of the everyday life of the municipalities. It is precisely the (in)coherence of an intangible and at times volatile relationship that promises to meet the communities' economic hopes and also speaks of alignment to, and respect for, the cultural field that we address through our concept of "affective indeterminacy."

As outlined, the sense of social license entails the requirement to be sensitive to cultural expression and ways of living around which mining revolved alongside economic needs. Thus, the social license implies a sensitivity grounded in a politics of recognition of the multiple layers of community identity. If anything, companies have to address not just the materiality of rich industrial pasts but also the symbolic residues that mining communities endeavor to keep alive through the remembering and performance of labor heritage (Smith 2006; Coupland and Coupland 2014). As research has shown, while there are evident signifiers of decay and de-investment in landscapes of industrial decline these are increasingly coupled to attempts to prevent the "collective forgetting" of industrial pasts and rather to use such signifiers as a "stepping stone to the future" (Harfst 2015, 218). Consequently, buildings, infrastructures, skills, traditions, and human knowledge have been considered as bearing "post-mining potential" to be preserved in the form of museums, exhibitions, art, and tourism. Add to this the symbolic potential of mining-related events, traditions, and mining identity. These are not to be thought of purely as confined to the "past" but rather utilized in the strengthening of local identities, income generation, educational purposes, and a general fostering of tourism.

Attempts to utilize the Valley's labor heritage in the reinvigoration of the local economy were pursued unsuccessfully in various forms from the 1990s mostly due to financial mismanagement. Nevertheless, the Valleys have three active mining museums which have received regional funding. Their display of artifacts takes the form of a practical consciousness where living and interrelating continuity with the past is represented through the staging of narratives conveyed through the museums' explanatory notes. The cultural heritage of mining is also visible in its continued performances. For example, annual mining processions and religious mass to honor Santa Barbara, the patron saint of mining, and various mining fetes and commemorations organized throughout the year.

In this sense, the structure of feeling carries the emotional bonds that link personal to social history and in turn points to the narratives of affect that shaped and continue to structure a communal sense of cautious optimism toward a future with *Altamin.* As we explain in the next section, the importance of ensuring that initiatives of heritage preservation and valorization remain active was a key demand that local authorities in the Valleys made from the outset when interest in the mines was first shown.

Precarious Hopes, Uncertainties, and the Arrival of "the Australians"

When *Energia Minerals* made its appearance in the Valleys and was granted exploration licenses in 2013, the locals responded with cautious expectancy. As commented in local newspapers, some feared that the reopening of the mines might have potentially negative environmental impacts and dreaded a return to high-risk jobs. Others saw an invaluable opportunity that could

Figure 8.2 A Restored Artifact: The Minecart Was Recovered and Given Back to the Community by the Australian Company. Photo by Robin West.

bring employment and economic opportunities to the Valleys (Filisetti 2013a), finally putting an end to depopulation in the area (Ghisalberti 2013). Local mayors reportedly viewed the possible development with prudent hope and emphasized ensuring that the Valleys' mining heritage and touristic potential would have to expand together with any possible revival of extraction activities (Filisetti 2013b).

The organization of a symposium in Gorno in March 2014 on the valorization and development of mining in the Valleys with the participation of representatives from *Energia Minerals*, and the commencement in August 2014 of the company's trading on the Australian Securities Exchange, injected new hopes on the feasibility of a contemporary life for the mines. *Energia Minerals* named the "Gorno zinc project" as its flagship project and over the following months, both in its marketing material and in communications with local newspapers, issued continued reassurances over its commitment to the mines' reopening while respecting the environment, the local community, and the Valleys' mining heritage (Ghisalberti 2014a).

As the months passed, the local media painted a picture of surging enthusiasm on the part of local authorities. Local mayors had been reassured that the company was committed to the environment and would leave one of the drifts available for touristic activities (Ghisalberti 2014b). Villagers were also presented as mostly supportive of the project; they had started to believe in it and were waiting to see it happen (Filisetti 2014; Furia 2014). The wait during these first stages was not long as *Energia Minerals*' plans unfolded promptly. In December 2014, it awarded a rehabilitation and decline contract to an Italian company to refurbish old portals and begin its exploration program. In January 2015, it opened an office on the main street of one of the villages making its presence even more visible. At least 400 job applications were sent to the company (Ghisalberti 2015) and "expectations, hope and moderate optimism"[5] (Irranca 2015) remained in place according to local media, while local administrators asserted the need to be kept informed on progress and to have their requests fulfilled (ibid.).

The "Community" pages of *Energia Minerals*' (now closed[6]) website praised the support and collaboration of the local communities and authorities whose trust, it claimed, the company was happy to have secured. Local knowledge and hard work had greatly benefitted the development of the project, the website also stated, and the company was excited to have contributed to the region's economic growth. Videos and statements by local mayors reinforced the conveyed message of a strong, unfaltering, and trustful relationship between *Energia Minerals* and the local community.

However, not everyone appeared to have so unquestioningly embraced new developments. Already in September 2013 someone had mown the message "Australia, no thank you" in 2-meter high letters on a local meadow (Filisetti 2013c). In September 2014, hostile graffiti appeared on a wall in one the villages: "Hands off the mines!," "Uranium is close and it's alluring," "Jobs for a few, cancer for all" (*L'Eco di Bergamo* 2014). Around the same time, a local resident published a controversial article in the local media implying that the local population had never been consulted by "the Australians"—as people referred to the company—and had been duped with the promise of new jobs without knowing that the now highly technological mining industry requires specialized skills that most locals lacked (Epis 2014). The article also claimed that the company's planned works would erase the old mining tracks and structures, over- and underground, and hinting at the graffities' messages concluded that the company's historical interest in uranium should alert locals of the possible real aim of the project: to get closer to uranium mines in the adjacent valleys should Italy return to extracting the mineral. In January 2017, another article in a local newspaper advanced the theory that the Australian company, owned by an investment fund, is not really interested in reopening the mines, but is merely conducting explorations to inflate the value of its shares (Araberara 2017a). The relationships between *Energia Minerals* and the local community were thus not as "unfaltering and trustful" as the company's marketing material suggested.

Indeed, some of the arguments presented in the articles were also shared with us by locals during our visits in 2016 and 2017. Those we spoke with were not only hopeful that the mines would eventually reopen but also uncertain and anxious about when this might happen and with what implications for local infrastructures and heritage. They were also concerned with the rumored short duration (ten to fifteen years) of the extraction plan and what might happen to the Valleys in the longer term. Our conversations also revealed a less-than-optimal relationship between the locals and *Energia Minerals*, revealing a discrepancy with the ideal strong bond of trust publicized by the company. In September 2016, for example, one former miner explained to us that he had attempted to organize a community event with "the Australians" to allow them to share information about project developments. The company had confirmed their willingness, albeit not compelled by any regulation to do so, but never committed to anything. Consequently, he told us that in reality "they are not really that available." Similarly, another former mine worker reported that "the Australians" seemed to have no interest in communicating with the locals. Old miners, he said, were particularly keen to know what was going on in "their mines" but were not relayed any information apart from the occasional article in local newspapers. He added that the publicity posted by *Energia Minerals* on their website, in local media and various tourist information brochures, and even their office on the village main street, were mostly aimed at showing off to potential investors the company's integration into the community. He perceived these initiatives as a "façade" that did not reflect a genuine and trusting relationship with the locals. When we met in June 2017, a local administrator was keen to emphasize that *Energia Minerals* had established a dialogue with the local authorities from the very beginning. However,

> after a year or more of positive conversations, in concrete they have not written down a single word. [. . .] Today we are a little disappointed. [. . .] They made videos, they wrote to the newspapers [. . .], it's part of the game, when you deal with an international market you have to show positivity [. . .] that's important to know that there are no conflicts with the local populations. [. . .]. But we need to know what they want to do. [. . .] We cannot really understand their intentions.

Conversations held in 2017 in particular signalled frustration over the uncertainty surrounding the project and need to be understood in the context of developments that unfolded in the first half of 2017. As an Australian mining magazine reported, while "at the end of 2016 Alta [Zinc] appeared to have everything going for it—a strong zinc price environment, a robust resource at its Gorno zinc project in Italy and positive share price movement," complexities emerged with the revising down of resource estimates combined with the delaying of a feasibility study in spring 2017 which ultimately "curtailed that momentum" leaving the company "devoid of market support" (Piper 2018,

79). When we visited the Valleys in the summer of 2017, the company was "gathering itself" (ibid.) and was therefore possibly less keen to engage with the locals before knowing how to respond to unfavorable market conditions.

In September 2017, having been informed of the setback, a local mayor commented in a local newspaper: "We can only wait and see what the company is going to decide, hoping that soon a project will be presented formally to the region, something that has not yet happened in the last years" (Araberara 2017b). When the decision to continue with the Gorno project albeit on a smaller scale-staged approach was communicated to the local authorities, the response reported in local media revealed increasing frustration. As one mayor commented: "They are talking about a possible beginning of the works in 2020 but it's all up in the air, in essence there is nothing concrete [. . .] They keep postponing and postponing and in the meantime time goes by, months and years" (Araberara 2017c). In September 2018, a public meeting was held in one of the villages and the project re-developments presented to the community, but without a feasibility study yet compiled the overall sense of uncertainty over the future of the project was intensified (Araberara 2018a).

While the website of the Australian company continues to be regularly updated with news of further exploratory drillings, quarterly reports and updates to shareholders, the decreasing coverage in local media on the "never ending story of the mines" (Araberara 2018b) emphasized feelings that also emerged from our direct contact with locals. Most prominent was the frustration and powerlessness of being kept in the dark over the future of their own territory and having to wait indeterminately for external "others" to make decisions based on forces and calculations that have little to do with their lives. Some of the locals we remained in touch with suggested that many people believed the company had now left and felt disappointed that, without being able to fully understand why, once more their hopes for the rebirth of their mines remain unfulfilled.

"Cruel Optimism?" The Ethereal Promise of Subjectivity and Security

Mezzadra and Neilson (2017) make the case that, under the logic of contemporary capitalism, there are now "multiple frontiers of extraction" that lay beyond the literal sense of "mining," and are now being transformed into resources ripe for extractive exploitation. This suggests potential for the appropriation of once-relatively autonomous cultural and affective characteristics being subsumed into the business and marketing strategies of companies such as *Altamin.* As shown throughout this chapter, many in the Valleys' communities have optimistically embraced the promises and potentials that have emerged following the arrival of "the Australians." As the previous section outlined, this level of optimism has been challenged and weakened through a neglect of the early-established relations that underpinned the intangible nature of the SLO and the lack of concreteness regarding the actual course of the mines' reopening. This exposes vulnerable elements that point to the instrumental reality of a formulaic interpretation of "social license" in promotional material. Even though the closing of the mines meant the loss of a once secure economic base, today, the Valleys remain an "affective space" (Reckwitz 2012) with regard to the mining experience that has lived on following the closures. Despite the oft-conflicting views held among the Valleys' population, this affective space generates a collective subjectivity forged through the highs and lows of the historic mining experience: a subjectivity that *Altamin*'s outreach to the communities promised to make sustainable. As the Mayor of Oltre il Colle states, albeit prematurely, in a promotional video for the company: "They have brought a windfall to our community" citing the pregnant promise of employment opportunities (Alta Zinc 2019a).

Berlant (2010, 2011) describes optimism as an "affective form" in which "the subject leans towards promises contained within the present moment" (2010, 93). Optimism relates to desire, whereby "desire" speaks of embedded promises and endurance in our supposed relationship with the world. In short, optimism and desire amount to the affective sense of continuity and investment connecting subjects to their life. If the Valleys' "structure of feeling" provides for a relatively assured subjectivity refreshed by promises for the Valleys' future, then we argue by way of conclusion that this may indeed be a "cruel optimism" (Berlant 2010, 2011).

We have seen above that the fate of the Valleys' communities has been subject historically to the combined force of politics and capital. For the reasons set out above, the promises of job security and future cultural sustainability have become, we argue, ethereal at this stage of the mining project perhaps betraying the Valleys' trust and goodwill. This has resulted in a state of what we term "affective indeterminacy" as the sense of certainty that characterized early responses to the project no longer corresponds to the cultural (and optimistic economic) structures of feeling due to the loss of transparency in the promised future. This is compounded by the loss of any sense of control over when and if that promised future might materialize, and a shared feeling of exclusion from knowledge of the present situation.

The nature of some sections of the extractive industries today seems to be, as we have suggested, one of "surfing the peaks and troughs" of global demand and increasingly hypersensitive to the precarity of the market. The sun may have risen for zinc, but it will likely set. At the turn of 2019, and amid volatile share prices, zinc resources were insufficiently defined preventing the company from setting clear production and financial targets (*Mining Journal* 2019). We argue further then, although cautiously, that the sense of "cruel optimism" takes shape with regard to the promise of jobs: as Macdonald (2018, 591) puts it, the reality of today's extractive industry is that "[l]ocal benefits in terms of new employment opportunities frequently disappoint in terms of their magnitude." This in the case of the Riso and Parina Valleys has seemed to be an understatement from the start. Equally, the increase in extractive companies demonstrating their "positive contributions to their host communities" (ibid., 511) as reflected in the concept of the SLO also raises questions as to future intentions of *Altamin* in relation to the incorporation of local heritage into its strategy. The performance of heritage within the communities facilitates the expression of autonomy, acting as a "cipher" for collective experience tied to the locality by bringing to the surface affective and subjective contexts as a measurement of the past's value for the future (Dicks 2008). However, the desired future has become indeterminate and ambiguous preventing what many in the Valleys had hoped to be an evolution that would guarantee both collective subjectivity and financial security.

NOTES

1. *Energia Minerals* changed its name to *Alta Zinc* in 2017 and to *Altamin* in 2021.
2. Observations and informal conversations with locals took place during two visits to the Valleys in September 2016 and in July 2017. The documentary sources analyzed include local newspapers, online news, and other relevant websites, as well the Australian company's websites (*Energia Minerals'* and *Alta Zinc*'s: both now unavailable, having been substituted by *Altamin*'s website). Moreover, one of the authors has good knowledge of the area and the Valleys having spent many summers vacationing there since the 1980s.
3. Whereas Williams' original analytical use of the structure of feeling was related to the literary representation of material culture, Highmore (2016) shifts the concept into the realm of everyday shared living practices and the "sensual forms" of objects. "Feelings" are therefore taken to be form-giving social forces.
4. Owen (2016, 102–103) argues that in reality, there is a paucity of empirical evidence of where SLOs have "demonstrated a sustainable future-orientated program for host communities" and that conceptually the SLO is a "pseudo-permitting instrument." As our argument can only briefly cover here, *Alta Zinc*'s promotional literature makes much of its relationship to the Valleys' communities and recognition of mining history. However, this perhaps has much to do with image management and is instrumental in ensuring a positive reading of social performance.
5. All translations from Italian into English are by the authors.
6. www.energiaminerals.com/community

REFERENCES

Addison, Tony, and Alan Roe, eds. 2018. *Extractive Industries: The Management of Resources as a Driver of Sustainable Development*. Oxford: Oxford University Press.

Alta Zinc. 2019a. "Media. Video gallery." Accessed August 17, 2019. http://www.altazinc.com/media/video-gallery.

Alta Zinc. 2019b. "Italy Gorno." Accessed August 5, 2019. http://www.altazinc.com/projects/gorno-zinc-project'.

Araberara. 2017a. "E se gli australiani fossero un bluff? 'Hanno già investito troppi soldi...'" January 27, 2017. https://www.araberara.it/archivio_testate_pdf/2017/2017_01_27.pdf.

Araberara. 2017b. "Energia Minerals Italia: ridimensionato il progetto per le miniere e il titolo crolla in Borsa." September 22, 2017. https://www.araberara.it/archivio_testate_pdf/2017/2017_09_22.pdf.

Araberara. 2017c. "Una miniera di promesse." November 3, 2017. https://www.araberara.it/archivio_testate_pdf/2017/2017_11_03.pdf.

Araberara. 2018a. "MINIERE: ancora rinvii e la società degli Australiani cambia ancora nome." September 28, 2018. https://www.araberara.it/archivio_testate_pdf/2018/2018_09_28.pdf.

Araberara. 2018b. "'La storia infinita' delle miniere in un buco nero." November 9, 2018. https://www.araberara.it/archivio_testate_pdf/2018/2018-11-09.pdf.

Berlant, Lauren. 2010. "Cruel optimism." In *The Affect Theory Reader*, edited by Melissa Gregg and Gregory J. Seigworth, 93–117. Durham: Duke University Press.

Berlant, Lauren. 2011. *Cruel Optimism*. Durham: Duke University Press.

Bice, Sara. 2016. *Responsible Mining: Key Principles for Industry Integrity*. Abingdon: Routledge.

Bice, Sara, and Kieren Moffat. 2014. "Social licence to operate and impact assessment. Impact Assess." *Project Appraisal* 32 (4): 257–262.

Broad, Robin. 2014. "Responsible mining: Moving from a buzzword to real responsibility." *The Extractive Industries and Society* 1 (1): 4–6.

Coupland, Bethan, and Nikolas Coupland. 2014. "The authenticating discourses of mining heritage tourism in Cornwall and Wales." *Journal of Sociolinguistics* 18 (4): 495–517.

Dicks, Bella. 2008. "Performing the hidden injuries of class in coal-mining heritage." *Sociology* 42 (3): 436–452.

L'Eco di Bergamo. 2014. "«Giù le mani dalle miniere» Oltre il Colle: «Scritte stupide»." September 4, 2014. https://www.ecodibergamo.it/stories/Cronaca/giu-le-mani-dalle-miniere-oltre-il-colle-scritte-stupide_1076487_11/.

Epis, Roberto. 2014. "Le miniere tradite: gli australiani riaprono le miniere a spese dei... gornesi." *Araberara*, September 12, 2014. https://www.araberara.it/archivio_testate_pdf/2014/2014_09_12.pdf.

Filisetti, Andrea. 2013a. "Cacciatori australiani di argento e oro Altre 5 ricerche nella Val del Riso." *L'Eco di Bergamo*, August 22, 2013, page 26.

Filisetti, Andrea. 2013b. "Febbre dell'oro Gli eredi dei minatori stanno cauti." *L'Eco di Bergamo*, August 23, 2013. http://www.orobievive.net/20_montagna/miniere/miniere%20riso%2023%20agosto%202013.pdf

Filisetti, Andrea. 2013c. "Gorno, nel prato spunta la scritta «No Australia»." *L'Eco di Bergamo*, September 7, 2013, page 47.

Filisetti, Andrea. 2014. "'Riapriamo le miniere e porteremo lavoro'. Estrazione per il 2016." *L'Eco di Bergamo*, August 26, 2014, page 28.

Furia, Luigi. 2013. *Le miniere di piombo e di zinco della bergamasca*. Bergamo: Bolis Edizioni.

Furia, Luigi. 2014. "Gli Australiani: quasi 8 milioni di tonnelate di zinco in Val del Riso. Investimenti per 8 millioni di dollari." *Arabeara*, December 19, 2014.

Ghisalberti, Giovanni. 2013. "Metalli hi-tech e zinco nell'Arera. Gli ex minatori pronti a tornare." *L'Eco di Bergamo*, August 26, 2013, page 24.

Ghisalberti, Giovanni. 2014a. "Riapre la miniera, maxi investimento." *L'Eco di Bergamo,* August 24, 2014, page 34.

Ghisalberti, Giovanni. 2014b. "Il sindaco: chiediamo di usare le nostre imprese e garantire la viabilità." *L'Eco di Bergamo*, August 26, 2014, page 29.

Ghisalberti, Giovanni. 2015. "Zinco a Oltre il Colle Dalle prime ricerche riemergono i minerali." *L'Eco di Bergamo*, October 31, 2015, page 45.

Grossberg, Lawrence. 2009. "Raymond Williams and the absent modernity." In *About Raymond Williams*, edited by Monika Seid, Roman Horak, and Lawrence Grossberg, 18–33. London: Routledge.

Harfst, Jörn. 2015. "Utilizing the past: Valorizing post-mining potential in Central Europe." *The Extractive Industries and Society* 2 (2): 217–224.

Harvey, Charles, and Jon Press, eds. 1990. *International Competition and Industrial Change: Essays in the History of Mining and Metallurgy 1800-1950*. Abingdon: Routledge.

Heffron, Raphael J., Lauren Downes, Oscar M. Ramirez Rodriguez, and Darren McCauley. 2021. "The emergence of the 'social licence to operate' in the extractive industries?" *Resources Policy*. 74: 101272.

Highmore, Ben. 2016. "Formations of feelings, constellations of things." *Cultural Studies Review* 22 (1): 144–167.

Humphreys, David. 2015. *The Remaking of the Mining Industry*. Basingstoke: Palgrave Macmillan.

Irranca, Franco. 2015. "Riapriranno nel 2017 le vecchie miniere La «calamina» porterà nuova ricchezza." *L'Eco Di Bergamo*, July 19, 2015, page 37.

Litmanen, Tapio, Tuija Jartti, and Eero Rantala. 2016. "Refining the preconditions of a social licence to operate (SLO): Reflections on citizens' attitudes towards mining in two Finnish regions." *The Extractive Industries and Society* 3 (3): 782–792.

Locatelli, Lucio. 2012. "L'industria dello zinco nella provincial di Bergamo (1869-1946)." Tesi di Laurea, University of Milan.

Macdonald, Catherine. 2018. "The role of participation in sustainable community development programmes in the extractive industries." In *Extractive Industries: The Management of Resources as a Driver of Sustainable Development*, edited by Tony Addison and Alan Roe, 591–610. Oxford: Oxford University Press.

Mercer-Mapstone, Lucy, Will Rifkin, Kieren Moffat, and Winnifred Louis. 2017. "Conceptualising the role of dialogue in social licence to operate." *Resources Policy* 54: 137–146.

Mezzadra, Sandro, and Brett Neilson. 2017. "On the multiple frontiers of extraction: Excavating contemporary capitalism." *Cultural Studies* 31 (2–3): 185–204.

Mining Journal. 2019. "Alta seeks to grow Italy zinc project." January 29, 2019. https://www.mining-journal.com/base-metals/news/1355207/alta-seeks-to-grow-italy-zinc-project.

Moffat, Kieren, and Airong Zhang. 2014. "The paths to social licence to operate: An integrative model explaining community acceptance of mining." *Resources Policy* 39: 61–70.

Owen, John R. 2016. "Social license and the fear of Mineras Interruptus." *Geoforum* 77: 102–105.

Piper, Dominic. 2018. "Alta rebalance nearly complete." *Australia's Paydirt*, September 1, 2018. http://www.altazinc.com/assets/20180927%20Paydirt%20PD264_Sept2018_Mag_Final%2079-b389e031367471c464dee05cc791534aca1745a45aeda88c7e34433 875d2d4c5.pdf

Reckwitz, Andreas. 2012. "Affective spaces: A praxeological outlook." *Rethinking History* 16 (2): 241–258.

Rodon, Thierry, and Aude Therrien. 2015. "Resource development & land claim settlements in the Canadian Arctic: Multilevel governance, subsidiarity and streamlining." *Arctic Yearbook* 2015: 1–13.

Sanderson, Henry. 2015. "Zinc traders dig in as price rebounds." *Financial Times*, May 12, 2015. https://www.ft.com/content/83a025a8-f492-11e4-9a58-00144feab7de.

Smith, Laurajane. 2006. *Uses of Heritage*. Abingdon: Routledge.

Tiainen, Heidi, Rauno Sairinen, and Olga Sidorenko. 2015. "Governance of sustainable mining in Arctic countries: Finland, Sweden, Greenland & Russia." *Arctic Yearbook* 2015: 132–157.

Tiess, Guenter. 2010. "Minerals policy in Europe: Some recent developments." *Resources Policy* 35 (3): 190–198.

Todeschini, Marta. 2016. "Le scoperte nell'Arera fanno sognare Gorno «Pernoilevaturistica»." Accessed October 26, 2022. https://valserina.blogspot.com/p/val-serina-e-val-del-riso.html

TuttItalia. 2021a. "Statistiche demografiche Oltre il Colle." Accessed September 16, 2023. https://www.tuttitalia.it/lombardia/54-oltre-il-colle/statistiche/popolazione-andamento-demografico/.

TuttItalia. 2021b. "Statistiche demografiche Gorno." Accessed September 16, 2023. https://www.tuttitalia.it/lombardia/38-gorno/statistiche/popolazione-andamento-demografico/.

Veltmeyer, Henry, and James Petras. 2014. *The New Extractivism: A Post-Neoliberal Development Model or Imperialism of the Twenty-First Century?* London: Zed.

Waterton, Emma, and Laurajane Smith. 2010. "The recognition and misrecognition of community heritage." *International Journal of Heritage Studies* 16 (1–2): 4–15.

Williams, Raymond. 1961. *The Long Revolution*. Harmondsworth: Penguin.

Williams, Raymond. 1977. *Marxism and Literature*. Oxford: Oxford University Press.

Zhang, Airong, Kieran Moffat, Justine Lacey, Junxiu Wang, Roberto González, Kathleen Uribe, Lijuan Cui, and Yan Dai. 2015. "Understanding the social licence to operate of mining at the national scale: A comparative study of Australia, China and Chile." *Journal of Cleaner Production* 108 (Part A): 1063–1072.

Part V

LARGE-SCALE/SMALL-SCALE

Artisanal and Small-Scale Mining

Kuntala Lahiri-Dutt

Social studies of extraction generally portray people and communities as victims who are being crushed by the multifaceted impacts of large-scale operations or highlight their agency in resisting these operations. This literature has created a main binary: of large-scale mining as opposed to and strictly different from small-scale mining (Hilson et al. 2020). A core implication of this binary is the erasure of informal, artisanal, and small-scale mining from our understanding of extractivism. Not only is artisanal and small-scale mining (ASM) presented as the opposite of "Large-scale mining" (LSM), it is also considered a smaller version or a "sector" of industrial-scale mining. This attention to scale, and the "sectoral" view that it entails, distracts the attention away from the processes that are impacting the rural countryside and isolates ASM. However, extending my previous arguments (Lahiri-Dutt 2004, 2019), I contend against this epistemic form of exclusion and for making visible this genre of extractivism, practiced primarily by the poor in the countries of the Global South. Throughout this introduction, I show that the binary of exclusion/inclusion that this edited volume critiques is unfit to capture the complexities of ASM. To do so, this chapter first defines informal ASM carefully as part of a broader objective to make it visible in our understanding of extractivism. I then show that informal ASM must be seen as more than a mere survival strategy for the excluded and the poor but must be situated within the ongoing intensification of agrarian distress, the decay of peasantry, and the transformation of agrarian/rural/forest/grazing-based societies in these countries. The third point that I developed in this introduction is about casting ASM as dangerous, illegal, and disorganized, and placing it in opposition to large-scale mining (LSM) as formal, official, legal, and organized. I reveal how first, informal ASM finds its roots in other frames of understanding of property rights (including customary rights), and second that LSM and ASM often occur hand in hand, overlap, and cohabit sometimes with noticeable tensions and conflicts. Finally, I look at ASM as part of the burgeoning informal sectors that sustain the innumerable poor and show that efforts for formalization (inclusion) sometimes produce their own forms of exclusions. The two chapters in this section by Hilson et al. and Kurniawan et al. highlight some of these complexities.

Making visible informal ASM by defining its boundaries is a first step to challenging the epistemic exclusion of ASM from our understanding of extractivism. I classify this broad genre of mining "informal, artisanal, and small-scale mining" to highlight the fact that it is part of a sprawling informal economy. Scholars such as Behzadi (2019) and Geenen (2014) highlight how resistance to broader forms of resource development may also be articulated through local populations' involvement in ASM. Often exploiting marginal deposits, ASM is characterized by a high degree of labor intensity, low compliance to labor (or environmental) regulations, and poor occupational health and safety standards (Adler Miserendino et al. 2013; Buxton 2013; Chaparro Ávila 2003; Hentschel et al. 2002; Hinton 2006; Mutemeri et al. 2016). Other features—which are highly variable and depend on the context—include relatively low capital investment, and lower levels of machine or technology use resulting in lower production or recovery of minerals than industrial mines.

The many extractive practices are carried out by local or migrant groups or "lone wolf" operators, as well as by small- and medium-sized enterprises owned by local entrepreneurs who hire laborers in extremely exploitative conditions. This is rudimentary and artisanal mining carried out using traditional or more contemporary methods. The extractive practices vary from individuals panning for gold or digging for precious stones along riverbanks or in the tailings of industrial operations, to relatively large and organized operations using heavy machinery such as excavators and drilling machines (Buxton 2013; Collins and Lawson 2014; Hinton 2006; Veiga et al. 2006). The commodities in this extractive landscape also vary: from high-value or semi-precious gemstones, to minerals and fuels, to low-value construction materials including various stones, gravel, and even sand. While contemporary work often looks at current dynamics in the Global South, ASM also has a long history of socio-political significance all over the world—an example of this being the Californian gold rush in the USA.

Understanding informal ASM as part of broader contemporary agrarian transformations is essential to add complexity to conventional readings of ASM as a mere survival strategy. Today, throughout the mineralized tracts of the Global South, thousands of peasants are moving out of agriculture-related livelihoods and into informal extractive practices (Lahiri-Dutt 2018). The reasons for this are complex and the process has far-reaching implications for the governance of extractive industries. This unprecedented and widespread shift from agrarian to informal mineral extractive economies is driven by six overlapping factors: (1) the agricultural sector is unsustainable and characterized by poor productivity; (2) economic reforms carried out in developing countries primarily liberalize land markets and help foreign direct investment; (3) the state undertakes initiatives to earn revenue from mineral extraction; (4) policy-makers equate expansion of extractive industries with "development," which has established an extractive model that favors large corporatized operators, ignoring local communities' claims; (5) environmental degradation at the local level and uncertainty regarding precipitation and temperatures have enhanced peoples' vulnerability and encouraged them to take up extractive livelihoods; and finally, (6) cyclic global commodity price increases have incentivized not only large corporations but also the rural poor to earn additional cash or to supplement their existing incomes through seasonal mining.

Perhaps where ASM has a major impact on the economy is in supporting subsistence during noncrop seasons, and the opportunity to earn cash incomes. However, no accurate estimate can ever be given for such a diverse, seasonal, and widely scattered activity. Recent estimates put the number of people directly involved in ASM at around 20–30 million in over eighty countries around the world (Buxton 2016; Villegas et al. 2012; World Bank 2012) compared to 13 million reported in 1999 (ILO 1999). In 2016, the International Institute of Environment and Development (IIED) estimated the number of people supported by ASM-related activities to be 100–150 million (Buxton 2016), up from 80 to 100 million people in 1999 (ILO 1999).

Academics and international donors now concede that ASM is a poverty-driven activity (Gamu et al. 2015; Hilson and Banchirigah 2009; Hilson and Garforth 2012). That said, in many areas, ASM "is predominantly a highly important and deeply rooted livelihood improvement activity" and "not a mere survival strategy to which people turn to, in times of distress or conflict" (Geenen 2013, 208–209, cited in Maclin et al. 2017). The potential of ASM for "wealth creation" (Hilson and Hilson 2015, 6; Verbrugge 2016, 113) is often acknowledged by scholars; Hilson (2016a), for example, has consistently advocated for the need to understand the links between ASM and smallholder farming and claims that the two activities are "inseparable." His work (e.g., Hilson 2011, 2016b; Hilson and Garforth 2012; Hilson and Van Bockstael 2012; Okoh and Hilson 2011) and that of other authors (e.g., Cartier and Bürge 2011; Maconachie and Binns 2007) argue that ASM complements and supports agriculture by providing income in the off-season for the purchase of fertilizers and other agricultural inputs. However, as smallholder agriculture stagnates, and large numbers of previously rural, farm-based communities are forced into nonfarm jobs (Lahiri-Dutt 2017), ASM provides an alternative or supplementary source of income for farmers struggling due to agricultural reforms that have made smallholder farming unfeasible. As Verbrugge (2016) argues, ASM is both a product and catalyst of rural transformation.

This view of informal ASM as a survival strategy for the poor and excluded also obscures the fact that ASM supply chains are global and involve many actors beyond those conventionally seen as the "poor," including local traders, city-based financiers, government elites (including the military), and even criminal networks. Associating ASM only with informality and casting it as a survival strategy for the poor is therefore inadequate (Mkodzongi and Spiegel 2019). Actions to upgrade and "clean" artisanal supply chains must take into consideration the political economy driving decisions by miners, buyers, traders, and other key stakeholders.

The relationship between ASM and property rights is a third significant aspect of informal ASM that challenges simplistic binaries that drive readings of this type of extractivism. The casting of informal ASM as disorganized and illegal often overlooks the fact that informal ASM frequently involves elaborate informal or customary property systems. These reflect the plurality of actors, kinds of capital and administrative landscapes that define informal ASM, and the often blurry boundaries between legality/illegality or formality/informality (Spiegel and Veiga 2010). On one level are the complex relationships between ASM interests and those of surface land tenure holders. Verbrugge et al. (2015) argue that the interactions between ASM (largely informal) and surface land claimants are not underscored by antagonism but rather negotiations over mutually beneficial arrangements. Meanwhile, evidence from sub-Saharan Africa suggests that insecure land tenure/lack of property rights offers little incentive to miners to formalize their activities, build their businesses and infrastructure or undertake environmentally responsible practices (Mitchell 2016; O'Faircheallaigh and Corbett 2016). On the administrative level, different kinds of authorities assert sometimes conflicting rights to guarantee property ownership. In addition to the state, traditional authorities also assert control over rural lands. For instance, in Ghana, where customary land tenure practices put ASM interests into conflict with those holding formal (state-sanctioned) land rights (Banchirigah 2008; Dube et al. 2016; Hilson and Yakovleva 2007; Nyame and Blocher 2010). Two scenarios occur frequently: traditional authorities give permission for ASM operators to work on lands without legal/official permits or licenses, or the government provides licenses for miners to work on tracts of land without the permission of traditional authorities (Dube et al. 2016).

The conventional opposition between ASM/LSM also obscures the relationship and overlaps between the two. In resource scholarship as well as in policy circles, conflict between ASM and industrial mining is an area of increasing attention and concern. ASM may occur on mining leases; in fact, the presence of ASM frequently provides a geological indicator or "target selection criteria" for large-scale exploration activities (Aubynn 2009; ICMM 2009, 16). In some countries, government policies promoting large-scale mining have created a situation whereby enormous tracts of land are under the concession of large mining companies but, in many cases, ASM existed prior to the arrival of the big company. Cases have been reported where large companies have been offered land once the ASM license (which is typically short in duration) expires. At other times, the presence of the large company instigates an influx of miners either to work on the outskirts of the concession or in the tailings of the large-scale mines. In this section, Kurniawan et al. illustrate how the inclusion and exclusion of artisanal oil miners in two villages have blurred the boundaries between legality and illegality, and how power relations are constantly being negotiated and renegotiated. They show that the Indonesian state's attempts for inclusion and exclusion have been a continuous process, one where a given situation is a result of particular local context and time.

From the perspective of industrial-scale mining companies, ASM poses potential financial liabilities and reputational risks, given its "illegal" nature, poor health and safety practices, use of child labor, and environmental impacts that may be mistakenly attributed to large companies' activities (ICMM 2009). In addition, ASM can undermine large companies' assets, either directly through vandalism and other acts of resistance by miners, high-grading, or other forms of encroachment on large-scale activities. Military-type tactics to evict artisanal and small-scale miners are frequently used. Around the world, security forces employed by either companies or governments to protect industrial mines assets have been accused of human rights abuses, including gender-based violence and sexual assault. Violence in ASM sites are also, in some instances, related to the fact that when ASM focuses on high-value minerals, ASM sites become attractive to criminal networks that seek to capture and control these resources. This could occur in two ways: First, it might involve capturing state power and thus controlling the sites. Second, it could entail capturing the site directly and engaging the state for control. High-value commodities (such as diamonds and gemstones) are able to finance this kind of conflict and are known as conflict minerals. In recent years tin, tantalum, tungsten, and gold (3TG) have been the key focus due to the conflict they have been fuelling in the Great Lakes region in Africa, cobalt is also becoming a source of concern as demand rises (Faber et al. 2017).

The last point that this introduction develops and which challenges the binary of ASM/LSM, as well as the broader exclusion/inclusion binary that this edited collection critiques, relates to the discussion on the benefits of regulation and formalization. Much has been—and is being—done with the aim of improving governance of ASM and enhancing its contribution toward achieving the Sustainable Development Goals. A number of initiatives have been proposed and these include broader processes and approaches (such as formalization), as well as individual initiatives (Weldegiorgis 2016). The Economic Commission for Africa (ECA) has argued for adequate allocation of funding to local centers of innovation and adaptation of technology as well as the dissemination of tested models of equipment hiring, payback, or hire-to-pay schemes. Experts are increasingly advocating for bottom-up approaches that directly engage miners (Salo et al. 2016). The need for a better understanding of the context-specific and nuanced characteristics of the practices and the people involved (including organizational structures and labor hierarchies) is also frequently highlighted (Hilson 2009). Hilson et al. show in their chapter in this volume that despite the challenges and tensions between LSM and ASM, Mali has established the architecture for what he calls "autonomous coexistence" between the two. They surmise that this model can be replicated and "scaled up" elsewhere in sub-Saharan Africa.

Formalization is one key strategy for regulation, increasing the sustainability of activities, creating benefits for communities, and minimizing negative impacts as illustrated in much of the literature on ASM (Banchirigah 2008; Dube et al. 2016; Maconachie and Hilson 2011; Salo et al. 2016; Siegel and Veiga 2009; Smith et al. 2017; Spiegel 2015; Verbrugge and Besmanos 2016). Formalization refers not only to the presence of legislation but also to "the activation and enforcement of it by authorities and the extent of their success" (Hilson et al. 2007, 276–277). The UN Environment Programme highlights some of these efforts (2012) and, in practice, many countries have formalized their ASM sector in the last few years. Recently, there has been a refocus on formalization through organizations such as the London-based research institute International Institute for Environment and Development (IIED) and international treaties such as the Minamata Convention on Mercury to protect human health and the environment from anthropogenic emissions of mercury and mercury compounds. They require that, in countries where ASM is "more than insignificant," a National Action Plan must be developed, containing, *inter alia*, steps to facilitate the formalization or regulation of the sector.

These efforts to formalize ASM are, however, not always successful, and sometimes lead to contradictions and new forms of exclusions. Lahiri-Dutt et al. (2021) argue, for instance, that in countries like Mongolia, these attempts to formalize are overridden by the explicit support given by the government to large, industrial mining. Dube et al.'s (2016) analysis of the institutional frameworks governing ASM in Zimbabwe identifies some common barriers to miners' formalization including "high registration and compliance fees; limited knowledge of formal institutional frameworks; limited access to the formal market and opaque nature of a registration process that often breeds corruption" (1091). Spiegel (2015, 544) points out that formalization policies are potentially misused by elites to "consolidate power rather than to distribute 'development' opportunities."

The ambiguous nature of efforts to formalize ASM runs also through Kurniawan et al.'s chapter in this volume. The authors show how artisanal miners in East Java, Indonesia, have perceived the state's alternating efforts to include artisanal mining in the formal economy and exclude local people from accessing and controlling local oil resources. The state has portrayed its interventions as legitimate means to securing decent working conditions, local economic development, and a healthy environment. Yet in practice, formalization has further blurred the boundaries between legality/illegality in the village, shaping new power relations and sometimes new forms of exclusions.

In conclusion, informal ASM, with its tremendous internal diversities, can be imagined as inhabiting one end of the extractive practices spectrum; not the binary "other" of large-scale, industrial mining. As Hilson et al. show in their chapter in this section, they coexist and recognition of this coexistence might be the key to better governance. Such a view would be able to intervene in the ongoing tendency to exclude ASM from debates about how people relate to nature and resources. We need to recognize that the diversity of ASM is produced by its highly context-specific nature, and scholarly debates over its origin and policies to regulate or formalize its practices must increasingly take a more holistic approach. In the future, ASM will continue to come with environmental and other externalities, including conflict; yet, as long as poverty and inequality exist, ASM will survive. Moreover, like everything else in our economy that occurs "under the radar," it is likely to remain low-tech, risky, and exploitative. As long as the minerals extracted informally can be traded locally, nationally, and even globally as an essential requirement for important industries such as construction, jewelry, and electronics, ASM, with all its contradictions, will continue to baffle scholars and administrators.

REFERENCES

Adler Miserendino, Rebecca., Bdidget A. Bergquist, Sara E. Adler, Jean-Remi Davée Guimarães, Peter S.J. Lees, Wilmer Niquen, P. Colon Velsasquez- López, and Marcello M. Veiga. 2013. "Challenges to measuring, monitoring, and addressing the cumulative impacts of artisanal and small-scale gold mining in Ecuador." *Resources Policy* 38(4): 713–722. https://doi.org/10.1016/j.resourpol.2013.03.007.

Aubynn, Anthony. 2009. "Sustainable solution or a marriage of inconvenience? The coexistence of large-scale mining and artisanal and small-scale mining on the Abosso Goldfields concession in Western Ghana." *Resources Policy* 34(1–2): 64–70. http://doi.org/10.1016/j.resourpol.2008.04.002.

Banchirigah, Sadia Mohammed. 2008. "Challenges with eradicating illegal mining in Ghana: A perspective from the grassroots." *Resources Policy* 33(1): 29–38. http://doi.org/10.1016/j.resourpol.2007.11.001.

Behzadi, Negar E. 2019. "Women miners' exclusion and Muslim masculinities in Tajikistan: A feminist political-ecology of honor and shame." *Geoforum* 100: 144–152. https://doi.org/10.1016/j.geoforum.2019.01.001.

Buxton, Abbi. 2013. *Responding to the challenge of artisanal and small-scale mining: How can knowledge networks help?* London: IIED. http://pubs.iied.org/16532IIED.html.

Buxton, Abbi. 2016. *Transforming mining through dialogue*. London: IIED. http://pubs.iied.org/pdfs/G04081.pdf.

Cartier, Laurent E., and Michael Bürge. 2011. Agriculture and artisanal gold mining in Sierra Leone: Alternatives or complements? *Journal of International Development* 23(8): 1080–1099. https://doi.or/doi:10.1002/jid.1833.

Chaparro Ávila, Eduardo. 2003. *Small-scale mining: A new entrepreneurial approach*. UN ECLAC. http://www.eclac.org/publicaciones/xml/1/13901/Lcl1834i.pdf

Collins, Nina., and Lynda Lawson. 2014. *Investigating approaches to working with artisanal and small-scale miners: A compendium of strategies and reports from the field*. IM4DC Action Resarch Report. http://im4dc.org/wp-content/uploads/2013/09/Collins-ASM-FR-Completed-Report.pdf.

Dube, Nqobizitha, Funa Moyo, Mkhokheli Sithole, Gracsious Ncube, Peter Nkala, Nevel Tshuma, Mandlekonsi Maphosa, and Clifford Mabhena. 2016. "Institutional exclusion and the tragedy of the commons: Artisanal mining in Matabeleland South Province, Zimbabwe." *The Extractive Industries and Society* 3(4): 1084–1094. https://doi.org/10.1016/j.exis.2016.08.006.

Gamu, Jonathan, Philippe Le Billon, and Samuel Spiegel. 2015. "Extractive industries and poverty: A review of recent findings and linkage mechanisms." *The Extractive Industries and Society* 2(1): 162–176. https://doi.org/10.1016/j.exis.2014.11.001.

Geenen, Sara. 2013. "'Who seeks, finds': How artisanal miners and traders benefit from gold in the Eastern Democratic Republic of Congo." *The European Journal of Development Research* 25(2): 197–212. https://doi.org/10.1057/ejdr.2012.19.

Hentschel, Thomas, Felix Hruschka, and Michael Priester. 2002. *Mining, minerals and sustainable development (MMSD) project: Global report on small-scale mining*. Working Paper No.70. London: IIED. http://pubs.iied.org/pdfs/G00723.pdf

Hilson, Gavin. 2009. "Small-scale mining, poverty and economic development in sub-Saharan Africa: An overview." *Resources Policy* 34(1–2): 1–5. https://doi.org/10.1016/j.resourpol.2008.12.001.

Hilson, Gavin. 2011. "Artisanal mining, smallholder farming and livelihood diversification in rural Sub-Saharan Africa: An introduction." *Journal of International Development* 23(8):1031–1041. https://doi.org/10.1002/jid.1829.

Hilson, Gavin. 2016a. *Artisanal and small-scale mining and agriculture: Exploring their links in rural sub-Saharan Africa*. IIED Issue Paper March 2016. London: IIED. http://pubs.iied.org/pdfs/16617IIED.pdf.

Hilson, Gavin. 2016b. "Farming, small-scale mining and rural livelihoods in Sub-Saharan Africa: A critical overview." *The Extractive Industries and Society* 3(2): 547–563. http://doi.org/10.1016/j.exis.2016.02.003.

Hilson, Gavin, and N. Yakovleva. 2007. "Strained relations: A critical analysis of the mining conflict in Prestea, Ghana." *Political Geography* 26(1): 98–119. https://doi .or/10.1016/j.polgeo.2006.09.001.

Hilson, Gavin, and Sadia Mohammed Banchirigah. 2009. "Are alternative livelihood projects alleviating poverty in mining communities? Experiences from Ghana." *The Journal of Development Studies* 45(2): 172–196. https://doi .org /10 .1080 /00220380802553057.

Hilson, Gavin, and Chris Garforth. 2012. "'Agricultural poverty' and the expansion of artisanal mining in Sub-Saharan Africa: Experiences from southwest Mali and southeast Ghana." *Population Research and Policy Review* 31(3): 435–464. https:// doi.org/10.1007/s11113-012-9229-6.

Hilson, Gavin, and Steven Van Bockstael. 2012. "Poverty and livelihood diversification in rural Liberia: Exploring the linkages between artisanal diamond mining and smallholder rice production." *Journal of Development Studies* 48(3): 413–428. https://doi.org/10.1080/00220388.2011.604414.

Hilson, Gavin, and Abigail Hilson. 2015. *Entrepreneurship, poverty and sustainability: Critical reflections on the formalisation of small-scale mining in Ghana.* IGC Working paper April 2015. The IGC.org/publications: International Growth centre. http://www.theigc.org/wp-content/uploads/2015/04/Hilson-Hilson-2015-Working -Paper.pdf.

Hilson, Gavin, Christopher J. Hilson, and Sandra Pardie. 2007. "Improving awareness of mercury pollution in small-scale gold mining communities: Challenges and ways forward in rural Ghana." *Environmental Research* 103(2): 275–287. https:// doi.org/10.1016/j.envres.2006.09.010.

Hilson, Gavin., Titus Sauerwein, and John Owen. 2020. "Large and artisanal scale mine development: The case for autonomous co-existence." *World Development* 13: 1-190. https://doi.org/10.1016/j.worlddev.2020.104919.

Hinton, Jennifer. 2006. *Communities and small scale mining: An integrated review for development planning.* Report to the World Bank. http://www.eisourcebook.org /cms/June%202013/CASM,%20an%20Integrated%20Review%20for%20Development%20Planning.pdf.

ICMM. 2009. *Working together: How large-scale miners can engage with artisanal and small-scale miners.* ICMM. Commdev.org http://www.icmm.com/document/789.

ILO. 1999. *Social and labour issues in small-scale mines. Report for discussion at the Tripartite meeting on social and labour issues in small-scale mines.* 17–21 May. Geneva: ILO (International Labour Organization). http://www.ilo.org/public /libdoc/ilo/1999/99B09_35_engl.pdf.

Lahiri-Dutt, Kuntala. 2004. "Informality in mineral resource management in Asia: Raising questions relating to community economies and sustainable development." *Natural Resources Forum* 28(2):123–132. https://doi .org /10 .1111 /j .1477 -8947 .2004.00079.

Lahiri-Dutt, Kuntala. 2017. "Resources and the politics of sovereignty: The moral and immoral economies of coal mining in India." *South Asia: Journal of South Asian Studies Association of Australia* 40(4): 792–809. https://doi.org/10.1080/00856401 .2017.1370211.

Lahiri-Dutt, Kuntala, ed. 2018. *Between the plough and the pick*. Canberra: ANU Press.

Lahiri-Dutt, Kuntala. 2019. "Extractive peasants: Reframing informal, artisanal and small-scale mining debates." *Third World Quarterly* 39(8): 1561–1582. https://doi.org/10.1080/01436597.2018.1458300.

Lahiri-Dutt, Kuntala, Emily Crawford, Jonathan Ratcliffe, and Michael Rose. 2021. "Resource politics in Mongolia: Large- and small-scale mines in collision." *Resources Policy* 73: 1-8 https://doi.org/10.1016/j.resourpol.2021.102137.

Maclin, Beth. J., Jocelyn T.D. Kelly, Rachel Perks, Patrick Vinck, and Phuong Pham. 2017. "Moving to the mines: Motivations of men and women for migration to artisanal and small-scale mining sites in Eastern Democratic Republic of the Congo." *Resources Policy* 51:115–122. https://doi.org/10.1016/j.resourpol.2016.12.003.

Maconachie, Roy, and Tony Binns. 2007. "'Farming miners' or 'mining farmers'?: Diamond mining and rural development in post-conflict Sierra Leone." *Journal of Rural Studies* 23(3): 367–380. https://doi.org/10.1016/j.jrurstud.2007.01.003.

Maconachie, Roy, and Gavin Hilson. 2011. "Safeguarding livelihoods or exacerbating poverty? Artisanal mining and formalization in West Africa." *Natural Resources Forum* 35(4): 293–303. http://doi.org/10.1111/j.1477-8947.2011.01407.

Mitchell, James.2016. "Pulling the rug out from under: The land tenure dynamics of mining concessions in sub-Saharan Africa." *The Extractive Industries and Society* 3(4):1117–1129. https://doi.org/10.1016/j.exis.2016.10.003.

Mkodzongi, Grasian, and Samuel Spiegel. 2019. "Artisanal gold mining and farming: Livelihood linkages and labour dynamics after land reforms in Zimbabwe." *The Journal of Development Studies* 55(10): 2145–2161. https://doi.org/10.1080/00220388.2018.1516867.

Mutemeri, Nellia., Joshua. Z. Walker, Nancy Coulson, and Ingrid Watson. 2016. "Capacity building for self-regulation of the artisanal and small-scale mining (ASM) sector: A policy paradigm shift aligned with development outcomes and a pro-poor approach." *The Extractive Industries and Society* 3(3): 653–658. https://doi.org/10.1016/j.exis.2016.05.002.

Nyame, Frank K., and Joseph Blocher. 2010. "Influence of land tenure practices on artisanal mining activity in Ghana." *Resources Policy* 35(1): 47–53. https://doi.org/10.1016/j.resourpol.2009.11.001.

O'Faircheallaigh, Ciaran, and Tony Corbett. 2016. "Understanding and improving policy and regulatory responses to artisanal and small scale mining." *The Extractive Industries and Society* 3(4): 961–971. https://doi.org/10.1016/j.exis.2016.11.002.

Okoh, Godfrieh, and Gavin Hilson. 2011. "Poverty and livelihood diversification: Exploring the linkages between smallholder farming and artisanal mining in rural Ghana." *Journal of International Development* 23(8): 1100–1114. https://doi.org/10.1002/jid.1834.

Salo, Matti, Juha Hiedanpää, Teemu Karlsson, Luciano Cárcamo Ávila, Juha Kotilainen, Pekka Jounela, and Roger Rumrrill García. 2016. "Local perspectives on the formalization of artisanal and small-scale mining in the Madre de Dios gold fields, Peru." *The Extractive Industries and Society* 3(4): 1058–1066. https://doi.org/10.1016/j.exis.2016.10.001.

Siegel, Samuel, and Marcello M. Veiga. 2009. "Artisanal and small-scale mining as an extralegal economy: De Soto and the redefinition of 'formalization'." *Resources Policy* 34(1–2): 51–56. https://doi.org/10.1016/j.resourpol.2008.02.001.

Smith, Nicole M., Jessica M. Smith, Zira Q. John, and Benjamin A. Teschner. 2017. "Promises and perceptions in the Guianas: The making of an artisanal and small-scale mining reserve." *Resources Policy* 51: 49–56. https://doi.org/10.1016/j.resourpol.2016.11.006.

Spiegel, Samuel J. 2015. "Shifting formalization policies and recentralizing power: The case of Zimbabwe's artisanal gold mining sector." *Society and Natural Resources* 28(5): 543–558. https://doi.org/10.1080/08941920.2015.1014606.

Spiegel, Samuel J., and Marcello M. Veiga. 2010. "International guidelines on mercury management in small-scale gold mining." *Journal of Cleaner Production* 18(4): 375–385. https://doi.org/10.1016/j.jclepro.2009.10.020.

Teschner, Benjamin. 2013. "How you start matters: A comparison of Gold Fields' Tarkwa and Damang Mines and their divergent relationships with local small-scale miners in Ghana." *Resources Policy* 38(3): 332–340. https://doi.org/10.1016/j.resourpol.2013.03.006.

United Nations Environment Programme. 2012. *Analysis of formalization approaches in the artisanal and small-scale gold mining sector based on experiences in Ecuador, Mongolia, Peru, Tanzania and Uganda.* https://wedocs .unep .org /20 .500 .11822/31429.

Veiga, Marcello M., Peter A. Maxson, and Lars D. Hylander. 2006. "Origin and consumption of mercury in small-scale gold mining." *Journal of Cleaner Production* 14(3–4): 436–447. https://doi.org/10.1016/j.jclepro.2004.08.010.

Verbrugge, Boris. 2016. "Voices from below: Artisanal- and small-scale mining as a product and catalyst of rural transformation." *Journal of Rural Studies* 47(Part A): 108–116. https://doi.org/10.1016/j.jrurstud.2016.07.025.

Verbrugge, Boris, and Beverly Besmanos. 2016. "Formalizing artisanal and small-scale mining: Whither the workforce?" *Resources Policy* 47: 134–141. http://dx .doi.org/10.1016/j.resourpol.2016.01.008

Verbrugge, Boris, Jeroen Cuvelier, and Steven Van Bockstael. 2015. "Min(d)ing the land: The relationship between artisanal and small-scale mining and surface land arrangements in the southern Philippines, eastern DRC and Liberia." *Journal of Rural Studies* 37: 50–60. https://doi.org/10.1016/j.jrurstud.2014.11.007.

Villegas, Cristina, Ruby Weinberg, Estelle Levin, and Kirsten Hund. 2012. *Artisanal and small-scale mining in protected areas and critical ecosystems programme (ASM-PACE): A global solutions study.* www.asm-pace.org. WWFP Report, September 2012. http://www.sidalc.net/repdoc/A10263I/A10263I.PDF.

Weldegiorgis, Fitsum. 2016. *Talking sustainable development in the troubled world of ASM.* IIED, October 25, 2016. IEED. IEED.org. http://www.iied.org/talking -sustainable-development-troubled-world-asm.

World Bank. 2012. *Artisanal mining in critical ecosystems: A look at Gabon, Liberia, and Madagascar.* http://www .profor .info /sites /profor .info /files /docs /ASM -brochure.pdf.

9

Artisanal and Large-Scale Mine Relations

Laying the Groundwork for Autonomous Coexistence in Sub-Saharan Africa

Gavin Hilson, John Owen, Titus Sauerwein, Massaran Traore, and Éléonore Lèbre

Over the past four decades, capital-intensive large-scale mineral exploration and mining activities have expanded rapidly across continental sub-Saharan Africa. During this period, numerous mining companies have flocked to the region, lured by enabling legislation and reforms designed to attract foreign investment. The scale of investment has been impressive, particularly with regards to gold mining, new and rehabilitated.[1]

Yet, the mining sector reforms implemented in sub-Saharan Africa have not, as Gajigo, Mutambatsere, and Ndiaye (2012) put it, "been problem-free" (16). Most were implemented largely under the guidance of the World Bank, under the assumption that "foreign investment in the sector was sufficient to ensure the sector's contribution to economic development." This belief is an oversight which led to there "initially [being] less emphasis on the fair distribution of economic rents or shoring up the required capacity of governments to improve their share of those rents, but more stress on reducing regulations" (16). Even officials at the World Bank now concede that in the case of mining-led development in sub-Saharan Africa, the "positive effect of revenue windfalls is underpinned by several assumptions," foremost "that local politicians are responsive to the preferences of the broad population, local institutions function well, and local bureaucracies have the technical capacity to provide these public goods and services" (Chuhan-Pole et al. 2017, 27).

One of the most significant impacts of these reforms and the rent-seeking behavior they have spawned has been their effect on people who depend on artisanal and small-scale mining (ASM)[2] for their livelihoods. From the evidence, ASM is the most important rural non-farm activity in sub-Saharan Africa, providing income to tens of millions of people. The sector generates revenue used by hundreds of thousands of the region's families to purchase fertilizers and the tools they need to cultivate crops for consumption and/or market. This type of mining also spawns an unrivalled collection of downstream and upstream economic activities which generate income for millions more people. Additionally, throughout sub-Saharan Africa, ASM is largely "poverty-driven," providing a lifeline to people with few, if any, alternative sources of income.

These economic impacts, however, are routinely ignored because most of the region's ASM activities are carried out informally. Efforts made to date to address this problem in sub-Saharan Africa, principally through legalizing and supporting the sector, have been woefully ineffective. They have yielded regulatory and policy frameworks which offer few navigable routes for individuals who fall into the category of "poverty-driven" operator. These individuals have struggled to secure licenses and viable plots to work in landscapes that are now mostly controlled by large-scale mineral exploration and mining companies. Mining sector reforms have rather established an interface between Indigenous artisanal and small-scale groups on the one hand, and the management and executives of mineral exploration and mining companies—many of which are foreign—on the other hand. Within the region's gold-rich territories in particular, tensions between these miners have escalated in recent years, often culminating in violent conflict. In most cases, it has been the encroachment of artisanal miners on to concessions that have been awarded to large-scale mineral exploration and mining companies that has triggered such conflict.

As Table 9.1 reveals, across sub-Saharan Africa, there are now numerous (sizeable) concessions in the hands of mining companies, a situation which has led to diminished opportunities for prospective small-scale licensees. But rather than conceding that reforms are, indeed, problematic and yield policy frameworks which are incapable of facilitating the formalization of ASM, donors and industry organizations have elected to portray the interface between operators as an economic opportunity for mining companies. Officials at the International Council on Mining and Metals (ICMM),[3] for example, are calling for these operators to forge partnerships, maintaining that "for LSM [large-scale mining] engagement with ASM to be successful, consideration also needs to be given to the business case for ASM engagement with LSM" (IFC 2011, 12). Similar claims were made in a World Bank report, *Working Together*, which calls on companies to engage and support ASM operators, going as far as supplying a "Tentative List of Actions" on how to achieve this. The idea that multinationals should be encouraged to

Table 9.1 Mine Claims Data from Selected Countries in Sub-Saharan Africa

Country	*Total Area of Country (km²)*	*Number of Mining Claims*	*Total Area, Mining Claims (km²)*	*Average Mine Claim Size (km²)*
Ghana	227,540	1,229	68,208.38	55.5
Tanzania	885,800	78,486	109,599.47	1.4
Namibia	823,290	5,100	593,277.33	116.33
Madagascar	581,800	4,998	213,029.37	42.62
DR. Congo	2,267,050	3,826	272,051.11	71.11
Cote d'Ivoire	318,000	2,383	156,232.76	65.56
Sierra Leone	72,180	1,286	7,168.73	5.57
Burkina Faso	273,600	1,129	120,428.52	106.67
Uganda	200,520	799	59,738.77	74.77
Mali	1,220,190	601	49,714.98	82.72
Guinea	245,720	528	109,461.36	207.31
Mauritania	1,030,700	310	212,719.76	686.19
Liberia	96,320	200	25,850.22	129.25
Niger	1,266,700	192	61,238.84	331.02
Zimbabwe	386,850	182	123,123.26	676.50
Cameroon	472,710	165	128,057.07	776.10
Senegal	192,530	89	37,520.06	421.57

Source: Data obtained from S&P Global (2022) and the World Bank (2021).

support local ASM operators and provide services to buy their gold at sites has also been proposed.

It is unlikely, however, that the business models of mining companies operating in gold-rich sections of sub-Saharan Africa were designed to take into account the possibility of partnering with and supporting local artisanal operators. Senior managers and executives are mostly based in Toronto, Perth, Vancouver, Denver, Johannesburg, or London—major mine investment hubs. Far removed from the realities of the sections of rural sub-Saharan African Africa where gold extraction takes place, these managers are often reluctant to implement programs to engage artisanal operators and unreceptive to the idea of forging working agreements with them. Over the years, there have been only rare instances in sub-Saharan Africa where companies have relinquished sections of their concessions to government, with the intention of handing these lands over to ASM operators.

Companies seem to prefer to take no action until absolutely necessary, or, as referred to in Damang, Ghana—the location of one of the most widely publicized cases of ASM-large-scale mining working agreements forged in sub-Saharan Africa to date—to adopt a "Live and Let Live" approach (Aubynn 2009). The tendency of company officials and site managers to *tolerate* ASM activities on sections of their concessions has routinely been misinterpreted over the years by donors and various experts as cohabitation underpinned by a camaraderie between operators, and therefore mistakenly viewed as an opportunity to facilitate further engagement (see, e.g., Hentschel et al. 2002; World Bank 2009; IFC 2011). This approach neither provides ASM operators with the security of tenure they covet, nor creates conditions conducive for innovation and "scaling up" their production.

Rarely are mining companies required by law to commit financial resources to support the activities of ASM operators. Historically, they have used artisanal and small-scale operators as "pathfinders" (Luning 2014; IGF 2017) to pinpoint the locations of viable gold deposits. In these instances, the decision by management to *tolerate* activities onsite may be deliberate. This chapter argues that in the case of sub-Saharan Africa, mining sector reforms should be designed with the goal of facilitating the autonomous *coexistence* of, and equal opportunities for, miners operating at different scales. This includes fostering a policy environment conducive to achieving this. As it stands, the region's ASM sector is not afforded the same legal or financial privileges as those enjoyed by companies engaged in capital-intensive mineral exploration and extraction. In sub-Saharan Africa, the ASM sector must be put in a position where it can evolve autonomously and securely in settings where host governments and the management of companies in possession of sizable concessions have little interest in pursuing their own interests. Ultimately, this approach to the coexistence of ASM and large-scale mining complicates perspectives on how institutions, governments, and companies have so far engaged with the idea of "including" the former in development, partly through formalization. It rather reveals the limits of past approaches which—as described above and as will be examined further—have at times led to new forms of exclusion.

The chapter engages critically with this conundrum, offering guidance on how to achieve a state of autonomous coexistence in sub-Saharan Africa. In doing so, it revisits the theme of ASM "corridors" (or "blocked out areas" for ASM), concluding with a case study of Mali, one of the region's largest gold producers.

ARTISANAL AND LARGE-SCALE MINE PARTNERSHIPS IN SUB-SAHARAN AFRICA: CRITICAL REFLECTIONS

With the African mineral exploration budget for gold remaining buoyant throughout the COVID-19 pandemic (Table 9.2), conflicts between miners are bound to intensify in the absence of an approach which fosters *autonomous* coexistence. Calls for large-scale operators to engage and partner with artisanal miners risks oversimplifying complex and unpredictable situations.

Table 9.2 Gold Exploration Budget in Africa, 2019–2021

Year	*Amount (in US$ millions)*
2019	615.9
2020	590
2021	689.4

Source: Data obtained from S&P Global (2022).

Proponents of cohabitation routinely oversimplify or ignore three crucial issues in their assessments, a detailed understanding of which casts light on why this strategy should be avoided.

Issue #1: The ASM Communities Themselves

Presumably, proponents (of these partnerships) see large-scale miners being in a position to establish buying arrangements with artisanal groups, irrespective of the challenges the former may face reconciling such a strategy with a business model linked to capital-intensive, export-led, mineral production. This was the logic behind the "Live and Let Live" strategy conceived at Gold Fields' Damang site in Ghana in the mid-1990s by the property's previous owner, Ranger Minerals. But the intervention, seen as ground-breaking at the time of its launch, eventually deteriorated because of the quantity of resources and logistics needed to keep it functional.

A more significant oversight is the assumption that a mining company can absorb, wholesale, artisanal, and small-scale groups, without any consequences at the community level. In the case of gold, the only similarity between ASM and large-scale mining is the mineral being extracted. The supply chains they are a part of, the equipment used, the markets they serve, and even the deposits they tend to work are all different. Throughout the developing world, and in sub-Saharan Africa in particular, capital-intensive resource extraction typically takes place in "enclaves" (Ferguson 2005); large-scale gold mining is no exception (Côte and Korf 2018; Hilson et al. 2019). Capital *hops* as opposed to *flows* into these "enclaves," which, consequently, are poorly integrated into local economies.

Conversely, the region's ASM activities *create* income-earning opportunities, even in an informal or quasi-legal state, and are multilayered with networks that spread across vast territory. Even the earliest studies conducted in the region capture these dynamics.[4] The assumption has been that mining companies operating in sub-Saharan Africa can absorb these structures and reorganize them as tributary systems. This, however, would lead to gold purchasing structures changing markedly. Many of the nodes and layers found in the labor pyramids of, and supply chains linked to, ASM, would disappear altogether, which would have enormous economic repercussions in local communities.

Today, the dialogue on ASM is primarily shaped by negative perceptions of the sector's operations. This includes intermediaries, who many scholars, NGO officials, and policy-makers have claimed over the years manipulate unlicensed operators' precarious negotiating positions, and the perceived inefficiency of the sector's operations due to an overdependence on rudimentary equipment. But it is these very attributes that *create* the multilayered labor structures that engage so many people. Setting boundaries for ASM groups and asking them to follow rules designed for large-scale companies operating within an enclave and asking them *not* to expand the networks they depend on for their success is bound to cause partnerships to deteriorate. This likely accelerated the collapse of both iterations of the "Live and Let Live" strategy adopted at Damang: the initial program instituted by Ranger Minerals and the more recent resuscitated Gold Fields' version.

Issue #2: The Unpredictable Behavior of Multinational Gold Mining Companies

The unpredictable strategies of gold mining companies make forging a robust and enduring partnership with local artisanal operators near-impossible. Hilson, Sauerwein, and Owen (2020) attribute this behavior, and ultimately why gold mining companies are so reluctant to embrace a strategy of cohabitation, to fluctuations in the gold price and the volatility of project ownership.

Partnering initiatives are most feasible for companies when the gold price is low, corresponding to times when a mine is scaling *down* its production or has suspended operations altogether because of high operating costs. Such was the case in Damang (Ghana), which again, authored the "Live and Let Live" approach (Aubynn 2009). In 2002, shortly after Gold Fields acquired the property from Ranger Minerals, its management committed, for several years, to dialoguing with local chiefs to "persuade ASM operators to voluntarily vacate the sites" whenever "[Gold Fields'] pit development was to take place at areas occupied by ASM" (Teschner 2013, 335). The same could be said about North Mara in Tanzania, which, for three decades, has been a conflict hotspot. In August 1993, a local company, Winani Mining, was issued a lease to work in this area, where mining had taken place informally since 1987. The lease included five villages, each of which had been issued separate mine titles in 1991. Winani Mining sold its rights, in October 1994, to another company, EAGM/Afrika Mashariki, which immediately commenced extensive drilling works, including in areas covered by claims belonging to the five villages, without incident (Lange 2008). Proponents of cohabitation have singled out cases such as these but fail to mention that these developments took place when the price of gold was low.

Conversely, when the gold price is high, corporate attitudes toward ASM operators change noticeably. In Damang, by 2008, relations between Gold Fields' staff and ASM operators began to deteriorate, after company managers decided to resume operations; this involved removing small-scale operators from the site. This was met with resistance because, by this time, "ASM had become highly lucrative as the world gold price increased and innovations in the ASM sector increased the sector's production capacity" (Teschner 2013, 335). Similarly, in Tanzania, tensions between artisanal and large-scale miners intensified in the mid-1990s in a number of major gold-producing areas, including Bulyanhulu, North Mara, and Geita. This followed a controversial decision made in favor of Sutton Resources, which was awarded a lease to work in Bulyanhulu. The "eventual resolution favouring large-scale foreign companies," it is believed, "set the tone for the subsequent [large-scale mining] investment boom" (Cooksey and Kelsall 2011 19). Consistent with how, "At different times, the Tanzanian state has recognised and banned local mining" (Cooksey and Kelsall 2011, 22), companies began showing hostility towards local artisanal mine operators, viewing them less as neighbors and more so as "invaders."

As was the case at Damang and other sites in Ghana in the 1990s and early-2000s, such as Prestea Underground (under Bogoso Gold Ltd.) and New Abirem-Ntronang (under Newmont Gold Mining), mining companies in Tanzania generally ignored ASM operators during the exploration phase but moved swiftly to evict these "invaders" once plans for production were approved. In both Ghana and Tanzania, the management of large-scale mining companies have resisted formalizing any type of partnership with artisanal operators, even when production has dipped in response to softening gold prices. In Ghana, this was perhaps best epitomized by Damang, where it could be argued that the dialogue and "Live and Let Live" approach taken by the company kept the community engaged during the period when the gold price was low, and management could not mine the areas where artisanal groups could work profitably.

The reluctance of gold mining companies to commit to partnering with local artisanal groups, however, may have more to do with the volatility of the sector's ownership structures. As explained by Hilson, Sauerwein, and Owen (2020), there are two dimensions of the ownership issue that must be considered. The first concerns the frequent changes in ownership of mining assets, a function of the numerous mergers and acquisitions that take place in the mining sector. In 2019, there were 1,973 major mining company mergers and acquisitions, a figure which rose to 2,141 in 2020.[5] Even the world's largest mining houses are not immune to acquisition, key examples being Colorado-based Newmont Gold Mining's acquisition of Normandy Mining; the establishment, in 2004, of AngloGold Ashanti, the result of the merger of giants Ashanti Goldfields and AngloGold; the US$18.3 billion merger of Randgold and Barrick Gold, in December 2018; Barrick's earlier acquisition, in 2006, of the Vancouver-based giant, Placer Dome; and Newmont's more recent merger with Goldcorp Inc., in January 2019.

Significantly, these mergers and acquisitions culminate in shifts in corporate strategy and ultimately spawn sweeping changes in how interactions with local people are managed at the site level. This leads to the second dimension, which is the anatomy of gold mining multinationals. These companies typically operate through a series of subsidiaries that are headquartered and/or listed on stock exchanges in different countries. The dynamics of mine ownership are in constant flux with continuous restructuring of regional divisions of parent companies and establishment of new subsidiaries. For community-based groups such as ASM operators, these changes surely obfuscate communication with site managers.

Simple questions such as "Which individuals make decisions on agreements?" and "How long would such agreements last and under which circumstances can they be forged?" become increasingly challenging to answer. For example, consider once again the case of Bulyanhulu, which Sutton Resources acquired from the Government of Tanzania in August 1994. In 1999, Barrick Gold acquired Sutton Resources for CAN$500 million, and Pangea Goldfields Inc. (which operated the Buzwagi mine and Tulawaka mine, the latter of which closed in 2013) the following year, and in 2006, when it took over Placer Dome, gained control of the North Mara Mine. Barrick Gold then established, in 2010, African Barrick Gold, which had its name changed to Acacia Mining in 2014 and was subsequently *reacquired* by Barrick Gold in 2019 (Government of Tanzania 2008; Cooksey 2011).[6]

Issue #3: Disengaged Host Governments

In sub-Saharan Africa, mining sector reforms have been implemented mostly with one objective in mind: to facilitate increased large-scale mineral extraction and exploration activity. This is because, once established, individual large-scale mines generate significant quantities of revenue for host governments (see Table 9.3). This large-scale mining "bias," however, has stifled the formalization of artisanal and small-scale activities. This is due in part to there being a shortage of gold-mineralized land—as most is in the hands of large corporations. Moreover, and as hinted at the beginning of the chapter, licensing schemes are incompatible with the needs of the ASM sector (Hilson 2019).

Most governments in sub-Saharan Africa have established "designated areas"[7] for (licensed) ASM operators, which, ironically, is the first step toward facilitating their *autonomous* coexistence. These "designated areas" *should* be easier for governments to monitor and regulate, provide a platform for

Table 9.3 Revenue Generated for Government by Selected Large-Scale Gold Mines in Sub-Saharan Africa

Corporation	*Country*	*Project*	*State Revenue (US$ millions)*	*2017*	*2018*	*2019*	*2020*
Barrick Gold	Tanzania	North Mara	Royalties	32.7	34.3	17.7	20.5
			Taxes	53.6	45.3	39	77.9
	Mali	Syama	Royalties	22.2	15.3	21.9	27.3
			Taxes	2.9	7.7	7.7	52.3
Kinross	Ghana	Chirano	Royalties	15.5	14.4	14	18.1
			Taxes	0	0	0	0
	Mauritania	Tasiast	Royalties	17.1	15.9	27.2	35
			Taxes	0	0	0	0
IAMGOLD	Burkina Fasso	Essakane	Royalties	22.2	24.5	24.4	25.3
			Taxes	18.5	24.2	3.9	51.6
AngloGold Ashanti	Guinea	Siguiri	Royalties	32.7	24.9	23.9	24
			Taxes	46.2	26.1	8.2	17.8

Source: Data obtained from S&P Global (2022).

collecting revenue from operators, and facilitate the adoption of practical environmental management strategies (Corbett et al. 2017). The problem, however, is that so few of these initiatives have progressed beyond what World Bank officials refer to as "paper formalization" of ASM (World Bank 2015). This is due to governments failing to ensure that these "designated areas" provide ASM operators with basic security of tenure. The most extreme case is DR Congo, where the government established, in the early-2000s, "Artisanal Exploitation Zones," claiming that these areas were put aside for ASM specifically. But as Geenen (2012) explains, the law allows for the government to close a zone if "the factors justifying its creation ceased to exist" or if "a new deposit necessitating large-scale exploitation has been discovered" (325).

The constraining factor in Côte d'Ivoire is the productivity of the areas put aside for ASM.[8] These areas are, as Van Bockstael (2019) explains, the "'left-overs' that remain after foreign investors have rushed in to secure relatively cheap exploration permits covering the most geologically appealing areas" (908). In Ghana, it seems to be precisely the opposite: here, the government is now *deblocking* locations for large-scale mining listed on its *Designated Areas for Small Scale Gold & Diamond Mining in Ghana* (Government of Ghana 2009), a database which it started assembling in 2008. The idea was to identify areas, ascertain their gold content—using mostly monies supplied by donors such as the World Bank—and to block them out for *licensed* ASM operators. The Government of Ghana, however, has failed to prospect these designated areas; it has also failed to issue many licenses to small-scale operators who have expressed interest in working them. Its strategy has instead been to allow informal ASM to emerge in these locations, and once enough geological information is obtained from studying their activities, to remove operators and then "deblock" and award these lands as concessions to corporations.

SINGLING OUT THE MALI CASE

One possible exception is Mali. It offers valuable lessons on how countries can tackle each of these three issues and lay the bedrock for autonomous coexistence. Mali is a landlocked nation in West Africa and is one of the most impoverished countries in the world. It performs poorly on most development indicators, ranking 184 out of 189 countries on the UN's most recent Human Development Index.[9] Malians have endured lengthy episodes of political instability, brought about by a string of *coup d'états*, the most recent taking place in 2020 and 2021. At the time of writing, West African countries had imposed severe sanctions on Mali after its ruling military junta announced a delay in the elections it had pledged to hold in February 2022.[10] Mali, therefore, seems the most unlikely of places from which to draw inspiration for the design of forward-looking government policies.

Significantly, however, Mali's "achievement" has come about inadvertently: it is the product of a series of interventions made by the government. In combination, these interventions have given rise to a situation in which both small and large-scale mining activities are able to coexist relatively autonomously. Ironically, it may be the rent-seeking behavior of the government, specifically its focus on promoting the growth of large-scale gold mining at all costs, that has facilitated this outcome. While not perfect, the arrangement that these changes have spawned provides a more stable foundation for coexistence over the long term than the unsustainable partnerships many donors and private sector bodies continue to advocate for.

The Government of Mali

Unlike other governments in sub-Saharan Africa, the authorities in Mali have done very little to shield incoming companies from local artisanal groups. Elements drawn from four separate mining codes (1970, 1991, 1999 and 2012) are used by the government to regulate the sector, and licenses are awarded on a "first come first served" basis by the Ministère des Mines, de l'Energie et de l'Eau du Mali.[11] The country's artisanal gold mining activities are confined to the traditional economy, "regulated" and monitored by a suite of local actors, including land owners (*dugutigui*), site owners (the *damantigui*), and mine police (the *tomboloma*). The latter[12] award a *Gold Washer's Card* and in some *Cercles*,[13] such as Kolondiéba, have moved their offices from the town hall to mine sites, where they reportedly rent machines and hire their own people to work deposits (Human Rights Watch, 2011).

As specified in the 1999 Mining Code (099-255), individuals possessing a *Gold Washer's Card* are permitted to work in designated zones referred to as artisanal gold mining corridors (*couloirs d'orpaillage*).[14] After they identify these corridors and oversee their transfer to the relevant devolved local authorities, government officials in the country capital of Bamako have little direct engagement with the country's artisanal gold mining sector. Significant criticism has emerged from the NGO community about how vast quantities of gold are being smuggled from these corridors across the Sahel to international destinations such as Dubai (Martin and de Balzac 2017; Raineri 2020).

Officials in Bamako are clearly focused on the revenue generated from the country's thirteen industrial gold mines, which amounted to US$850 million in 2020, up 13 percent from 2019, buoyed by higher global prices during the COVID-19 pandemic.[15] A case could be made, however, that it is because of Bamako's detachment that artisanal mining flourishes across the country. Specifically, while selling points for gold may be cross-border, artisanal mining activities are—at least in theory—being regulated and monitored by mayors and other local actors, and are only permitted to be carried out by nationals who are in possession of a *Gold Washer's Card.*

What differentiates Mali's system of *couloirs d'orpaillage* from most of the other ASM delineation strategies is that by law, the government is required to move artisanal gold miners to another designated area should the corridor in which they are working be parceled out as part of a large-scale concession. The Mining Code states that "In the event that the artisanal mining corridor is granted as exploration permit, exploitation permit, small mine or large mine, at the request of the community concerned, the administration in charge of mines will make available to the populations beforehand, in the limits of the available areas, another corridor [for] artisanal mining."[16]

The Orpaillage

Leaving the administration of the *Commune* to manage artisanal mining or *orpaillage* has helped to shield it from incoming commercial large-scale activity while preserving some of its traditional character. As indicated, security of tenure is crucially important for artisanal miners. The *couloirs d'orpaillage* strategy ensures that the burgeoning groups of gold washers found heavily concentrated in the south of Mali can access areas to work. This move has by no means been intentional, nor carried out with the welfare of these gold washers in mind. It has rather been the preoccupation of senior government officials with promoting and nurturing large-scale mining activity because of a fixation on extracting revenue from this sector that has led to the management and regulation of the *orpaillage* economy being consigned to the level of the *Commune,* and to mayors in particular. The impact of this move has been profound, galvanizing local administrative structures in Mali around artisanal gold mining.

It is likely for this reason that in Mali, ASM has been able to retain much of its traditional character (i.e., why it remains unmechanized). It begins with the *Gold Washer's Card,* which facilitates access to *couloirs d'orpaillage* and is the centerpiece of the quasi-formal ASM economy found in the country today. Technically, these cards are issued by the National Director for Geology and Mines to Malian nationals and entitle holders to exploit gold using nonmechanized methods, typically for one year (Keita 2001). In practice, however, it is administered at the *Commune*, specifically by the corresponding mayor.

The artisanal gold mining which has emerged in rural Mali is nestled within a village management structure that has remained largely intact for generations. A case can be made that by detaching themselves from *orpaillage* almost entirely, government officials in Bamako have inadvertently empowered the key actors found within this village management structure. Aside from mayors, these are, as already explained, the following (Keita 2001): (i) The land owner (*dugutigui*), who is the heir of the village land;

(ii) Site "owner" or central authority of the gold-washing area (*damantigui*); and (iii) The *tomboloma*, or "mine police," who are a group of young villagers tasked with peacekeeping, observing regulations, resolving disputes, and dialoguing with public administrators.

Ethnographic work carried out by Grätz (2007) in the commune of Mandé, captures in detail the power and influence these actors wield. For example, the *tomboloma* who police the area on behalf of the *damantigui* were observed to be more than just a vigilante group. They enjoy considerable autonomy, self-determination and in some cases, it is they who influence the *damantigui* and not the opposite. Moreover, mayors are often immersed in mining activities to the point where they frequent sites for the sole purpose of sponsoring operations and extracting taxes (Human Rights Watch 2011).

Mining Companies

Autonomous coexistence is a potential long-term strategy in Mali because mining companies are *forced* to be accommodating. To claim that mining company officials are "locked in" and committed to identifying and demarcating ASM corridors would be an exaggeration. But it is likely a small price to pay, given the potential backlash from local-level actors and the economic gains to be made through fostering a model of autonomous coexistence. Holders of a *permis d'exploitation* (the equivalent of a large-scale mining license) in Mali have traditionally only paid the standard 3 percent royalty on profits; are offered stabilization clauses which make it possible to freeze fiscal and/or legal provisions for the duration of a thirty-year lease; and have benefited from a three-year Value Added Tax exemption, introduced in the 2012 Mining Code, although the government removed this in 2019 (Schwartz 2019).[17] In Mali, annual investment in gold exploration climbed steadily between 2015 and 2020, from US$53.9 million to US$94.6 million.[18] Gold production has, unsurprisingly, increased over the same period, from 41 tons in 2015, to 42 tons in 2017, to 50 tons in 2021.[19]

Companies operating in Mali are now working with World Bank officials to identify and "block out" more ASM corridors. The Bank launched its US$35.7 million *Mali Governance of Mining Sector* project in 2019, pledging US$14 million of the budget to work in tandem with companies such as Barrick Gold and Hummingbird Resources to prospect "geological ground works in 20 areas designated as ASM corridors" based on geological mapping of high-potential localities (World Bank 2019, 20).[20] Progress thus far has been slow, but this could be attributed to the pandemic causing disruption to the World Bank's plans and, indeed, those of mining companies. Nevertheless, developments at Barrick Gold Mining's Loulo-Gounkoto mine in Kayes, and Hummingbird Resources' Komana mine in Yanfolila, are encouraging.

At the former, company officials have shown remarkable adaptability, moving on from contentious issues in Tanzania, headlined by legacy concerns at Bulyanhulu, and embracing the corridor strategy in Mali. At both sites, changes in ownership have had little impact on ongoing work aimed at "locating" ASM corridors on concessions. For Barrick, work on the Loulo Corridor was initiated by its predecessor, Randgold, and has continued undeterred since the merger of the two companies in December 2018. In the case of Hummingbird Resources, preliminary work on the Girin Corridor at the Komana Mine commenced during the exploration phase, when the property was under the control of Gold Fields. Hummingbird Resources acquired an 85 percent share of the mine in 2014.

CONCLUSION

Mali offers inspiration to any government or donor committed to ensuring that two very different segments of the mining sector can coexist. Officials at certain industry bodies and donors tend to get overexcited about situations where companies appear to be accommodating or ignoring outright the artisanal miners operating on their concessions. It is led to calls for cohabitation and the management of large-scale mining companies to formalize partnerships with artisanal mining groups. Proponents of this strategy, however, have oversimplified what are clearly far more complex environments.

The Mali case affirms that if the goal is to ensure that both artisanal and large-scale mining groups operate side-by-side relatively peacefully, a policy of autonomous coexistence is possible. All parties seem to be on board with ASM corridors as a strategy for facilitating different types of gold mining activities flourishing in the same landscapes. These corridors very importantly provide the security of tenure artisanal mine operators covet. While the Government of Mali appears content with permitting mayors and the administration at the *Commune* generally to monitor and regulate these operators' activities, mining companies have been engaged under the World Bank's *Mali Governance of Mining Sector Project* to ensure that efforts to identify more corridors continue.

The Mali setup is by no means perfect. Those operating in these corridors, for example, are not permitted to use machinery, which undoubtedly stifles production. Moreover, the government has been extremely slow in "approving" the corridors identified by mining companies, including those of Barrick Gold and Hummingbird Resources. Tensions between artisanal miners and company staff have escalated at times, principally because the former still feel that their production is constrained and too easily influenced by the latter. In May 2018, for example, at the Yanfolila mine, the police acted when it emerged that residents of the village of Dola were working one of Hummingbird's targets. This led to clashes, resulting in the deaths of three locals. Similarly, in July 2012, clashes between military officials and artisanal gold miners took place following the encroachment of the latter on to African Gold Mining Group Mali's Kobadani concession. Here, one person lost their life.

Despite these challenges, Mali has established the architecture for autonomous coexistence. As this has been accomplished in a challenging political environment, there is no reason why it cannot be replicated and "scaled up" in virtually any other setting in sub-Saharan Africa.

NOTES

1. For example, in Tanzania, investment in the mining sector rose from US$12 million in 1992 to US$184 million in 1997, and between 2005 and 2013, the government netted an average of US$137 million in revenue annually; in Mali, during the period 1998–2000, just four gold mining projects accounted for 87.5 percent of Foreign Direct Investment, generating an average of US$363 million annually for government; and in Ghana, where during the period 1983–1998 alone, US$4 billion of private capital was injected for mineral exploration, establishing new projects and rehabilitating existing operations, the majority of which were linked to gold (Aryee 2001; Boocock 2002; Chuhan-Pole et al. 2017).
2. Throughout this chapter, "small-scale mining" and "artisanal mining" are used interchangeably.
3. Claims data obtained from S&P Global Market Intelligence Database.
4. Key examples being the work of Amegbey et al. (1997) and Appiah (1998) in Ghana, and parallel analysis presented by Maponga (1993, 1994) on the organization of Zimbabwean operations.
5. Data obtained from S&P Global Market Intelligence Database.
6. "African Barrick is history, changes name to Acacia Mining," www.mining.com/african-barrick-is-history-changes-name-to-acacia-mining-31334/ (Accessed 4 December 2021).
7. Also commonly referred to as areas that are "blocked out" for ASM or (ASM) "corridors."
8. Articles 52 and 64 of the Mining Code 2014 laws established a framework for identifying zones in which only Ivorian nationals are permitted to mine (Sauerwein 2020).
9. "Latest Human Development Index Ranking," hdr.undp.org/en/content/latest-human-development-index-ranking (Accessed 5 December 2021).
10. "Mali's neighbours impose sanctions over election delay," www.ft.com/content/f4525017-eb6f-47ee-b05e-d381e1b05407 (Accessed 12 January 2022).
11. "Mali," https://eiti.org/mali#tax-and-legal-framework (Accessed 3 January 2022).
12. Mali's municipalities are composed of several villages and/or groupings or neighborhoods and are managed by a council which elects a mayor, who occupies

office for five years (Coulibaly and Hilhorst 2004). This setup emerged from an ambitious decentralization program that the country launched in the 1990s, under which thousands of villages were reorganized into 703 municipalities, each headed by a mayor (van Vliet 2013). These *communes* are the third layer of administration in Mali, behind *regions* and *cercles*.

13. Mali's second-level administrative unit.

14. See Ordonnance n°2019-022/P-RM du 27 septembre 2019 portant sur le code minier en République du Mali, titre ii : de la recherche et de l'exploitation des substances minérales soumises au régime des mines chapitre i: des dispositions communes aux titres miniers, sub-section 1 (de l'exploitation artisanale), Article 46. "Orpaillage" is a label used in francophone countries for "gold panning" and/or "gold washing."

15. "Mali gold mining revenue rises 13% in 2020 due to higher prices," www.reuters.com/article/mali-gold-revenue-idUSL5N2NN06E (Accessed 18 January 2022).

16. Translated by authors, from Article 46, Ordonnance N°2019-022/P-RM du 27 septembre 2019 portantsur le code minier en République du Mali

17. "Mali's new mining code ends tax exemptions, other p.rotections," www.reuters.com/article/mali-mining-idUSL5N25H5XY (Accessed 4 January 2022).

18. Data extracted from S&P Market Intelligence Database, 2021.

19. "Production volume of gold in Mali from 2005 to 2021," www.statista.com/statistics/1167250/production-volume-of-gold-in-mali/ (Accessed 14 January 2022).

20. Namely Kayes/Bakel, Kankossa, Yelimane, Sandare, Bafoulabe, Bafing-Makana, Nioro, Diema, Kita, Sirakoro, Kolokani, Banamba, Dioila, Koutiala, and Sikasso.

REFERENCES

Amegbey, Newton A., J.B.K. Dankwa, and S. Al-Hassan. 1997. "Small scale mining in Ghana - Techniques and environmental considerations." *International Journal of Surface Mining, Reclamation and Environment* 11 (3): 135–138.

Appiah, H. 1998. "Organization of small scale mining activities in Ghana." *The Journal of the South African Institute of Mining and Metallurgy* 98 (7): 307–310.

Aryee, Benjamin N. A. 2001. "Ghana's mining sector: Its contribution to the national economy." *Resources Policy* 27 (2): 61–75.

Aubynn, Anthony. 2009. "Sustainable solution or a marriage of inconvenience? The coexistence of large-scale mining and artisanal and small-scale mining on the Abosso Goldfields concession in Western Ghana." *Resources Policy* 34 (1–2): 64–70.

Boocock, Colin Noy. 2002. "Environmental impacts of foreign direct investment in the mining sector in Sub-Saharan Africa." Paper Presented at the OECD Global Forum on International Investment Conference on Foreign Direct Investment and the Environment Lessons to be Learned from the Mining Sector, February 7–8, 2002. Paris: OECD.

Chuhan-Pole, Punam, Andrew L. Dabalen, and Bryan Christopher Land. 2017. *Mining in Africa: Are Local Communities Better Off?* Washington, DC: The World Bank.

Cooksey, Brian, and Tim Kelsall. 2011. "The political economy of the investment climate in Tanzania." Research Report 1. London: UK Aid.

Corbett, Tony, Ciaran O'Faircheallaigh, and Anthony Regan. 2017. "'Designated areas' and the regulation of artisanal and small-scale mining." *Land Use Policy* 68: 393–401.

Côte, Muriel, and Benedikt Korf. 2018. "Making concessions: Extractive enclaves, entangled capitalism and regulative pluralism at the gold mining frontier in Burkina Faso." *World Development* 101: 466–476.

Coulibaly, Amadi, and Dorothea Hilhorst. 2004. "Implementing decentralisation in Mali: The experiences of two rural municipalities in southern Mali." Issue Paper No. 127. London: International Institute for Environment and Development (IIED).

Ferguson, James. 2005. "Seeing like an oil company: Space, security, and global capital in Neoliberal Africa." *American Anthropologist* 107 (3): 377–382.

Gajigo, Ousman, Emelly Mutambatsere, and Guirane Ndiaye. 2012. "Gold mining in Africa: Maximizing economic returns for countries." Working Paper No. 147. Abidjan: African Development Bank Group.

Geenen, Sara. 2012. "A dangerous bet: The challenges of formalizing artisanal mining in the Democratic Republic of Congo." *Resources Policy* 37 (3): 322–330.

Government of Ghana. 2009. *Designated Areas for Small Scale Gold & Diamond Mining in Ghana*. Accra: Minerals Commission.

Government of Tanzania. 2008. "Report of the Presidential Mining Review Committee to advise the government on oversight of the mining sector." Volume 2. Dar es Salaam: Government of Tanzania.

Grätz, Tilo. 2007. "Vigilante groups and the state in West Africa." In *Order and Disorder: Anthropological Perspectives*, edited by Keebet von Benda-Beckmann and Fernanda Pirie, 74–89. New York: Berghahn Books.

Hentschel, Thomas, Felix Hruschka, and Michael Priester. 2002. "Global report on artisanal and small-scale mining." London: Minerals Mining and Sustainable Development (MMSD) and International Institute for Environmental Development.

Hilson, Abigail, Gavin Hilson, and Suleman Dauda. 2019. "Corporate social responsibility at African mines: Linking the past to the present." *Journal of Environmental Management* 241: 340–352.

Hilson, Gavin. 2019. "Why is there a large-scale mining 'bias' in sub-Saharan Africa?" *Land Use Policy* 81: 852–861.

Hilson, Gavin, Titus Sauerwein, and John Owen. 2020. "Large and artisanal scale mine development: The case for autonomous co-existence." *World Development* 130: 104919.

Human Rights Watch. 2011. *A Poisonous Mix Child Labor, Mercury, and Artisanal Gold Mining in Mali*. New York: Human Rights Watch.

Intergovernmental Forum on Mining, Minerals, Metals and Sustainable Development (IGF). 2017. *Global Trends in Artisanal and Small-Scale Mining (ASM): A Review of Key Numbers and Issues*. Winnipeg: IISD.

International Finance Corporation (IFC). 2011. *Working Together: How Large-Scale Mining Can Engage with Artisanal and Small-Scale Miners*. www.icmm.com: The International Finance Corporation.

Keita, Seydou. 2001. "Study on artisanal and small-scale mining in Mali." Report No. 80. London: Mining, Minerals and Sustainable Development (MMSD) Project and International Institute for Environment and Development (IIED).

Lange, Siri. 2008. "Land tenure and mining in Tanzania." CMI Report, Chr. Michelsen Institute, Bergen.

Luning, Sabine. 2014. "The future of artisanal miners from a large-scale perspective: From valued pathfinders to disposable illegals?" *Futures* 62 (Part A): 67–74.

Maponga, Oliver. 1993. *Small-Scale Mining Operations in Zimbabwe.* Ottawa: International Development Research Centre.

Maponga, Oliver. 1994. "Small scale mining and the environment in Zimbabwe: The case of alluvial gold panning and chromite mining." In *Mining on a Small Scale*, edited by Ajoy K. Ghose, 185–211. The Netherlands: A.A. Balkema.

Martin, Alan, and Hélène Helbig de Balzac. 2017. *The West African El Dorado: Mapping the Illicit Trade of Gold in Côte d'Ivoire, Mali and Burkina Faso*. Ottawa: Partnership Africa Canada.

Raineri, Luca. 2020. "Gold mining in the Sahara-Sahel: The political geography of state-making and unmaking." *The International Spectator* 55 (4): 100–117.

Sauerwein, Titus. 2020. "Gold mining and development in Côte d'Ivoire: Trajectories, opportunities and oversights." *Land Use Policy* 91, Art 104323.

Schwartz, Brendan. 2018. "Revising Mali's mining code: Three key areas for improvement." London: International Institute for Environment and Development (IIED).

S&P Global Capital IQ Pro. 2022. New York: Thomson Reuters.

Teschner, Benjamin. 2013. "How you start matters: A comparison of Gold Fields' Tarkwa and Damang Mines and their divergent relationships with local small-scale miners in Ghana." *Resources Policy* 38 (3): 332–340.

Van Bockstael, Steven. 2019. "Land grabbing 'from below'? Illicit artisanal gold mining and access to land in post-conflict Côte d'Ivoire." *Land Use Policy* 81: 904–914.

van Vliet, Martin. 2013. "Family matters: The interplay between formal and informal incentives for accountability in Mali's local communities." In *Accountable Government in Africa: Perspectives from Public Law and Political Studies*, edited by Danwood Chirwa and Lia Nijzink, 235–254. Tokyo: United Nations University Press.

World Bank. 2009. *Mining Together: Large-Scale Mining Meets Artisanal Mining*. Washington, DC: The World Bank.

World Bank. 2015. "Tanzania sustainable management of mineral resources project." Project Paper. Washington, DC: The World Bank.

World Bank. 2019. "World Bank Mali governance of mining sector project." Project Appraisal Report. Washington, DC: The World Bank.

World Bank. 2021. "Land Area (sq. km)." Accessed December 1, 2021. https://data.worldbank.org/indicator/AG.LND.TOTL.K2.

10

Negotiating Inclusion and Exclusion in Artisanal Oil Extraction

The Case of Two Villages in East Java, Indonesia

Nanang Indra Kurniawan, Päivi Lujala, and Ståle Angen Rye

In recent decades, artisanal and small-scale mining (ASM) has grown to involve 15–30 million people, produce 15–20 percent of the total global mineral output (Buxton 2013; Verbrugge 2015; World Bank 2012) and contribute substantially to local economies and peoples' livelihoods (Banchirigah 2008; EARF 2018; Ghose and Roy 2007). In many countries, governments seek to regulate ASM activities (Hilson and Maconachie 2017). To validate the attempts to regulate and formalize ASM, governments often portray artisanal activities as destructive and even illegal, producing narratives wherein artisanal miners are people who deliberately pollute the environment and evade the law (Banchirigah 2008). Concerns about the safety and health effects of artisanal practices are also used to justify state intervention. Although governments often justify their regulatory activities on the grounds that they are acting in local communities' best interests, such processes often lack an understanding of the complex socio-political realities of local artisanal production and are thus often met with resistance by local communities (e.g., Cartier and Bürge 2011; Fisher 2007; Salo et al. 2016; Echavarria 2014; Matti 2010).

This chapter discusses how artisanal mining communities can (re)negotiate or, alternatively, resist attempts to include and exclude them from formal resource production and economy. The aim is to show how such processes of inclusion and exclusion are related to shifting boundaries between legality and illegality (Hilson and Maconachie 2017). Furthermore, this chapter shows how existing and new power relations at the extraction site and beyond form the negotiations that (re)establish notions of informality. Following Peluso (2018), who views small-scale extraction sites as entangled territories with their own history and complex arrangement, we also aim to highlight the dynamic process of how the state's attempts for inclusion and exclusion in artisanal mining are embedded in particular local conditions and (historical) processes at particular times. As Hook (2019) suggests, understanding exclusion and inclusion in extraction requires us to capture the complex and multidimensional interaction between social and economic dimensions.

We develop our discussion by drawing on the history of the state's efforts to regulate and formalize artisanal and small-scale oil production in Wonocolo and Hargomulyo, two villages in Bojonegoro Regency, East Java Province, Indonesia, since the late nineteenth century. We outline the historical case of these villages based on secondary sources (e.g., academic texts, institutional reports, and documents). For the analysis of the more recent situation, we collected materials through observation, focus group discussions, and interviews during five field visits to Wonocolo and Hargomulyo villages and Bojonegoro city between 2017 and 2019 (July and November 2017, June and September 2018, and April 2019). We conducted focus group discussions with local government officials and government-owned companies, while our interviews involved twelve artisanal miners. During each visit, we had informal discussions with oil miners and refiners in Wonocolo and Hargomulyo villages, and we observed artisanal oil mining and refining activities.

The chapter is structured as follows: First, we outline the historical development of artisanal mining and oil culture in the two villages. In the next section, we present and discuss how the central and local governments have sought to formalize and regulate artisanal oil mining in the two villages since the 1970s. The chapter ends with an analytical discussion on the national and local governments' alternating attempts to include and exclude artisanal oil miners from the production area and the local oil economy as they have sought to gain control over oil production.

A BRIEF HISTORY OF ARTISANAL OIL MINING AND OIL CULTURE IN WONOCOLO AND HARGOMULYO

The villages of Wonocolo and Hargomulyo, with a total of approximately 3,000 inhabitants, are located in a remote highland area northwest of Bojonegoro city, the capital of Bojonegoro District (see figure 10.1). These villages are close to the Cepu Oil Block, which straddles three regencies in two provinces: Bojonegoro and Tuban in East Java and Blora in Central Java.[1] The Wonocolo oil field is located under Wonocolo and Hargomulyo villages

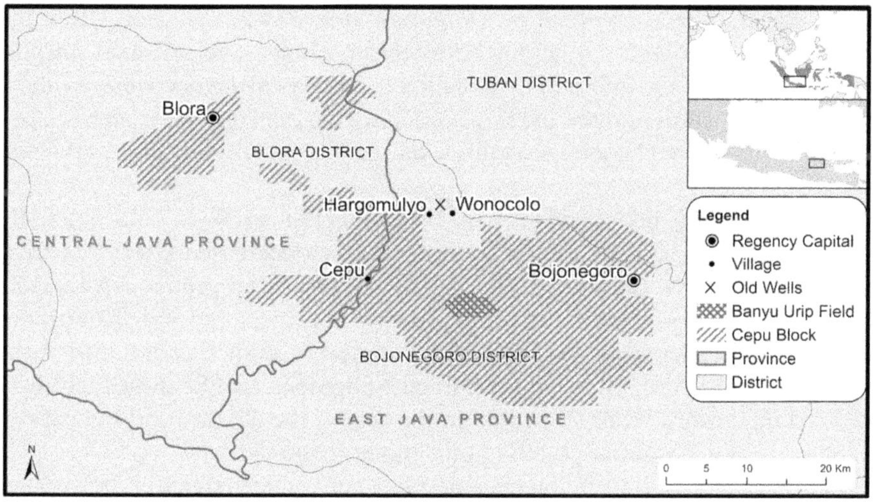

Figure 10.1 The Study Area. Map created by Päivi Lujala.

and is seen as a part of the larger Cepu Block, which is one of the largest sources of oil production in Indonesia. Although PT Pertamina EP Asset IV, a subsidiary of the state-owned oil company Pertamina, controls the Wonocolo field, the villagers have used oil wells dating from the Dutch colonial era for artisanal oil production since the 1940s.

Oil production in the two villages goes back to the Dutch colonial period when the Dutch colonial government granted permission to Adrian Stoop, an engineer who had found oil in the area, to exploit the reserves at an industrial scale. Stoop's company *Dordsche Petroleum Maatschappij* (DPM) drilled oil in the Wonocolo field for the first time in 1894, and the company controlled the field until 1925, when it was taken over by *Bataafsche Petroleum Maatschappij* (BPM). Between 1894 and 1940, 225 wells were drilled in the area (*Radar Bojonegoro* 2018) with well depths ranging between 50 and 784 meters. It is the history of these 225 "old wells," and the control over them, that we will discuss in the remainder of this chapter.

In the colonial era, agriculture was the villagers' main source of livelihood, and they were only involved in formal oil production to a limited degree, being employed as laborers. However, the villagers collected oil that leaked from the pipelines that crossed the villages, and they also illegally tapped them. Villagers not only used this oil for domestic purposes, such as lighting and cooking but also refined it into kerosene and sold it to limestone powder industries and residents of surrounding areas. These activities, the beginning of the small-scale oil industry in the two villages that has continued until today, were opposed by BPM. To curb the stealing of oil, the company paid the village chiefs a considerable amount of money annually on the condition that villagers would not steal oil (*Radar Bojonegoro* 2018). Thus, from the beginning, the complex relations between large-scale and artisanal oil production and among international, national, and local actors were evident. These relationships and the local oil economy have mixed with, been shaped by, and shaped local social structures and culture over generations (PolGov 2018).

After an initial period of growth, oil production in Wonocolo gradually declined in the 1930s as the production volumes decreased after the extensive oil lifting that occurred in the 1920s. BPM ceased production in 1938 and abandoned its oil wells, covering them with cement and soil. After Indonesia's independence in 1945, villagers started to reopen abandoned wells and began extracting oil using traditional equipment. Unlike the Dutch, who utilized the newest technology within the sector, the artisanal oil industry in Wonocolo and Hargomulyo relied on simpler methods.

Oil lifting and refining soon became a key economic activity in the villages and was organized based on existing social structures and power networks. This growing artisanal oil industry, however, had limited legal standing, as after independence, colonial land had been surrendered to the central government, which was legally entitled to all natural resources below the surface. This implied that local miners were operating on land and utilizing resources that were not legally theirs but the central government's. Nevertheless, in 1950, the regent of Bojonegoro issued an oral decree that allowed villagers to produce oil (*Radar Bojonegoro* 2018). As demand for kerosene also grew in the 1950s, traditional oil production in these two villages expanded significantly.

Crude oil was extracted from wells using pulleys and ropes, and oil lifting relied on manual labor until the 1990s. Extraction from one well required eight to twelve people, depending on the depth of the well. Each well produced an average of ten drums (the volume of one drum is 210 liters) of crude oil per day (Brata et al. 2018, 753). Since mechanization began in the early 1990s, with old trucks being used for extraction, each well has produced between forty and sixty drums a day (Brata et al. 2018, 753).

In addition to extracting oil for sale to contractors, the villagers in Wonocolo and Hargomulyo have refined oil and distilled it into kerosene and low-quality diesel. After the crude oil is extracted, it is manually separated from water and moved into drums that are boiled in a furnace (called a *pawonan*, literally "kitchen"). The resulting steam is cooled down and reliquefied by channeling it through pipes into a small water pool. The petrol that emerges from the cooling pipes at the beginning is called *bensin* by the workers and is the highest quality gasoline that can be produced by a *pawonan*; the low-quality petrol is categorized as kerosene and diesel fuel. In 2015, there were approximately 500 furnace chambers refining oil in Wonocolo and Hargomulyo (*Solopos* 2015).

In recent years, distilled oil has been distributed by *peréngkék* (sellers) using modified motorcycles that can carry up to 245 liters of gasoline in seven jerrycans (see figure 10.2). In 2015, there were approximately 225 transporters coming from not only Wonocolo and Hargomulyo but also neighboring villages and regencies (*BeritaBojonegoro.com* 2015). *Peréngkék* sells gasoline in districts and cities in East Java and Central Java as fuel for trucks, traditional boats, and small- and medium-size enterprises.

Over the years, the extraction, refinement, and distribution of oil from the old Dutch wells in these two villages have developed into a distinct and intricate system that involves a large part of the population. Moreover, it has become a major source of income. As argued by Peluso (2018), small-scale mining territories such as Wonocolo and Hargomulyo need to be understood as places governed without the presence of the formal state but rather through the "shadow state" with its different representations. According to Peluso (2018), control of an extraction site should be understood as a result of complex relations between the materiality of the resources and the social relations and institutions that have developed over time. The same kind of complex relations between people, resources, and society have been identified in other places (see, e.g., Hook 2019; Hilson et al. 2017; Salman 2016;

Figure 10.2 Artisanal Oil Mining and Refining Activities in Wonocolo and Hargomulyo. Photo by the authors.

Verbrugge 2015). Corresponding to these findings, the mode of production that we have described above for the case of Wonocolo and Hargomulyo is highly embedded in the two villages' social and economic lives, and oil and oil-related activities have become part of villagers' cultural identity and way of life. The core of this development was captured by an oil miner in Wonocolo, who said the following:

> Wonocolo has long been known as an oil village. The Dutch came here because this place is rich in oil. Our grandfathers taught us the skills to produce oil. People in Wonocolo have worked with oil for generations. I can say that oil is our life, and without oil, we would not be people of Wonocolo.

Importantly, villagers not only consider oil to be an essential part of their history but also believe that access to oil will determine their future. As another worker noted, "it will be very hard for us to live without oil."

As we will see in the next section, this reliance on oil for their livelihoods and way of life has made villagers suspicious of any government interventions and attempts to regulate oil-related activities. According to the villagers, the intervention and formalization attempts are an effort to regulate production and other oil-related activities, and these measures effectively double as attempts to regulate their way of life. To further explore this issue, in the next section, we highlight how the state has attempted to regulate and formalize the artisanal oil industry in Wonocolo and Hargomulyo, in particular, the 225 "old wells," during the last fifty years and how artisanal miners have met such formalization attempts with negotiation and resistance.

CURTAILING ARTISANAL OIL MINING THROUGH EXCLUSION AND INCLUSION

The national government, through the Ministry of Mining, first attempted to gain control of artisanal mining and remove workers from the old wells in 1977. The Ministry claimed that such artisanal activities were illegal, as they did not meet safety standards and harmed the environment—particularly the teak forests (PPT MIGAS 1994, 132). Extraction activities, however, continued. A second attempt was made in 1985 when the Ministry of Domestic Affairs requested the Governor of East Java to halt artisanal mining to protect young teak trees in the surrounding areas (PPT MIGAS 1994, 135). This time, the government argued that artisanal oil extraction was illegal because it occurred in an area controlled by Perhutani, the state-owned forestry company. This attempt also failed, as it faced resistance from communities that had relied on oil mining and refining for their livelihoods for decades. In addition, local political elites and military officials—who profited from their involvement in artisanal mining—backed the miners, making it also politically difficult to stop production (Pojok Pitu 2016).

As neither of these attempts worked, the government changed its strategy. Starting in the mid-1980s, it began to conduct negotiations, holding a series of meetings with local people in Wonocolo and Hargomulyo. The state sought to include artisanal oil extraction in the formal oil production chain and thereby incorporate it firmly into the legal economy. Following meetings between the government and local residents, a contract between the Bogasasono Village Cooperative (KUD Bogasasono)[2] and Pertamina was signed in 1988. Pursuant to this contract, KUD Bogasasono collected the oil extracted by artisans in Wonocolo and Hargomulyo and delivered it to Pertamina (PPT MIGAS 1994). KUD Bogasasono received compensation for this service, which was distributed to the oil miners. Miners chose to join KUD Bogasasono because it gave them legal protection for their artisanal activities, including a legal channel to sell the oil they produced. Notably, however, although the contract between the village cooperative and Pertamina legalized artisanal oil extraction, the villages' traditional refinement activities remained illegal because these refineries were unlicensed. Moreover, the government forbade the direct transport and sale of gasoline to consumers in 1977 (PPT MIGAS 1994); as such, all oil produced by artisanal miners in Hargomulyo and Wonocolo was supposed to be delivered to Pertamina.

The 1988 contract obligated the oil miners to sell the crude oil to the cooperative at fixed prices and prohibited them from conducting refinement activities. Miners perceived this contract as transforming them from oil producers into contract laborers (Brata 2018). Although miners used their own capital and equipment, Pertamina treated them as extraction and transportation workers; this is exemplified in the company's use of the term "wage" to refer to payments made to oil miners through KUD Bogasasono (Brata 2018). Furthermore, owing to an article in the 1988 contract between Pertamina and KUD Bogasasono that specified that oil miners operated on Pertamina's property, the artisanal miners had a weak bargaining position. This implied that, as in the colonial era, oil resources were controlled by external actors, with the villagers being nothing but laborers.

As mentioned above, the government sought to regulate artisanal oil extraction in Wonocolo by arguing that traditional practices did not meet environmental protection and safety standards (PPT MIGAS 1994). However, the 1988 contract with KUD Bogasasono did not address these issues, showing that the state did not take the negative health and environmental effects of traditional extraction practices seriously. The contract was more about strengthening government control over oil resources, curtailing artisanal miners' refinement of crude oil, and ensuring that the oil produced by the artisanal oil miners was delivered to Pertamina. As a result, the inclusion of artisanal miners in the formal oil economy did not ease tensions, and, as will be seen below, the inclusion of local oil miners in the formal oil production chain was followed by other forms of exclusion and repression.

Over the years after the 1988 contract was signed, tensions between miners and KUD Bogasasono increased. Miners were unhappy with the situation, suspecting that the managers were buying crude oil at a low price and selling it to Pertamina at a high price (i.e., profiteering). The miners saw KUD Bogasasono as an extension of Pertamina and it being under the company's control. To improve their situation and to increase their incomes, they thus restarted artisanal refinery activities, selling the products directly to customers. However, these efforts were strenuous, as Pertamina worked continuously with the police and military to stop illegal oil refinement and trading activities.

Tensions peaked in 2006, when unhappy miners mobilized a demonstration in front of Pertamina's Cepu office, demanding that the company revise its 1988 contract and change the article that referred to them as workers of Pertamina. Brata (2018) describes, based on an ethnographic study of the miners' community, how the ensuing negotiations first occurred as a direct talk between miners and Pertamina at the Pertamina Cepu office. The negotiations were tense and ended in a deadlock. Then, the negotiations continued with the Bojonegoro and Blora regency police chiefs and military commanders functioning as mediators between the miners and Pertamina (Brata 2018, 199). Negotiations were still unsuccessful, and ultimately, military and police officials stated that they would "turn a blind eye" to oil refining and selling activities. This meant that they gave local miners informal permission to continue their illegal (from the state's perspective) activities. In practice, this meant that the state—through Pertamina and KUD Bogasasono—no longer had a monopoly over the oil from Wonocolo and Hargomulyo. Artisanal oil miners could now more easily improve their livelihoods by refining oil and selling the products on the black market as the police and military informally supported their efforts (Brata 2018, 200).

As a result of this "blind eye" agreement, KUD Bogasasono's role in oil mining diminished. Nonetheless, Pertamina continued to cooperate with the cooperative, signing new three-year contracts in 2006 and 2009. Between 2012 and 2015, rather than continuing to cooperate with KUD Bogasasono, Pertamina worked with two other village cooperatives (KUD Sumber Pangan and KUD Usaha Jaya Bersama) to collect and transport oil from the Wonocolo field. The cooperation between Pertamina and KUD Sumber Pangan and KUD Usaha Jaya Bersama was not successful. As the cooperatives were directly involved in the illegal drilling of new oil wells in the Wonocolo field, Pertamina decided to terminate its contracts with the two

KUDs as well (*Bisnis.com* 2015). Following the termination of the contracts, new and stronger efforts to exclude local miners from oil production have emerged.

In the 2010s, it became increasingly difficult for the military and police to continue to turn a "blind eye" to illegal activities in the Wonocolo field (Brata 2018). The mining area was assigned status as a National Vital Object, and President Joko Widodo demanded state institutions and state-owned companies to eradicate illegal activities in the oil and gas sector. Furthermore, Pertamina estimated that between 2013 and 2014, approximately 300 to 500 barrels of oil per day (bopd) were not handed over to the company, causing the national government to have an annual loss of approximately US$ 11 million (Boediwardhana 2015). In this context, it became possible for Pertamina to seek the support of the local military and police to help combat unlawful drilling, refining, and selling (*Geotimes* 2015). Consequently, in March 2015, the Commander of Indonesian Armed Forces, General Moeldoko, visited Wonocolo and Hargomulyo as part of Pertamina's and the military's plan to stop illegal oil activities in the area (Sindo 2015).

In July 2015, together with Pertamina, the military, police, and the Public Order Agency[3] established a task force to combat illegal oil production, drilling, and gasoline transportation. This task force, the *Satgas Pengamanan Sumur Tua* [Old Well Security Task Force], spent six months monitoring the area around Wonocolo and Hargomulyo. For this, it used checkpoints established by Perhutani, a state forestry company, as its bases of operation. In 2015, a total of 50 military, 27 police, 8 Pertamina Security, and 4 Public Order Agency personnel were deployed at these checkpoints (*BeritaBojonegoro.com* 2015).

Artisanal miners and local residents resisted these efforts to combat illegal oil mining. They were backed by "local strongmen," who were part of the local elite and wielded power outside formal political and government institutions. For example, the locals constructed a roadblock near the main entry to the oil extraction and refinery area, thereby preventing vehicles from entering the site. Responding to this open resistance, Pertamina sought to build new alliances. It started cooperating with two local artisanal oil mining associations, *Paguyuban Wonocolo* and *Paguyuban Wonomulyo*, which were led by local strongmen who had long been involved in artisanal oil mining and controlled the informal oil business (*Portal Indonesia News* 2015). The associations were established in 2015, after the termination of the KUD contracts, and differed from the earlier cooperatives in that they were legally recognized as social organizations rather than business entities. Oil miners accepted working together with Pertamina through these associations, which provided them with a legal basis to continue extracting and selling oil to Pertamina (thereby mitigating the problems caused by the military and task force). This scheme did, however, end in 2016, when Pertamina decided to sign a new three-year contract with KUD Sumber Pangan.

In recent years, village cooperatives have faced staunch competition from the Bojonegoro district government. The district government has used the continuous tension between Pertamina and artisanal miners as an opportunity to become directly involved in oil production in Wonocolo and Hargomulyo. As its legal foundation, it has used Energy and Mineral Resources Ministry Decree No. 1/2008 on old oil wells, which states that contractors (such as Pertamina) may establish partnerships not only with village-level cooperatives but also local government-owned enterprises. In 2016, PT Bojonegoro Bangun Sarana (BBS)—a local government-owned company[4]—sought to partner with Pertamina to operate the old wells and to deliver crude oil to Pertamina. After its application was approved by the Ministry of Energy and Mineral Resources in August 2017, the district government was allowed to compete with village cooperatives to serve as the entity delivering oil from the old wells to Pertamina.

To strengthen its position, BBS began actively garnering support among artisanal oil miners and other villagers (Nugroho 2019). One of its strategies was to appoint an influential young person from Hargomulyo as its middleman. BBS also portrayed itself as a better partner than village cooperatives for community and artisanal miners. It emphasized issues such as worker safety, environmentally friendly oil extraction, modern equipment, and worker insurance. As of 2020, BBS was, together with the other actors discussed above, a central player in the contest over oil resources in Wonocolo and Hargomulyo.

CONCLUSION

This chapter has shown how artisanal oil mining has developed for eighty years in Wonocolo and Hargomulyo, from the local take-over of Dutch-drilled wells in the mid-1940s to the current situation where state-owned and local government-owned companies have sought to control artisanal oil production in these villages. Oil extraction in the villages has been dominated by artisanal oil mining and refining, and over generations, oil has become villagers' main source of income and an integral part of their everyday life. Accordingly, villagers perceive external interventions and state regulation (most recently by BBS) not only as an economic threat but also as a threat to their way of life and their social, cultural, economic, and political roots. The regulation of oil-related activities is thus about more than regulating the economic sector and formalizing artisanal extraction. As also highlighted by Peluso (2018), Hook (2019), and Verbrugge (2015), understating small-scale and artisanal mining thus requires an integrated approach that takes into account the complex relations between production and social structures that have their own history in particular places.

Against this complex historical, social, and economic background, we have shown how the state's alternating attempts to include and exclude artisanal miners in the two villages have been shaped by negotiations and resistance. Informality in the local artisanal oil mining sector has at times enabled the "shadow state" to gain influence and indirect territorial control, and the miners have resisted state control through shifting alliances with various local power holders (see also Peluso 2018). At the same time, the state has made efforts to gain direct territorial control through its attempts to formalize artisanal oil production. These attempts for state regulation have been justified by the need to include community members in the formal national economy to secure decent jobs, promote economic development, and preserve the environment. However, the villagers involved in artisanal mining, refining, and selling have questioned this narrative, conceiving it as creating insecurity and loss of control by excluding them from their own labor and severing their access to local resources and their roots. As a consequence, miners and their communities have come to see government regulation and intervention efforts as a means of gradually taking over local oil resources and as a threat to their way of life.

Overall, the case of Wonocolo and Hargomulyo shows how inclusion and exclusion within small-scale mining are much more complex than an issue of a formal versus an informal economy, and it is about much more than state versus community interests (see also Hilson et al. 2017). Accordingly, this chapter contributes to the literature that seeks to understand the persistence of informality and the diverse causes and consequences of formalization attempts for the different stakeholders involved in ASM (Verbrugge 2015). In this way, we have addressed Verbrugge's (2015, 1024) call for a "more comprehensive understanding of the diverse causes and consequences of informality for the different (categories of) actors involved in the [ASM] sector." This is particularly relevant, he claims, when mining has over time developed into more complex production structures, such as in the case of Wonocolo and Hargomulyo.

For the two villages, we can see the contours of what Peluso (2018) calls the "shadow state" where territorial control is intertwined and where key actors such as the army and police are dynamic forces taking different positions toward miners at different times. In the Wonocolo and Hargomulyo case, it is evident that the competition over local resources and their extraction is much more complex than a simple question of insiders versus outsiders. We have also seen how struggles over inclusion and exclusion are driven by shifting alliances between various local actors and external actors.

As a result, the inclusion and exclusion of artisanal oil miners in these two villages must be seen in relation to the blurred boundaries between formality and informality, legality and illegality, and insiders and outsiders. These practices are not fixed but are created through power relations and shaped by negotiations and resistance with local artisanal miners, refiners, and sellers. As a result, inclusion and exclusion in the artisanal oil industry should be seen as never-ending processes and as always being in the making.

ACKNOWLEDGMENTS

We are grateful to the editors of this book and two anonymous reviewers for providing us with thoughtful comments on our chapter. We thank Primi Suharmadi Putri, Muhammad Djindan, and Indah Suryawardhani for their insightful thoughts and discussions. We especially thank Udin, who assisted us in the data collection processes in Bojonegoro. This work was supported by the Research Council of Norway under Grant 275387 and the Academy of Finland under Grants 309206 and 314143.

NOTES

1. Cepu Block is named after a small city, Cepu, in Blora District, which has been the center of oil production in the area since the end of nineteenth century.
2. The government initiated the establishment of KUD Bogasasono as part of its strategy to take over control of old oil wells in Wonocolo (*Tempo* 1988).
3. The Public Order Agency is a municipal police unit that assists the local government in enforcing its regulations.
4. BBS was established in 2010 to provide services for oil-related business activities in Bojonegoro following the development of the Banyu Urip Field (operated by Exxon Mobil Cepu Limited) and the Sukowati Field (operated by Petrochina).

REFERENCES

Banchirigah, Sadia Mohammed. 2008. "Challenges with eradicating illegal mining in Ghana: A perspective from the grassroots." *Resources Policy* 33 (1): 29–38.

BeritaBojonegoro.com. 2015. "Penertiban Perengkek, Menyelamatkan 45 Ribu Liter Minyak Yang Hilang." December 16, 2015. https://beritabojonegoro.com/read/2397-penertiban-perengkek-menyelamatkan-45-ribu-liter-minyak-yang-hilang.html.

Bisnis.com. 2015. "Pertamina EP Depak 2 KUD Pelanggar Kontrak Di Blok Cepu." May 20, 2015. https://ekonomi.bisnis.com/read/20150520/44/435390/pertamina-ep-depak-2-kud-pelanggar-kontrak-di-blok-cepu.

Boediwardhana, Wahyoe. 2015. "Illegal oil wells found in Cepu." *The Jakarta Post*, April 16, 2015. https://www.thejakartapost.com/news/2015/04/16/illegal-oil-wells-found-cepu.html.

Brata, Nugroho Trisnu. 2018. "Social conflict and military roles in public oil mining in Bojonegoro." *MIMBAR: Jurnal Sosial dan Pembangunan* 34 (1): 195–203.

Buxton, Abbi. 2013. *Responding to the Challenge of Artisanal and Small-Scale Mining. How Can Knowledge Networks Help?* London: IIED. https://pubs.iied.org/16532IIED/

Cartier, Laurent E., and Michael Bürge, M. 2011. "Agriculture and artisanal gold mining in Sierra Leone: Alternatives or complements?" *Journal of International Development* 23 (8): 1080–1099.

EARF. 2018. *Understanding the Economic Contribution of Small-Scale Mining in East Africa.* East Africa Research Fund (EARF). https://assets.publishing.service.gov.uk/media/5a3929b640f0b649cfaf86ce/Pact__DFID_EARF_Overarching_Synthesis_Jan2018VF.pdf

Echavarria, Cristina. 2014. *'What Is Legal': Formalising Artisanal and Small-Scale Mining in Colombia.* London: IIED. https://pubs.iied.org/16565IIED/

Fisher, Eleanor. 2007. "Occupying the margins: Labour integration and social exclusion in artisanal mining in Tanzania." *Development and Change* 38 (4): 735–760.

Geotimes. 2015. "Sumur Minyak Di Bojonegoro Dijaga Ketat 6 Bulan." July 8, 2015. https://geotimes.id/arsip/sumur-minyak-di-bojonegoro-akan-dijaga-ketat-6-bulan/.

Ghose, Mrinal K., and Surendra Roy. 2007. "Contribution of small-scale mining to employment, development and sustainability–An Indian scenario." *Environment, Development and Sustainability* 9 (3): 283–303.

Hilson, Gavin, and Roy Maconachie. 2017. "Formalising artisanal and small-scale mining: Insights, contestations and clarifications." *Area* 49 (4): 443–451.

Hilson, Gavin, Abigail Hilson, Roy Maconachie, James McQuilken, and Halima Goumandakoye. 2017. "Artisanal and small-scale mining (ASM) in sub-Saharan Africa: Re-conceptualizing formalization and 'illegal' activity." *Geoforum* 83: 80–90.

Hook, Andrew. 2019. "The multidimensionality of exclusion in the small-scale gold mining sector in Guyana: Institutional reform, landlordism, and mineral uncertainty." *World Development* 123: 104607.

Matti, Stephanie. 2010. "Resource and rent seeking in the Democratic Republic of the Congo." *Third World Quarterly* 31 (3): 401–413.

Nugroho. 2019. "100 Penambang Sumur Minyak Tua Bergabung BUMD Bojonegoro | SuaraBanyuurip.Com." *Suarabanyuurip.com*. May 26, 2019. https://www.suarabanyuurip.com/amp/migas/read/188695/100-penambang-sumur-minyak-tua-bergabung-bumd-bojonegoro.

Portal Indonesia News. 2015. "Penambang Tradisional Wonocolo : Lebih Nyaman Dengan Paguyuban." September 14, 2015. https://portalindonesianews.com/posts/view/2198/penambang_tradisional_wonocolo_lebih_nyaman_dengan_paguyuban.

Peluso, Nancy Lee. 2018. "Entangled territories in small-scale gold mining frontiers: Labor practices, property, and secrets in Indonesian gold country." *World Development* 101: 400–416.

PPT MIGAS. 1994. *100 Tahun Perminyakan di Cepu [100 years of oil mining in Cepu]*. Cepu: PPT MIGAS.

Radar Bojonegoro. 2018. "Dibor 1894, setahun setelah Cepu [Drilled in 1894, one year after Cepu]." https://radarbojonegoro.jawapos.com/read/2018/03/17/57827/dibor-1894-dilakukan-setahun-setelah-cepu.

Research Center for Politics and Government (PolGov). 2018. "Artisanal oil mining governance in Wonocolo, district of Bojonegoro." PolGov Policy Brief 1421/PB-PolGov/VIII/2018. Yogyakarta, Indonesia: Universitas Gadjah Mada. https://regina.polgov.fisipol.ugm.ac.id/publication?topic=artisanal-mining.

Salman, Ton. 2016. "The intricacies of 'being able to work undisturbed': The organization of alluvial gold mining in Bolivia." *Society & Natural Resources* 29 (9): 1124–1138.

Salo, Matti, Juha Hiedanpää, Teemu Karlsson, Luciano Cárcamo Ávila, Juha Kotilainen, Pekka Jounela, and Róger Rumrrill García. 2016. "Local perspectives on the formalization of artisanal and small-scale mining in the Madre de Dios gold fields, Peru." *The Extractive Industries and Society* 3 (4): 1058–1066.

Sindo, Koran. 2015. "Panglima TNI Tertibkan Sumur Minyak Tua." *SINDOnews.com.*, March 14, 2015. https://daerah.sindonews.com/berita/976384/151/panglima-tni-tertibkan-sumur-minyak-tua.

Solopos. 2015. "PERTAMBANGAN BOJONEGORO : Pertamina EP Janji Persuasif Kepada Perengkek Minyak." December 17, 2015. https://www.solopos.com/pertambangan-bojonegoro-pertamina-ep-janji-persuasif-kepada-perengkek-minyak-671808.

Tempo. 1988. "Raja Minyak Itu Telah Tiada." September 10, 1988. https://majalah.tempo.co/read/nasional/25441/raja-minyak-itu-telah-tiada.

Verbrugge, Boris. 2015. "The economic logic of persistent informality: Artisanal and small-scale mining in the southern Philippines." *Development and Change* 46 (5): 1023–1046.

World Bank. 2012. *Artisanal Mining in Critical Ecosystem.* http://www.profor.info/sites/profor.info/files/docs/ASM-brochure.pdf.

Part VI

INCLUSION/EXCLUSION

Precarious Resource Inclusions

Penda Diallo

The studies in this section use empirical evidence to illustrate how inclusionary mechanisms such as FPIC, IBAs, and other CSR tools can, in some contexts, be both a tool for inclusion and exclusion simultaneously—what this book refers to as "ambivalent and precarious inclusions." Often, local realities and the influence of different stakeholders are not understood in the broader context of extractive industries. Consenting to inclusion through these mechanisms also sometimes creates personal and collective tensions and struggles as the chapters in this section illustrate. These chapters support the need to understand the "complex social dimensions" relevant to mining which are key for the operationalization of inclusionary tools (Owen and Deanna 2014) and the understanding of their impacts. I will particularly develop my perspective based on my expertise on Free, Prior, and Informed Consent (FPIC) which is "today widely endorsed as a principle and process that must be applied when any company seeks to operate in lands used and occupied by Indigenous peoples" (Tomlinson 2019, 1). I have heard first-hand feedback from Indigenous and local communities on the challenges and opportunities facing inclusionary mechanisms like FPIC. In addition to my academic work on FPIC and corporate social responsibility (CSR), my expertise draws from over ten years of experience and engagement in mining, humanitarian aid, and international development. I have had the opportunity to exchange with various stakeholders on FPIC issues and representatives from Indigenous and local communities from Guinea, Ghana, India, Australia, Suriname, and Niger. I have also discussed mining and sustainable development issues with stakeholders from various sectors, including academia, NGOs, bilateral organizations, mining companies, and the media. In my current role, in additon to my research activities, I support the Aluminum Stewardship Initiative's (ASI) Indigenous People's Advisory Forum (IPAF), meeting with representatives from Indigenous and local mining communities affected by bauxite mining globally. Indeed, these communities are keen to move beyond debates on FPIC toward addressing the challenges with implementation, including how to ensure that FPIC is respected in a context where the state and mining companies undermine the voices and concerns of local communities. While there is a wide range of literature on FPIC, there are very limited examples of initiatives that can support the implementation of FPIC beyond the overwhelming debate on the subject. Therefore, this introduction also seeks to address this gap by drawing on the broader literature on FPIC, reviewing the chapters in this section, and providing a practical example supporting FPIC efforts in the long term.

FROM CORPORATE SOCIAL RESPONSIBILITY (CSR) TO FPIC

There seems to be a growing awareness of problems with image and reputation for the extractives industry. Mining companies have been criticized by stakeholders, nongovernmental organizations (NGOs), environmental activists, human rights activists, global institutions, and governments about the impact of resource extraction. Companies have recognized that CSR initiatives can assist in mitigating against resource nationalism, conflict with local communities, and provide them with the reputational capital enabling them to maintain their license to operate in the region. As a result, most oil and mining companies have broadened their corporate strategies to include CSR activities.

Today, "CSR involves a complex agenda including managing localised impacts, e.g., water and biodiversity impacts but also contributing to solving broader-scale issues such as national-level prosperity and international-scale problems including climate change" (Fordham and Robinson 2018, 2). CSR requires strong community participation in business activities, and it requires businesses to mitigate their negative social and environmental impacts by respecting human rights and mitigating environmental damages. To effectively do this, businesses need to collaborate with local/and Indigenous communities. Thus, the need for FPIC has become prevalent for the effective implementation of mining activities throughout the mining life cycle from exploration to closure.

The FPIC principle was introduced in the 2007 United Nations Declaration on Rights of Indigenous People (UNDRIP), one of the most comprehensive international instruments on the rights of Indigenous peoples. FPIC is based on the collective right of Indigenous peoples to self-determination concerning developments affecting them and their environment (U.N. Human Rights 2013; Owen and Kemp 2014; Tamang 2005, 5). Therefore, FPIC has become an important topic in discussions related to mining and societies, particularly the inclusion and exclusion of local and Indigenous communities in questions related to mining activities. In the context of inclusion and exclusion regarding participation in mining projects, FPIC can be the most important first step, which builds the foundation for the effective inclusion of Indigenous and local communities in mining projects.

FPIC is not the only existing right but is considered essential to the recognition and protection of other Indigenous peoples' rights. FPIC establishes a framework for Indigenous peoples' participation in decision-making on activities that affect their rights. The four elements of FPIC are interdependent, with "free," "prior," and "informed" setting the conditions for a "consent" based decision-making process. To engage in an effective FPIC process, all stakeholders (including communities, governments, and companies) must establish a shared understanding of these elements.

Despite the widespread implementation of FPIC, there remain major challenges around the definition of FPIC, the meaning of "consent," the legitimacy around representation, the gap between local legislation and international standards (Iseli 2020, 272; Tomlinson 2019; Owen and Kemp 2014; Foley Hoag 2010, 93). In addition, FPIC is being implemented in places where, for instance, the government does not respect the internationally enshrined rights of Indigenous communities, and where FPIC is not part of national laws (Tomlinson 2019). The lack of clarity and agreement on some FPIC elements presents a challenge for all stakeholders, including local and Indigenous communities and mining companies. FPIC has become a "tick box" to satisfy certain stakeholders such as lenders (including the IMF and World Bank) and/or to meet legal mandates, rather than in the spirit of UNDRIP to mandate mining companies to seek meaningful consent (Hanna and Vanclay 2013).

In 2021, during consultations as part of my work with Aluminum Stewardship Initiative (ASI), Indigenous and local community groups from bauxite mining regions, including Guinea, Suriname, India, Australia, and Ghana, all raised the violation of FPIC as a significant concern. Participants stated that their consent is often not sought, and when they are informed of a project, it is often not to seek consent but to be informed of the implementation of activities. These complaints are not unusual: neither FPIC nor the right to self-determination for Indigenous peoples are being respected globally (Hanna and Vanclay 2013; U.N. Human Rights 2013). However, "while FPIC is an important approach to empower [I]ndigenous peoples against companies and governments, the main weakness of FPIC lies in its implementation" (Tomlinson 2019, 3). Effective engagement of local and Indigenous communities during the implementation of FPIC processes remains uncommon.

FPIC works on the assumption that the decision-making process of Indigenous and local communities are "legitimate, inclusive, and nonauthoritarian" (Fontana and Grugel 2016, 257). However, in many cases, "the extractive industry faces particular challenges around inclusion" (Bebbington and Humphreys Bebbington 2018, 14). In practice, FPIC is not neutral and will not automatically lead to inclusion; in many cases, it can lead to unplanned consequences. Many cases illustrate how FPIC implementation can have unintended consequences, and inclusion of local and Indigenous communities is always embedded in existing social, cultural, and economic tensions (Fontana and Grugel 2016, 257). For instance, if selection of key stakeholders during negotiation is not "done carefully, a project can 'divide and conquer' communities" (Foley Hoag 2010, 93). Also, in many cases, "the illusion of free, prior and informed consent is achieved by the exclusion of the majority of community members from effective participation in decision-making" (Hanna and Vanclay 2013, 7).

In this section, Leah Horowitz invites readers to look beyond the illusion that FPIC and negotiated agreements (which operate through consent) translate into inclusion in the chapter, "Impact-Benefit Agreements and the Violent Exclusions Hidden by 'Consent.'". This chapter builds on feminist theories of consent, in its examination of how Impact-Benefit Agreements (IBAs) work in practice. As negotiated settlements, usually between extractive firms and impacted communities, these agreements spell out exactly what is expected from the company to ensure the "consent" of local communities and a "social license to operate." Using the case of New Caledonia, Horowitz illustrates how the result of companies' engagement and agreement with local communities can, on one hand, end "physical conflict" and, on the other hand, these same agreements may perpetuate other "less visible forms of violence." The chapter highlights the implications of inclusion by recognizing local realities and existing power asymmetries (along the lines of gender, age, race, among others) within specific contexts where "consent" can lead to unintended consequences and exclusions across multiple axes of power.

The second chapter in this section, "In the ebbs and flows of resource extraction, who is a stakeholder? Insights from west Greenland," explores the complexities related to stakeholder engagement. Nina Nikola Doering examines the importance of understanding the local history and context to help understand what shapes how local stakeholders engage or choose not to participate in stakeholder engagement initiatives in the extractive industry. Where FPIC and CSR emphasize and even assume participation, Doering focuses on a common occurrence: nonattendance in the rare public fora for communities to learn about and comment on extractive projects. An effort to understand the history, past, and present relationships within a place may be necessary to ensure effective stakeholder engagement in FPIC processes. While the study focuses on Greenland, the findings are similar to the challenges concerning stakeholder engagement in mining regions across Africa. I have personally observed these kinds of challenges to participation, including the cultural barriers that stop effective stakeholder engagement. These include the mismatch between how open forums may externally be viewed to resolve issues and beliefs that prevent participants from asking questions or challenging authority in an open forum.

As illustrated in both chapters, existing social, cultural, traditional, political, and economic networks are key to understanding how inclusionary tools like FPIC can still create ambivalent and exclusionary outcomes. Therefore, "longer histories of place and development are the context in which movement, and the management of movement, in mine-affected areas should be understood" (Bebbington and Humphreys Bebbington 2018, 441). FPIC and other inclusionary mechanisms need to be viewed based on the contextual realities where it is being implemented, the complexities of the political dynamics of Indigenous communities and the relationships among various stakeholders, including the state and mining companies.

Both chapters in this section highlight the assumption that making use of inclusionary tools like FPIC and IBAs can lead to meaningful agreements with stakeholders without considering the contextual realities and local specificities of power relations. To date, in many extractive communities, the lack of understanding or consideration for contextual realities are often not addressed before the implementation of inclusionary mechanisms (FPIC, CSR, IBAs), which ultimately leads to a failure of the processes in advance of implementation. Therefore, it is crucial for all stakeholders, particularly those from mining companies implementing these interventions, to build a deep knowledge of the socio-political, historical, and cultural contexts and "characterize its associated conditionalities where they are planning mining projects" (Owen and Deanna 2013). A failure to understand these realities "would put projects—and communities—at great risk" (Owen and Deanna 2014, 3). While the FPIC debate tends to focus on consent, "companies need to be ready to listen and to accept 'no' as an answer sometimes, as not every community will be agreeable to accept all development projects affecting them, despite the potential benefits they might receive" (Hanna and Vanclay 2013, 10).

EXAMPLE OF PRACTICAL INITIATIVES THAT CAN SUPPORT INDIGENOUS COMMUNITIES' PARTICIPATION IN MINING AND FPIC PROCESS—THE ALUMINUM STEWARDSHIP INITIATIVE'S (ASI)[1]—INDIGENOUS PEOPLES ADVISORY FORUM (IPAF)[2]

The Aluminum Stewardship Initiative (ASI) is a global nonprofit standard-setting and certification organization that brings together producers, users, and stakeholders in the aluminum value chain. As a voluntary sustainability standard initiative for the aluminum value chain, ASI is part of the inclusionary model discussed in this edited collection. In addition to the Certification Program, ASI has a formal engagement mechanism with Indigenous communities affected by or involved with the aluminum value chain called the Indigenous Peoples Advisory Forum (IPAF). IPAF is a platform for exchange, learning, support, and opportunities, which aim to ensure that the voices and desires of Indigenous communities in bauxite mining regions are effectively addressed. It has representatives from various areas around the world, who prior to COVID-19 attended a yearly face-to-face meeting to interact with, learn from each other, and work on practical actions to protect Indigenous communities' rights. IPAF continues to hold regular on-line meetings. For instance, at IPAF's last meeting, in India, participants had the opportunity to learn from Indigenous communities from Odisha, where there are clashes between external interest groups and local communities impacted by bauxite mining. This became a forum to draw comparisons and connections with representatives from other bauxite mining countries and regions.

What became apparent through the IPAF is that many Indigenous communities lack the technical and specialized means to protect their rights. For the effective implementation of FPIC, it is important to support Indigenous communities' capacity to participate "effectively in dialogue" (Mercer-Mapstone et al. 2019, 2) and decision-making.

While moving from debates to effective implementation of FPIC, it is important to note that "many Indigenous groups are not in principle opposed to extractive projects but simply want them on their terms" (Tomlinson 2019, 15). Therefore, an effective FPIC process can enable Indigenous and local communities to participate in decision-making about mining projects on their land, which is the initial premise of FPIC. This is where networks such as IPAF become critical in supporting the effective implementation of FPIC. Creating mechanisms that facilitate the meaningful participation of communities in FPIC processes has the potential to mitigate the possible risks of mining activities to the environment and the local communities.

CONCLUSION

Among the inclusionary tools in natural resource extraction, like corporate social responsibility, IBAs, and sovereign wealth funds, FPIC stands apart as a major change delineated in international law with on-the-ground impacts for communities in extractive zones. While FPIC has been adopted and is required to advance most large-scale mining projects, to date, the implementation of FPIC has not been successful. Local and Indigenous communities continue to voice their concerns regarding the negative impact of mining activities and the inadequacies of existing FPIC in practice. Yet, FPIC can still be an important step that can improve the engagement of local and Indigenous communities in mining projects. Therefore, academics, policy-makers, nongovernmental organizations, mining companies, bilateral organizations, and donors need to collaborate with local and Indigenous communities to improve the implementation of FPIC processes.

As illustrated in both chapters and the broader academic literature, the illusion that FPIC automatically leads to inclusion in most cases is not true. As described in the chapters in this section, FPIC processes can inadvertently foster exclusion. Recognizing the limitations of the binary thinking of inclusion/exclusion, we need to ask what is required to ensure that context-specific realities are taken into account to support the effective participation of Indigenous and local communities in implementing FPIC processes and other inclusionary tools.

This introduction has focused on the limits to the effective implementation of FPIC. The chapters outline why the lack of consideration for local, political, and historical issues and the lack of sensibility to social structures and traditional practices in specific regions and localities presents a major limitation to the effective operationalization of FPIC in practice. These gaps call for a deeper knowledge base in relation to social, cultural, historical, economic, and political realities of specific places where FPIC needs to be implemented (Owen and Deanna 2014). A deeper understanding of the contexts in which mining occurs may enable industry actors to engage more effectively with Indigenous communities. In addition, it is important to look at some of the innovative initiatives such as IPAF and explore how to scale such examples and start to discuss them in the academic literature as potential models to support the foundation of FPIC processes in the long term.

For stakeholders to collaborate effectively, external investors need to understand the historical context that has shaped the local social, economic, and political realities of countries and communities where they plan to intervene. With Indigenous communities' engagement in mining operations, mining activities can align with their values and respect for their environment. IPAF is one model that can empower Indigenous communities, strengthen their contribution to sustainable development and mining, and foster effective communication and engagement with various external stakeholders. An effective FPIC process can ensure that mining happens without losing sight of the social, economic, and environmental concerns and the cultural needs and aspirations of Indigenous and local communities.

NOTES

1. https://aluminium-stewardship.org/
2. See the ASI Governance Handbook https://aluminium-stewardship.org/wp-content/uploads/2019/09/ASI-Governance-Handbook-v2-September2019.pdf

REFERENCES

Bebbington, Anthony, and Denise Humphreys Bebbington. 2018. "Mining, movements and sustainable development: Concepts for a framework." *Sustainable Development* 26 (5): 441–449. https://doi.org/10.1002/sd.1888.

Fontana, Lorenza B., and Jean Grugel. 2016. "The politics of Indigenous participation through 'free prior informed consent': Reflections from the Bolivian case." *World Development* 77: 249–261. https://doi.org/10.1016/j.worlddev.2015.08.023.

Hanna, Philippe, and Frank Vanclay. 2013. "Human rights, Indigenous peoples and the concept of free, prior and informed consent." *Impact Assessment and Project Appraisal* 31 (2): 146–157. https://doi.org/10.1080/14615517.2013.780373.

Iseli, Claudia. 2020. "The operationalization of the principle of free, prior and informed consent: A duty to obtain consent or simply a duty to consult?" *UCLA Journal of Environmental Law and Policy* 38 (2): 260–275. https://doi.org/10.5070/L5382050112.

Lehr, Amy K., and Gare A. Smith. 2010. *Implementing a Corporate Free, Prior, and Informed Consent Policy: Benefits and Challenges.* Washington DC: Foley Hoag.

Mercer-Mapstone, Lucy, Will Rifkin, Winnifred Louis, and Kieren Moffat. 2019. "Power, participation, and exclusion through dialogue in the extractive industries: Who gets a seat at the table?" *Resources Policy* 61: 190–199. https://doi.org/10.1016/j.resourpol.2018.11.023.

Office of the High Commissioner for Human Rights (U.N. Human Rights). 2013. "Free, prior and informed consent of Indigenous peoples." Last modified September 2013. https://www.ohchr.org/sites/default/files/Documents/Issues/IPeoples/FreePriorandInformedConsent.pdf.

Owen, John R., and Deanna Kemp. 2014. "'Free prior and informed consent', social complexity and the mining industry: Establishing a knowledge base." *Resources Policy* 41: 91–100.

Tamang, Parshuram. 2005. "An overview of the principle of free, prior and informed consent and Indigenous peoples in international and domestic law and practices." *Australian Indigenous Law Reporter* 9 (2): 111–116.

Tomlinson, Kathryn. 2019. "Indigenous rights and extractive resource projects: negotiations over the policy and implementation of FPIC." *The International Journal of Human Rights* 23(5): 880–897. https://doi.org/10.1080/13642987.2017.1314648.

11

Contractual Violence

Impact-Benefit Agreements and the Violent Exclusions Hidden by "Consent"

Leah S. Horowitz

FROM REPRESSION TO "ENGAGEMENT"

Industrial capitalism is inherently violent. By its very nature, it involves overproduction, destruction of natural resources, and exploitation of human labor (O'Connor 1991). Community resistance—which can itself turn violent—may be sparked by concerns about environmental degradation, human health, or inequitable distributions of benefits and costs. When the agents of development encounter such resistance, they can usually rely upon support from local governments, who may send armed personnel to subdue protesters through forced relocation, repression of protests, or even military force (Frederiksen and Himley 2020). Foreign nations also play crucial roles in this violence, directly through military aid or training, or indirectly by creating political and economic conditions; local governments may feel obliged to "use all means necessary to protect resource extraction activities so as to meet their debt obligations, ensure continued foreign investment, and minimize conflict with more powerful nations and institutions" (Downey et al. 2010, 424). Thus, globalization has historically facilitated the exploitation of remote places and nourished the violence that this exploitation engenders.

However, the same forces of globalization that have promoted the transnational exchange of material and informational resources have also enhanced communication and networking abilities enabling local grassroots groups to resist multinational corporations' activities (Keck and Sikkink 1998). Nowadays, protest can delay, prevent, or vastly increase the costs of industrial projects (Kuecker 2007; O'Faircheallaigh 2008). Global information flow has also made industrial activities increasingly accessible to international public scrutiny while corporations' international reputations have become more important (Garsten and Hernes 2009). Meanwhile, the recent rise of a Socially Responsible Investment movement, with indicators such as the FTS-E4Good Index Series (FTSE 2011) and environmental and social standards imposed by institutions such as the International Finance Corporation (IFC 2011), places investor pressure on companies to behave ethically. Seeking to minimize any involvement in the violent conflict associated with resistance and its repression, corporations attempt to preempt problems through Corporate Social Responsibility (CSR), that is, voluntary self-regulation of environmental and social practices, including codes of conduct and benefit-sharing with local communities. An important tool within the CSR toolkit is the Impact and Benefit Agreement (IBA), a company-community contract that documents community acceptance of a project in exchange for various types of benefits that may include, inter alia, financial payments, employment quotas, support for local business development, environmental management, and cultural heritage preservation (see Gibson and O'Faircheallaigh 2010).

Ideally, such agreements would be negotiated under the conditions necessary for the achievement of Free, Prior, and Informed Consent (FPIC) for the project. As defined by the United Nations Economic and Social Council (ECOSOC 2005, 12), FPIC involves "no coercion, intimidation, or manipulation," is obtained "sufficiently in advance of any authorization or commencement of activities," and is predicated on provision to the community of sufficient information about the project. Of course, FPIC is only meaningful if it allows for the "choice to give or withhold consent" (Motoc and Tebtebba Foundation 2005, 15), although powerful funders such as the World Bank long required "a process of free, prior, and informed *consultation* with affected communities" (World Bank Group 2004, v, emphasis added), only embracing the language of "consent" in 2015 (World Bank 2017).

Often, corporations' willingness to negotiate with communities reduces the physical violence committed both in opposition to and in support of industrial activities. However, this obviation of direct confrontation begs the question of whether IBAs eliminate all violence associated with industrial expansion, or whether they may actually perpetuate other, less visible, forms of violence. This chapter addresses this question through a study of the various exclusions integral to negotiations toward an IBA between a multinational mining corporation and an Indigenous community in New Caledonia. This agreement, titled the "Pact for Sustainable Development of the Far South [of New Caledonia]," effectively ended a six-year protest movement that had included the burning of company equipment and confrontations with armed police in which several gendarmes were injured and a protestor's truck fired upon. However, I argue that while the Pact eliminated the physical violence associated with the mining project, it actually used a series of exclusions to further both ecological and social harm, particularly toward women and other marginalized social sectors, resulting in what I call "contractual violence." I examine this question through the framework of theories about different types of violence, and through the lens of feminist theory's insights about the meaning of consent.

VIOLENCE AND CONSENT

In a classic definition, Galtung defines violence as a condition under which "human beings are being influenced so that their actual somatic and mental realizations are below their potential realizations" (1969, 168, emphasis removed); in other words, their functional capacity is reduced. I agree with this definition but supplement it in two ways. First, to "somatic and mental" I add "social realizations": the ability or power to achieve goals within one's society. Second, I extend notions of violence to the ecological domain—as does Galtung in a later work (1990). Stretesky and Lynch (1999, 164) object to "the common view of pollution as 'negative externalities' or as natural by-products and consequences of production," arguing that it must be reconceptualized as "corporate environmental violence." Of course, ecological and social conditions are so intertwined that environmental damage necessarily impacts humans; the link is particularly direct in communities, many of which are Indigenous, that depend closely on natural resources for their subsistence, livelihoods, and/or spirituality.

Thus, types of violence may be categorized according to who or what is harmed. They may also be grouped by their mode of operation. Galtung (1969, 170–171, emphasis removed) distinguishes between "personal or direct" violence, committed by an actor, and "structural or indirect" violence, involving systemic harms such as unequal power and the uneven distribution of resources, which equate to "social injustice." Corporate environmental violence may involve direct harm to ecosystems but, like social violence, it is often indirect and structural, driven or supported by national- and international-scale systems including "a world-encompassing commercialization that makes the consequences non-visible to the perpetrators" (Galtung 1990, 294) as well as inadequate environmental legislation or enforcement thereof. Another form of indirect violence is cultural violence, comprised of belief systems "that can be used to justify or legitimize direct or structural violence" (Galtung 1990, 291), making harmful acts or conditions socially acceptable. Cultural violence may support ecological violence, such as through "[t]he buzzword 'sustainable economic growth'" (Galtung 1990, 294), which—rhetoric aside—simply cannot be achieved without large-scale exploitation of natural resources. It also often serves to rationalize violence against women, justifying both direct violence, such as by "blaming the victim" of rape (Suarez and Gadalla 2010), and structural violence, such as through the enculturation of inequitable gender roles (Meyers 2002).

Feminist theorists have highlighted the problematic nature of any contract that renders violence acceptable on the grounds that it is based in consent. Indeed, the "consent" that the contract putatively embodies often "masks coercion" by concealing the negotiation process, its social context, and the resultant "bargain" (Razack 1998, 350). As MacKinnon notes (1990, 4), "when fear and despair produce acquiescence and acquiescence is taken to mean consent, consent is not a meaningful concept." Similarly, apparent community acceptance of an industrial project may indicate "resignation" rather than true consent (Benson and Kirsch 2010). Even more problematic is consent granted by putative representatives on behalf of a wider group.

The existence of a contract between a community and a project proponent, then, rather than unquestionably justifying the agreement it contains, should raise at least three questions. First, does this apparent consent result from a real choice? Second, exactly whose "consent" does it really represent, and whom does it exclude? On a related note, does the contract involve true power-sharing and compromise, or does it simply represent a new—"contractual"—form of exclusionary violence? This chapter explores these questions through a case study from New Caledonia, a French possession in the South Pacific. Social structures within New Caledonia, particularly among its Indigenous populations, have historically been, and largely continue to be, highly patriarchal. Meanwhile, the nation possesses vast mineral wealth, yet suffers from outdated mining legislation that pays little heed to environmental impacts and none at all to social ones. Because of its attractiveness for multinational corporations, inadequate legislative framework, and inequitable power distributions, New Caledonia is a particularly appropriate location for the examination of issues surrounding contractual agreements between communities and corporations.

NEW CALEDONIA

A Pacific archipelago, New Caledonia's population is approximately 269,000 (ISEE 2015), with several ethnic groups, primarily Indigenous Kanak (40 %) and European-Caledonians (29 %) (Rivoilan and Broustet 2011). Within Kanak societies, the first clans that occupy an area possess customary (although not legal) rights to determine land use there, and have the highest customary social status (Bensa and Rivierre 1982; Naepels 1998). Administered by France since 1853, New Caledonia is now an "Overseas Country," but remains a French possession. An independence movement formed in the

1970s, and anti-colonial uprisings followed in the 1980s. Accords in 1988 and 1998 officially recognized Kanak identity, created an advisory Customary Senate, and promised redistribution of economic benefits to the Kanak.

New Caledonia possesses approximately one-sixth of the world's nickel reserves (USGS 2014). Mined since 1874, it currently hosts over thirty active mines (DIMENC 2012), some run by locally based entrepreneurs and others by multinational corporations (MNCs). Vale's "Southern Refinery", at first named Goro Nickel and now officially called Vale Nouvelle-Calédonie, is located at the southern tip of the main island (fig. 11.1). Due to low mineral content, the refinery uses hydrometallurgical technology, never before implemented in New Caledonia: Acid under pressure leaches nickel and cobalt from the ore, with effluent discharged into the sea. Operations commenced in 2010, with production increasing despite delays caused by acid leaks in 2009, 2010, 2012, and 2014 (two of which devastated freshwater ecosystems), the effluent diffuser's rupture in 2013, and plummeting nickel prices.

Rhéébù Nùù, an Indigenous protest group, formed in 2002 in response to the Southern Refinery. Predominantly a rhizomic structure (Horowitz 2016), it had a loosely defined membership but a charismatic leader, Gabriel (pseudonym). For six years, Rhéébù Nùù leaders initiated actions, including

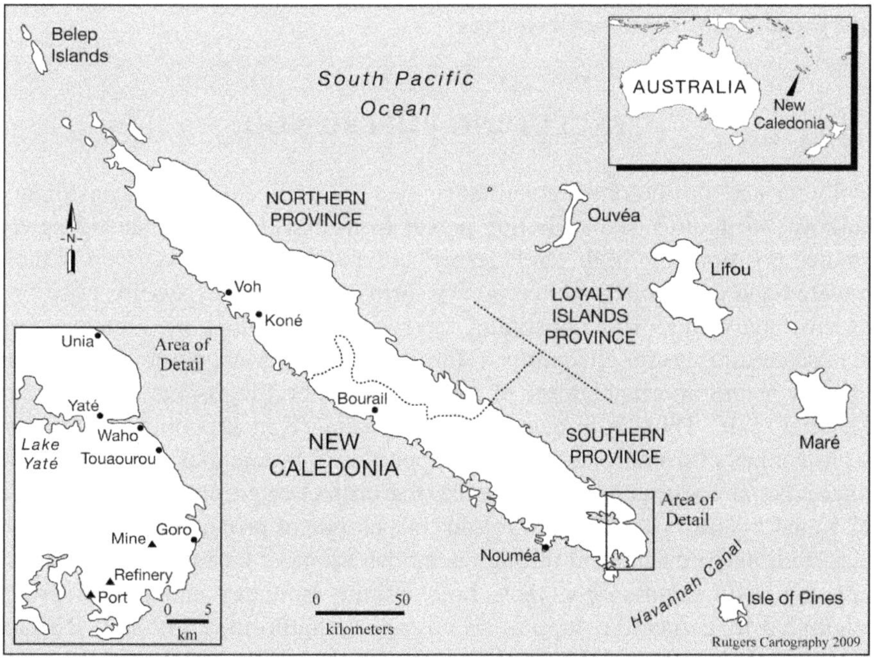

Figure 11.1 New Caledonia and the Southern Refinery Area. *Source*: Horowitz (2017a).

pamphlets denouncing company activities, public meetings, open letters to political leaders, legal action, and blockades of the construction site which turned into violent encounters with armed police. In 2008, Rhéébù Nùù leaders, customary authorities, and company representatives signed the Pact. The company committed to creating a Corporate Foundation to fund local development initiatives, training local youth as "environmental technicians," and establishing a reforestation program as well as a "Consultative Customary Environmental Committee" of customary authorities who could recommend further studies. In exchange, Rhéébù Nùù members committed to "assert their point of view not through violent or illegal actions, but by dialogue" (Vale Inco et al. 2008).

Methods

Between 2001 and 2012, I conducted face-to-face, semi-structured interviews with over two hundred stakeholders, including local residents, grassroots leaders, international nongovernmental organization (NGO) representatives, government officials, and company representatives. Additionally, I conducted telephone interviews with twenty NGO and grassroots leaders, lawyers, researchers, and company officials, based in Australia, Canada, France, and New Caledonia. Most interviews were in French; all translations are mine. I use pseudonyms for all interviewees.

EXCLUDING RHÉÉBÙ NÙÙ

Colonial and postcolonial governance systems radically altered customary authority structures, redistributing power from first-occupant clans to newly created positions of "high" and "lesser" chiefs (Naepels 1998). Even in their mutated and convoluted contemporary forms, however, respect for positions of customary authority constitutes a crucial component of contemporary Kanak cultural-political identity (Horowitz 2008). Meanwhile, Indigenous identity is an important source of political and moral legitimacy on an international scale. Thus, both customary legitimacy (Kanak people's sense that a project has customary authorities' support) and Indigenous legitimacy (the international community's acceptance of a project or group as representative of Kanak interests) are crucial for any development project in New Caledonia. Both may be achieved through demonstrations of customary authorities' approval. Not surprisingly, then, both mining company and protest group attempted to garner the support of customary authorities. At first, Rhéébù Nùù was more successful in publicly winning customary authorities' support for its activities and reinforcing its own legitimacy through this association.

While neither of its leaders was from a first-occupant clan, they had involved as large a number of customary authorities as possible from the beginning, by inviting customary authorities from the entire archipelago to participate in ceremonies celebrating the group's foundation (Horowitz 2009), if not in actual power-sharing. By mid-2006, however, that support was eroding. Customary authorities began to express mixed feelings about the protest group, viewing it not as the activist arm addressing their concerns but as an independent and increasingly unruly entity. Meanwhile, Vale was doing its best to position Rhéébù Nùù as a fringe element that represented neither customary authorities nor the general population (Horowitz 2012).

In June 2006, Rhéébù Nùù, company representatives, non-Indigenous environmental groups, and government officials sat down for a series of closed-door round table discussions organized by the Southern Province. Eventually, the government excluded the environmentalists and then found itself excluded as the discussions evolved into two-way talks between Vale and Rhéébù Nùù. The talks persisted but did not bring the parties any closer to a resolution. In early 2008, the company began laying the pipeline that would dump effluent into the marine environment, sparking fresh protests and blockades. At this point, Vale flew in a legal specialist from Brazil as the new lead negotiator. He took a different tactical approach, the distinguishing feature of which, and the basis of its "legitimacy," was that it was more "inclusive." Criticizing previous strategies that had "focused on talking to those who were most able to disrupt operations" (Benke 2010), the negotiator realized that exclusive engagement with Rhéébù Nùù had allowed the group to exert pressure on Vale. In attempting to achieve efficiency, the company had not seen a need to negotiate with other, less troublesome, community sectors. However, Vale could counterbalance and undermine their influence by capturing the deeply held cultural ideal of customary legitimacy. To do this, it needed to broaden its engagement strategy, including not the entire community, as many women and young people sympathized with the protest group, but the sector most sympathetic to the project: customary authorities, who were also, of course, at the top of the social hierarchy. Drawing from fashionable CSR discourses, the negotiator portrayed this strategy as "culturally sensitive." In his narrative, by reaching out to customary authorities, he was following culturally correct protocols by understanding "the customary leadership structures and interests" and engaging directly "with all legitimate customary representatives of all tribes related to the project" (Benke 2010). Thus, beyond taking the focus off Rhéébù Nùù, the company began working to undercut the group's claims to customary/Indigenous legitimacy and instead establish Vale's own claims to that powerful resource. Customary authorities' increasing disillusionment with the protest group provided just that opportunity. The tripartite negotiations that ensued effectively marginalized Rhéébù Nùù; if the group was, as it claimed, representing the customary authorities, the presence of the latter at the table rendered the protestors redundant, and silenced them.

In capturing customary legitimacy, Vale also captured the ideal of Indigenous legitimacy from the international perspective, leading to support from UN representatives (see Horowitz 2015). In 2008, ironically (and misleadingly), the UN further undercut Rhéébù Nùù's movement when UNESCO inscribed sections of New Caledonia's coral reef on its World Heritage list (Horowitz 2016). Facing multiple pressures, Rhéébù Nùù leaders chose to sign the Pact, although they had not achieved their environmental or economic goals. In summary, by capturing the customary authorities and the multiple forms of legitimacy they represented, Vale's Pact excluded Rhéébù Nùù's original aims. In doing so, it also excluded the group's adherents, many of whom were youth and women.

EXCLUDING WOMEN

In precolonial Kanak societies, senior men held all positions of authority; the colonial period exacerbated and racialized Kanak women's oppression (Berman 2006). This history informs current social relations in New Caledonia; an intersectional analysis reveals that young Kanak women with little formal education and low earnings are the most vulnerable to domestic violence (Hamelin and Salomon 2004). Over the last sixty years, however, Kanak women—especially urban residents—have been challenging male dominance (Salomon and Hamelin 2008). Meanwhile, Kanak women increasingly turn to options provided by contemporary institutions, such as the market economy or the French legal system, for some autonomy from husbands or community leaders, and have also made great strides politically (Salomon 2010). In rural areas, though, customary social norms still prohibit women and youth from adopting leadership roles. Thus, Rhéébù Nùù's main leader, Gabriel, was a senior man who, although not from a first-occupant clan and unable to determine land-use activities per se, could organize resistance efforts. Women and youth were restricted to following Gabriel. Later, local residents, many of whom had participated in direct actions at great personal risk, bitterly condemned Gabriel for signing the Pact, suspecting him of having received bribes and being a self-serving traitor. Young men brandished machetes and hurled insults at Gabriel when he dared visit their villages. However, women and youth sank into despondency as protest actions, without a leader to organize them, died out. The fact that women and youth who wanted to resist the refinery had no choice but to follow Gabriel made their subsequent perceptions of his betrayal all the more painful.

Although women's social position did not allow them to lead the entire community, they could fight for recognition of their own needs. However, Vale committed what I term "retrogradation" (Horowitz 2017b), ignoring Kanak women's recent social gains in order to justify sidelining women's concerns and excluding them from negotiations. When I asked company representatives why no women had been invited to negotiate or sign the Pact, they replied—with scoffing chuckles—that this was "normal," due to Kanak "cultural values" (Luc and Gilles September 9, 2010). Marc, a former Vale employee who had helped negotiate the Pact, lamented that "unfortunately, it's Custom that's like that," adding, "personally, it shocks me, the place of women in Custom." Nonetheless, he insisted, company representatives had "talked about" and "shown" the Pact to women's associations, who found it "very good," and whose comments "fed into" the document's drafting (September 10, 2010). Similarly, another Vale representative of European ancestry insisted that she often lamented women's exclusion from company–community interactions. She explained that Community Relations staff were eager to include women, but due to "historical" factors, "it's not easy" (Flavia July 28, 2010).

Customary authorities thought it perfectly normal that only senior men were chosen as representatives; nonetheless, several noted that they would not have objected had Vale held separate meetings with local women to address their concerns. Even had the customary authorities objected to negotiating with women, referencing "Custom" is of questionable morality. Similar arguments have been applied to "explain and excuse" Aboriginal men's behavior in Canadian courts, where "culture" has been used to reinforce "judicial tendencies to minimize the harm of rape" (Razack 1994, 899). Arguably, it is inappropriate to accept violence against women, personal or structural, in the name of "cultural relativism" (Gibson and Kemp 2008).

Vale failed even to identify women's concerns; the company's Socio-Economic Impact Study (Goro Nickel 2007) neglected to separate survey responses by gender. However, an earlier, independent study had shown that beside concerns about fishing grounds, women worried that cash flow to men (the majority of employees) would entail increased alcohol consumption, hence more domestic violence and more car accidents, and other social disruptions (Holbein and Marakovic 1996). Addressing such concerns would have necessitated extensive social programs.

Ultimately, Kanak women engaged strongly with the Southern Refinery project, despite hegemonic ideologies that militated against challenging male authority. However, Vale committed retrogradation, using outdated interpretations of the social context to justify negotiating exclusively with senior men, primarily customary authorities, who were more in favor of the projects and less concerned about their impacts.

EXCLUDING THE GOVERNMENT

In addition to reinforcing inequitable power structures within the community for the company's advantage, the Pact allowed Vale to take greater control over its environmental performance by directing regulatory power away from the government, which, as noted above, was excluded from negotiations surrounding the Pact. It also channeled community members' environmental observations, concerns, and objections into a form that the company could control and that did not threaten company activities.

Rhéébù Nùù had earlier made significant gains through the courts, such as the revocation in June 2006 of the MNC's permit to operate its refinery due to an insufficient environmental impact assessment, forcing Vale to invest significant time and money in a new study. To preempt legal action, the Pact specified that any future disagreements should be resolved through "discussion"—or, failing an "amicable agreement," through a group of third-party arbitrators selected by each party—rather than through the courts (Vale Inco et al. 2008). According to a local magistrate, this "strategy" aimed to "extinguish any possibility of referral to a judge" (Auguste September 3, 2010). In contrast to earlier gains, the Pact did not reduce the project's environmental impacts. One component of its Environmental Program was the reforestation of areas affected by projects prior to Vale's. While seemingly positive, this activity served to distract attention from the project's own impacts (e.g., its effluent, the potential—later realized—of acid spills, carbon emissions from its coal-fired power plant), displacing blame for ecological damage onto Vale's predecessors. Additionally, the mining company committed to creating a "Consultative Customary Environmental Committee" (CCCE) and to recruiting and training local "environmental technicians," and it also provided half the funding for l'Œil, an "observatory" charged with monitoring the project. With all three groups charged with "watching over" the mining project, a government official quipped, "I think one can't be more watched-over than that" (Nicole June 19, 2012). However, the technicians (who received a one-year training program organized by Vale through the local university) and the CCCE (ten customary authorities and five Rhéébù Nùù representatives) could only "recommend" environmental studies, provide "opinions" on project activities, and "elaborate a strategy for showcasing Kanak traditional knowledge within the operations" (Vale Inco et al. 2008). They were provided no capacity to alert the regulatory authority to any concerns. The only other environmental provision was a commitment to "put into place corrective mechanisms so as to maintain conformity with the environmental standards applicable to the project" (already legally required) "in light of the relevance" (determined, of course, by Vale) of the committee's recommendations. Indeed, there was no provision for pressuring Vale, in any meaningful way, to change its environmental management procedures. Meanwhile, there was in fact a lesser chance than before the Pact's signature that the company would alter these procedures, given that there was now a lesser risk of public exposure. Confident that they had managed to channel the expression of community concerns away from the courts, international nongovernmental organizations, and the media, Vale began reducing their on-site Community Relations personnel. While providing a false sense of participation in environmental decision-making, the Pact—an instance of "legal enclosure" (Szablowski 2019)—left the community without a channel to effect real change.

Some customary authorities worried that they had been co-opted, assimilated into company-created structures while their "demands, [. . .] grievances that were formulated since the beginning of the project" had "never been really responded [to], listened to" (Loïc July 21, 2011). Furthermore, they wondered whether they could remain independent critics of Vale's activities through structures such as the CCCE or l'ŒIL, as both were funded by the MNC (and both presided by Gabriel): "I wonder if we can go against Vale even if it's Vale that pays for everything" (Oscar July 13, 2012). In any case, they were acutely conscious of having been manipulated in order to build support for the Pact. Some expressed regret and wondered whether they had "signed too quickly" (Guy September 14, 2010), especially after recurrent acid spills in subsequent years. Many also noted the fury of other community members upon learning of the signature, particularly women and youth who had participated in blockades. However, they felt that the Pact had negated their ability to protest; they had pledged "no more actions against Vale" (Eugène August 7, 2010).

CONCLUSIONS

Building on feminist theories of "consent," this chapter has demonstrated that, while IBAs may reduce the physical violence associated with both resistance to industrial development projects and its suppression, they may also involve both ecological and social violence. In the case examined here, the Pact enabled the performance of ecological violence by allowing the company's plans for the construction and operation of the refinery and pipeline to continue with no alterations whatsoever. While company representatives often complained about the "concessions" they had made, these were purely financial (in the form of relatively insignificant amounts of monetary support for the community) rather than technical. By providing the community with direct monetary benefits, while promising further financial benefits through employment, the Pact implicitly justified the project's environmental damages through an appeal to the virtues of economic development, committing cultural violence by legitimizing harm through reference to social discourses and belief systems. Beyond enabling ecological violence, the Pact also not only enabled but also actually performed social violence, both cultural and structural, by excluding people from non-first-occupant clans, youth, and women, thus reinforcing—through reference to erstwhile ideologies—their subordinate social positions, that is, performing retrogradation. By also excluding the government from negotiations, and ultimately diverting regulatory power away from the state, it also cut off a potential source of protection from capitalism's excesses, disempowering community members and thus performing further structural social violence. Yet the Pact achieved all of these (indirectly) violent acts under the guise of officially verified "consent," embodied in a signed contract, earning praise even from the UN. Such contractual violence, largely invisible, does not threaten a company's reputation as does an association with direct physical violence. However, as this chapter demonstrates, it may also be harmful to local communities.

ACKNOWLEDGMENTS

I am deeply grateful to the residents of Goro, Unia, and Île Ouen for their hospitality, and to all interviewees for their time. Thanks to Lena Richter for research assistance. Research for this chapter was funded by the Australian National University, the Centre National de la Recherche Scientifique, the Centre National de Recherche Technologique "Nickel et son Environnement," Hawai'i Pacific University, the Ministère des Affaires Etrangères, the National Endowment for the Humanities, the University of Leeds, the University of Wisconsin-Madison, and the Vilas Trust Estate. Parts of this chapter have been reproduced, with permission, from Horowitz, L.S. "Culturally articulated neoliberalization: Corporate social responsibility and the capture of Indigenous legitimacy in New Caledonia," *Transactions of the Institute of British Geographers* 40 (1) (2015): 88–101 and from Horowitz, L.S. "'It shocks me, the place of women': Intersectionality and mining companies' retrogradation of Indigenous women in New Caledonia," *Gender, Place & Culture* 24(10) (2017): 1419–1440. Any errors are solely my responsibility.

REFERENCES

Benke, Rafael. 2010. "Perspective." Interview with BASESwiki, December 17, 2010. http://baseswiki.org/en/Vale_New_Caledonia,_Negotiation_of_a_Sustainable_Development_Pact_with_Local_Stakeholders,_Canada_2008.

Bensa, Alban, and Jean-Claud Rivierre. 1982. *Les Chemins de l'alliance : L'organisation sociale et ses représentations en Nouvelle-Calédonie (Région de Touho—Aire linguistique cèmuhî)*. Paris: SELAF.

Benson, Peter, and Stuart Kirsch. 2010. "Capitalism and the politics of resignation." *Current Anthropology* 51 (4): 459–486.

Berman, Alan. 2006. "Kanak women and the colonial process." *International Journal of Law in Context* 2 (1): 11–36.

DIMENC (Direction de l'Industrie des Mines et de l'Énergie). 2012. "Les sites d'activité du nickel en Nouvelle-Calédonie 2012." Accessed April 7, 2013. http://www.dimenc.gouv.nc/portal/page/portal/dimenc/mines_carrieres/mines.

Downey, Liam, Eric Bonds, and Katherine Clark. 2010. "Natural resource extraction, armed violence, and environmental degradation." *Organization & Environment* 23 (4): 417–445.

ECOSOC (United Nations Economic and Social Council). 2005. *Report of the International Workshop on Methodologies Regarding Free, Prior, and Informed Consent and Indigenous Peoples*. New York: UNESC.

Frederiksen, Tomas, and Matthew Himley. 2020. "Tactics of dispossession: Access, power, and subjectivity at the extractive frontier." *Transactions of the Institute of British Geographers* 45 (1): 50–64.

FTSE. 2011. "FTSE4Good index series." Accessed March 24, 2011. http://www.ftse.com/Indices/FTSE4Good_Index_Series/index.jsp.

Galtung, Johan. 1969. "Violence, peace, and peace research." *Journal of Peace Research* 6 (3): 167–191.

Galtung, Johan. 1990. "Cultural violence." *Journal of Peace Research* 27 (3): 291–305.

Garsten, Christina, and Tor Hernes. 2009. "Beyond CSR: Dilemmas and paradoxes of ethical conduct in transnational organizations." In *Economics and Morality: Anthropological Approaches*, edited by Katherine E. Browne, and B. Lynn Milgram, 189–210. Lanham, MD: AltaMira Press.

Gibson, Ginger, and Deanna Kemp. 2008. "Corporate engagement with Indigenous women in the minerals industry: Making space for theory." In *Earth Matters: Indigenous Peoples, the Extractive Industries and Corporate Social Responsibility*, edited by Ciaran O'Faircheallaigh, and Saleem Ali, 104–122. Sheffield: Greenleaf.

Gibson, Ginger, and Ciaran O'Faircheallaigh. 2010. *IBA Community Toolkit: Negotiation and Implementation of Impact and Benefit Agreements*. Toronto: Walter & Duncan Gordon Foundation.

Goro Nickel. 2007. *Étude d'Impact Socio-Économique du projet Goro Nickel*. Nouméa: Goro Nickel.

Hamelin, Christine, and Christine Salomon. 2004. "Parenté et violences faites aux femmes en Nouvelle-Calédonie. Un éclairage sur l'ethnicité différenciée des violences subies au sein de la famille." *Espace populations sociétés* 2: 307–323.

Holbein, Marguerite, and Grégory Marakovic. 1996. "Etude socio-géographique de perception et d'impact sur les populations d'un projet minier: le projet Goro-Nickel, Aspect environnemental et social." Master's diss., Université française du Pacifique.

Horowitz, Leah S. 2008. "'It's up to the clan to protect': Cultural heritage and the micropolitical ecology of conservation in New Caledonia." *The Social Science Journal* 45 (2): 258–278.

Horowitz, Leah S. 2009. "Environmental violence and crises of legitimacy in New Caledonia." *Political Geography* 28 (4): 248–258.

Horowitz, Leah S. 2012. "Translation alignment: Actor-network theory, resistance, and the power dynamics of alliance in New Caledonia." *Antipode* 44 (3): 806–827.

Horowitz, Leah S. 2015 "Culturally articulated neoliberalization: Corporate social responsibility and the capture of indigenous legitimacy in New Caledonia." *Transactions of the Institute of British Geographers* 40 (1): 88–101.

Horowitz, Leah S. 2016. "Rhizomic resistance meets arborescent assemblage: UNESCO World Heritage and the disempowerment of indigenous activism in New Caledonia." *Annals of the American Association of Geographers* 106 (1): 167–185.

Horowitz, Leah S. 2017a. "Indigenous by association: Legitimation and grassroots engagements with multinational mining in New Caledonia." In *Grassroots Environmental Governance: Community Engagements with Industry*, edited by L. S. Horowitz and M. J. Watts, 80–100. London: Routledge.

Horowitz, Leah S. 2017b. "'It shocks me, the place of women': Intersectionality and mining companies' retrogradation of Indigenous women in New Caledonia." *Gender, Place & Culture* 24 (10): 1419–1440.

IFC (International Finance Corporation). 2008. "Environmental and social standards." Accessed March 24, 2011. http://www.ifc.org/ifcext/sustainability.nsf/Content/EnvSocStandards.

ISEE (Institut de la statistique et des études économiques Nouvelle-Calédonie). 2015. "Recensement." Accessed February 23, 2015. http://www.isee.nc/population/recensement.

Keck, Margaret E., and Kathryn Sikkink. 1998. *Activists Beyond Borders: Advocacy Networks in International Politics.* Ithaca, NY: Cornell University Press.

Kuecker, Glen David. 1998. "Fighting for the forests: Grassroots resistance to mining in northern Ecuador." *Latin American Perspectives* 34 (2): 94–107.

MacKinnon, Catherine A. 1990. "Liberalism and the death of feminism." In *The Sexual Liberals and the Attack on Feminism*, edited by Dorchen Leidholdt, and Janice G. Raymond, 3–13. New York: Pergamon Press.

Meyers, Diana Tietjens. 2002. *Gender in the Mirror: Cultural Imagery and Women's Agency.* Oxford: Oxford University Press.

Motoc, Antoanella-Iulia, and Tebtebba Foundation. 2005. *Standard-Setting: Legal Commentary on the Concept of Free, Prior and Informed Consent.* New York: Expanded working paper commissioned by the United Nations Commission on Human Rights.

Naepels, Michel. 1998. *Histoires de terres kanakes.* Paris: Belin.

O'Connor, James. 1991 "On the two contradictions of capitalism." *Capitalism Nature Socialism* 2 (3): 107–109

O'Faircheallaigh, Ciaran. 2008. "Introduction." In *Earth Matters: Indigenous Peoples, the Extractive Industries and Corporate Social Responsibility*, edited by Ciaran O'Faircheallaigh, and Saleem Ali, 1–7. Sheffield: Greenleaf.

Razack, Sherene. 1994. "What is to be gained by looking white people in the eye? Culture, race, and gender in cases of sexual violence." *Signs* 19 (4): 894–923.

Razack, Sherene. 1998. "Race, space, and prostitution: The making of the bourgeois subject." *Canadian Journal of Women and the Law* 10 (2): 338–376.

Rivoilan, Pascal, and David Broustet. 2011. "Recensement de la population 2009." *Synthèse, Institut de la Statistique et des Études Économiques Nouvelle-Calédonie* 19: 1–4.

Salomon, Christine. 2010. "Vers un changement des normes de genre." In *La Nouvelle Calédonie, vers un destin commun?*, edited by Elsa Faugère, and Isabelle Merle, 203–224. Paris: Karthala.

Salomon, Christine, and Christine Hamelin. 2008. "Challenging violence: Kanak women renegotiating gender relations in New Caledonia." *The Asia Pacific Journal of Anthropology* 9 (1): 29–46.

Stretesky, Paul, and Michael J. Lynch. 1999. "Corporate environmental violence and racism." *Crime, Law & Social Change* 30: 163–184.

Suarez, Eliana, and Tahany M. Gadalla. 2010. "Stop blaming the victim: A meta-analysis on rape myths." *Journal of Interpersonal Violence* 25 (11): 2010–2035.

Szablowski, David. 2019. "'Legal enclosure' and resource extraction: Territorial transformation through the enclosure of local and indigenous law." *The Extractive Industries and Society* 6 (3): 722–732.

USGS (U.S. Geological Survey). 2014. *Mineral Commodity Summaries 2014*. Washington, DC: U.S. Geological Survey.

Vale Inco, and Conseil Coutumier de l'Aire Drubéa Kapume and Comité Rhéébù Nùù. 2008. "Pacte pour un développement durable du Grand Sud." Nouméa: Vale Inco Nouvelle-Calédonie.

WHO (World Health Organization). 2008. "Eliminating female genital mutilation: An interagency statement." UNAIDS, UNDP, UNECA, UNESCO, UNFPA, UNHCHR, UNHCR, UNICEF, UNIFEM, WHO. Geneva: WHO Press.

World Bank. 2017. *The World Bank Environmental and Social Framework*. Washington, DC: The World Bank.

World Bank Group. 2004. *Striking a Better Balance: The World Bank Group and Extractive Industries - The Final Report of the Extractive Industries Review*. Washington, DC: World Bank Group.

12

In the Ebbs and Flows of Resource Extraction, Who Is a Stakeholder?

Insights from West Kalaallit Nunaat (Greenland)

Nina Nikola Doering

Much can be learned about the present from the ways in which past extractive activities are remembered and talked about. The Danish-owned coalmine on Disko Island in west Greenland, active between 1924 and 1972, provides a powerful example. Half a century later, it is the closure of the mining town, Qullissat, and the forced relocation of the Greenlandic mine workers, their families, and all other inhabitants of the community, that remains most vivid in national memory. When several settlements in northwest Greenland were hit by a devastating tsunami in 2017, a woman who had been evacuated to Aasiaat in the south of Disko Bay told Greenland's national broadcaster KNR that she "could not help but think of the people who were forcefully resettled from Qullissat, it must have been terrible. At least that was the feeling I had on my way here" (Uldum 2017, translated). In both cases, people had no control over the events and decisions that forced them to abandon their homes. Working conditions in the coal mine undoubtedly were hazardous and strenuous. Yet memories of Qullissat come up rarely today in discussions about planned extractive projects in Greenland. Instead, these memories are evoked in the context of Greenland's history as a Danish colony and Danish policies in Greenland (see also Jørgensen 2019; Sejersen 2014, 44–46), highlighting that public concerns relating to extractive industries are focused, at least for the moment, primarily on questions of control and agency.[1] This emphasizes a need for meaningful public involvement in the decisions that determine extractive activities (see also Tejsner 2014). This chapter engages with the impacts of anticipated resource extraction as well as the exclusion of people and their views from decision-making processes in Greenland, which, the chapter illustrates, are also connected in several ways with the consequences of colonial practices.[2]

Current concerns shape the way that the past is evaluated and given meaning. Similarly, memories of the past and hopes for the future influence the present and affect extractive activities and public participation practices. Ongoing and planned resource operations in Greenland have been interpreted and narrated in political discourse in the context of the country's relationship with Denmark and are tied up in Greenlandic visions for future independence (Lindroth 2020). Following the 2009 Act on Self-Government, Greenland took home several policy areas that had remained under full or partial Danish control during the previous Home Rule arrangement, including extractive resource activities (Naalakkersuisut and Government of Denmark 2009; Inatsisartut 2009). The act recognizes Greenland's right to its subsurface resources, and the Government of Greenland has since viewed extractive development as a potential opportunity to diversify the economy and reduce financial and political dependence on Denmark (Naalakkersuisut 2014, 2020; Poppel 2018). To render Greenland's subsoil attractive to foreign investors, a "one-door" policy provides the hydrocarbon and mining industries with a single point of contact in the administration for all matters regarding licensing and monitoring (Inatsisartut 2009, Section 3). Guidelines on the preparation of social and environmental impact assessments (EIAs/SIAs) accompanying Greenland's Mineral Resources Act stipulate that companies are to "involve relevant stakeholders" as part of the licensing process prior to resource extraction in the mining sector and hydrocarbon exploration and extraction activities (Inatsisartut 2009, with amendments; Naalakkersuisut 2015; 2016a, 15, 2016b). However, the public meetings[3] carried out in fulfillment of this requirement over the past decade have been widely critiqued (e.g., Aaen 2012; Dahl and Hansen 2019; Siegstad and Fægteborg 2015).

While Greenland's early years of self-government were marked by high hopes for a hydrocarbon find, the focus has since shifted (back) from oil and gas to mining.[4] Yet, two small-scale mines, producing rubies/sapphire and anorthosite, are the only sites of active commercial extraction at the moment (Naalakkersuisut 2020, 8), as fluctuating resource prices and the high risks and costs of operating in the Arctic have slowed or ended several of the numerous other planned projects in recent years.[5] Greenland's history has long been marked by these ebbs and flows of outside interest in its mineral resources. This is exemplified by early Danish exploration (Nuttall 2013, 371–72), cryolite mined in Greenland for aluminum production in Canada during World War II (Berry 2012), the question of resource ownership during Greenland's struggle for increased autonomy in the 1970s (Poppel 2018) and a snap election in the spring of 2021 centered around political disagreement over a planned rare-earth elements and uranium mine in southern Greenland (McGwin 2022).

As interest shifts and levels of activity change in different parts of the country, some towns and settlements receive focused attention in national and international debates on resource extraction, before being relegated to the sidelines, as companies retreat temporarily or permanently. People living in the areas of planned projects are turned into publics for participation for periods of time, depending on official notions of geographic proximity and potential impacts of future resource extraction. This chapter illustrates the way that one such town, Aasiaat, has been affected over time. It shows that the anticipation of future hydrocarbon exploitation was given a material presence in Aasiaat and traces how local hopes and concerns were connected with community-level needs and past experiences, including the presence of the oil company. The chapter then explores the exclusion of people and their views from decision-making as a result of participatory procedures, which are often ill-suited to local and cultural contexts. In engaging with local resistance to the (un-)making of stakeholders,[6] the chapter challenges binary interpretations of nonparticipation as nonengagement or consent and concludes by pointing to other forms of engagement with extractive activities, such as street protests and art. The chapter is closely based on elements of a PhD dissertation and Master's thesis, for which observations and interviews were conducted in 2012 and 2015/16 in Aasiaat and other towns along Greenland's west coast.[7] This would not have been possible without the help and translation work of Karina Abelsen, Mette Abelsen, Inuaraq Gedionsen, Ane Siegstad Joelsen, and Hans Henrik Suersaq Poulsen, to whom I am deeply grateful.

ACADEMIC AND ACTIVIST ENGAGEMENT WITH PUBLIC PARTICIPATION

The right of Indigenous peoples to be involved in and have control over decisions affecting their lands, waters, and ways of life, fought for by Indigenous activists and organizations, has been recognized in international conventions and guidelines (e.g., International Labour Organization 1989; Secretariat of the Convention on Biological Diversity 2004; UN General Assembly 2007). Indigenous academic, activist, and political engagement with participation in resource management has formed part of broader movements for self-determination and sovereignty across domains, including scientific practice (e.g., Inuit Circumpolar Council [ICC] 2021; Inuit Tapiriit Kanatami 2018; Sámiráđđi 2019). In its charter, the Inuit Circumpolar Council states that "our right to self-determination must be confirmed and Inuit participation in policies and activities affecting our homeland assured" (ICC 2010). Yet, affected communities, a growing number of NGOs and interest groups, Indigenous rights organizations, and scholarly observers in Greenland have noted that insufficient, badly translated, biased and overly technical information, dismissal of Indigenous knowledge, one-way communication, incompatible expectations, untransparent decision-making procedures, and poorly timed meetings render meaningful participation impossible (Aaen 2012; Dahl and Hansen 2019; Graugaard et al. 2020; Hansen 2014; Hansen and Johnstone 2019; Langhoff 2013; Nuttall 2012a, b; Olsen and Hansen 2014; Siegstad and Fægteborg 2015; Wilson 2015). Their criticisms are echoed by those raised in other parts of the world and in debates across academic disciplines.

For decades, efforts to include members of the public in decision-making processes have been the center of contentious debate. In the 1990s, public participation emerged as "orthodoxy" (Cornwall 2006, 62) in the social sciences and in interventions carried out in the name of development by organizations such as the World Bank in diverse geographic locations, with far-reaching promises ranging from better project outcomes to better citizens (Cornwall 2002, 2006; World Bank 1996).[8] A broad body of literature, produced mainly in Western academic institutions, sought to identify appropriate methods for eliciting participation and compiled a wealth of evaluation criteria (e.g., Bishop and Davis 2002; Buchy and Hoverman 2000; Reed 2008; Rowe and Frewer 2000). In response, critical scholarship emerged to illustrate how participatory processes can serve to reinforce hegemonic power dynamics and disrupt culturally appropriate ways of reaching decisions (e.g., Cooke and Kothari 2001; Cornwall 2002), noting especially exclusionary and violent consequences of problematic participatory practices in the context of planned extractive projects on Indigenous lands (e.g., Fjellheim 2023; Lawrence and Larsen 2017; Mitchell et al. 2019). Approaches in science and technology studies draw attention to the dynamics by which relatively well-established public participation procedures—or "technologies of participation" (Chilvers and Kearnes 2016b, 42; Soneryd 2016, 145; which may include public hearings)—are made to travel across cultural and political contexts, producing both homogenizing and diversifying effects, and engage with the "organizational carriers" by which ideas about universally applicable "best practices" are transplanted (Soneryd 2016, 145, 150). Rather than taking "the public" as preexisting and available to be activated for participation, the construction, emergence, and enactment of publics and participation procedures as well as resistances to these processes form the subject of inquiry (Chilvers and Kearnes 2016a, b). This has opened new perspectives on the normative quest for greater inclusion and representativeness (ibid., 11–12).

In Greenland, recent years have also seen increasing academic and activist engagement with the continued consequences of Danish colonialism, disentangling and confronting the notion of Denmark as a "benevolent" colonial power (e.g., Boassen et al. 2022, 1; Boassen et al. 2022b; Kuratorisk Aktion 2006; Lynge 2011). New forms of public engagement and resistance are emerging in consequence. This debate continues to intensify and resonates with the following sections, which illustrate how specific public participation procedures have been put in place, including as a consequence of colonial structures and practices, and have clashed with local understandings of appropriate methods of decision-making.[9]

MATERIAL AND NONMATERIAL EFFECTS OF ANTICIPATED EXTRACTION

In 2010 and 2011, Scottish Cairn Energy[10] used Aasiaat, a town of about 3,100 (Statbank Greenland 2017, 9) in the south of Disko Bay, as regional supply base for the first offshore hydrocarbon drillings since Greenland had achieved full control over its extractive resources. Two years earlier, the United States Geological Survey (USGS) had made public estimates of large quantities of undiscovered offshore oil and gas in Greenlandic waters, which were circulated widely and reinforced the notion of Greenland as constituting a "resource frontier" (Nuttall 2012a; Poppel 2018). Following earlier unsuccessful attempts by other companies, Cairn Energy did not strike relevant quantities of oil or gas and gave up drilling after two summers. Nevertheless, the estimated volumes of oil and gas assumed a lasting material presence and created hopes, concerns, and expectations.

The Materialization of Expectations[11]

What lies (or is expected) below the surface influences socio-political developments above the ground. The anticipated future exploitation of extractive resources in Greenland has profound present implications and shapes how citizens interact with an evolving political and administrative system.[12] Recent studies engage with the social and political impacts and material effects of envisioned future resource exploitation in Greenland and other parts of the Arctic (e.g., Gad 2017; Hansen and Johnstone 2019; Wilson et al. 2017), and draw attention to the role of emotions and affects in these contexts, which may be evoked, employed, or silenced for specific political aims and actions (Bjørst 2020; Kangasluoma and Lempinen 2022; Lindroth and Lempinen 2021). In the context of participation procedures, attention may be paid to their existence and impacts as well as the reasons for when and why emotions and affects do not enter into or influence decision-making. Lindroth and Lempinen (2021) note that "emotional impacts start well before any concrete developments and do not just disappear if the envisioned developments never materialize" (6). What does materialize are maps, laws, reports, infrastructures, institutions, activist groups, and so on (Kangasluoma and Lempinen 2022, 12; Weszkalnys 2008). In the early years of self-government, Greenland resembled what Gisa Weszkalnys (2011) has named an "oil economy without oil" as a result of "anticipatory activities" (351). An institutional framework was set up to govern hydrocarbon operations. This included the establishment of a sovereign wealth fund for future oil and gas revenues. Greenland's small national oil company, Nunaoil, participates as a carried partner in all oil and gas exploration licenses (Nunaoil n.d.), and the government has promoted the country's extractive resource potential abroad (e.g., Naalakkersuisut 2014; Nuttall 2012b).

Preparations were also made at the local level. In 2011, a report published by Qaasuitsup Municipality singled out Aasiaat as the location that would best meet onshore requirements for servicing regional offshore operations (Qaasuitsup Kommunia 2011). The report envisioned scenarios for potential exploration, exploitation, and expansion phases. Maps and illustrations depicted large industrial and residential areas and an extension to the airport. Neighboring Tupilak Island was selected in the report as suitable location for a larger harbor, which, the plans indicated, could be made accessible from Aasiaat via a bridge (Qaasuitsup Kommunia 2011). With these maps and images, clear boundaries were drawn around imaginations of the town's future physical appearance. In addition, opportunities were created for residents to prepare for their own roles in this imagined future extractive economy: local authorities initiated training courses for jobs servicing the oil and gas industry. Cairn Energy, the oil company, supported a business simulation game to give high-school students hands-on experience with oil and gas activities (Simprentis 2013). Researchers (such as myself) and activists arrived in town to learn about local experiences and concerns or to draw attention to the risks associated with offshore drilling operations. For a period of time, the petroleum industry became part of everyday life and ceased to belong solely to an abstract vision of the future.

Hopes, Concerns, and Expectations

When I spent time in Aasiaat in 2012, people remembered the circulation of hopeful speculations and a general feeling of excitement prior to Cairn Energy's arrival. They recalled rumors of a large hotel that might be built, and some expected the town to eventually become rich. Expectations had been influenced and reshaped by the two summers (2010 and 2011) during which Cairn Energy had been present in town. Accounts of that time bore close resemblance to what Ferguson (2005) has referred to as the oil industry's "enclave" character: the company had placed a fence around the harbor (for which the market stand used by hunters and fishermen had to be relocated), accommodated foreign oil rig workers on a vessel offshore, and transported staff by helicopter from the airport to the oil rig and back. Excitement had soon started to fade as people felt the company had not kept its promise to create more than a few jobs. As one woman explained,

> Finally we thought that we were needed, that we could get work, and then they used other people from other countries. So we thought 'well, we were not needed'. Why did we have to prepare, if they were not going to use us anyway? (July 2012)

Nevertheless, some hope did endure. People in Aasiaat often remained of two minds over the oil and gas drilling activities. Only few people I spoke with supported exploration activities without any concerns, and none rejected them without also acknowledging potential benefits. Many worried about certain aspects of oil operations, yet also saw potential benefits. These local responses were tied up in past experiences and current needs and focused on labor market prospects, the pace of change (see also Wilson 2015, 14–15), environmental impacts, and the risk of culture loss.

Many referred to the period that followed the incorporation of Greenland into the Kingdom of Denmark in 1956. While officially the end of Greenland's colonial status, this new set-up marked the beginning of a substantial "modernization" programme carried out mainly by a Danish labor force, over which the Greenlandic population had little control (Petersen 1995). I was told of concerns that, as in the past, people would not be able to determine the pace or process of change and be bypassed on the labor market (see also Hansen et al. 2016).

At the same time, it was feared that contemporary challenges facing the community would create problems in the future. The low level of formal training, language barriers, and lack of relevant experience were expected to hinder local employment. Many also worried that oil and gas revenues would accrue mainly to the larger towns and bypass Aasiaat—a fear rooted in a recent municipal reform that had stripped Aasiaat of its status as municipal center. Natural resources and the surrounding environment have sustained culture and subsistence in Greenland for thousands of years. Threats to the environment took precedence among interviewees, who referred to the importance of natural resources for everyday life and culture.

Hopes and positive expectations were based more strongly on current needs. In light of a potential oil future, people in Aasiaat primarily hoped for economic development and infrastructure improvements. Many expected the town to grow and provide more labor opportunities and leisure activities, should oil or gas be discovered. The oil industry, it was hoped, could address the urgent need for economic diversification, create new employment opportunities, and stimulate incentives for better education. Needs identified by people in Aasiaat were primarily communal in nature and included better roads, additional medical staff, and youth leisure activities. Many also hoped that Greenland would develop closer ties with countries other than Denmark. Contrary to political and media discourse, political independence was rarely mentioned.[13]

While it is the stated aim of Greenland's Social Impact Assessment (SIA) guidelines "to involve in a meaningful manner affected towns, settlements and communities (individuals) that may be directly or indirectly impacted throughout [a] project by utilizing and respecting local knowledge, experience, culture and values" (Naalakkersuisut 2016a, 6), many of the insights, views and concerns held by individuals on oil and gas exploration activities never entered into decision-making processes. As a consequence, both frustration with and calls for improved public involvement have been growing (see also Nuttall 2013; Hansen and Johnstone 2019).

NONPARTICIPATION AND THE MAKING AND UNMAKING OF STAKEHOLDERS[14]

As parts of Greenland are turned into "resource spaces" (Nuttall 2016), some towns and settlements are turned into affected communities and their residents become stakeholders and potential publics for participation. This has involved political and industry narratives about the places of planned resource projects, which are also advanced in the public meetings organized by the administration and companies (ibid). However, all of these processes are met with resistance and consequently remain unstable.

Affected Communities

On a Sunday at the end of November 2015, representatives of the Greenlandic government, a member of the Danish Centre for Environment and Energy (DCE), the managing director of Australian mining company Ironbark, and a consultant from Sweco Danmark arrived in Ilulissat on Greenland's west coast for a public meeting concerned with Ironbark's plans to build a zinc and lead mine in Greenland's national park in the far northeast of the country. Stormy weather had delayed the group's visit by a week and it now coincided with the traditional lighting of the town's communal Christmas tree. Less than ten people filled a small proportion of the available seats in the auditorium, as the sounds of the Christmas market taking place in the sports hall next door floated into the room and competed for attention with the presentations.

With the use of colorful slides, an overwhelmingly positive picture of the project was painted. Ironbark highlighted expected financial benefits and employment opportunities and assured listeners that environmental impacts would be low, although the presentation remained vague on this point. The basis on which impacts were judged was equally obscure. For instance, the company claimed that animals in the area would not be affected in a significant way, as they were simply expected to move elsewhere. Representatives of the Department for Industry, Labour, and Trade and the DCE presented assessments of Ironbark's Environmental and Social Impact Assessment (EIA/SIA) reports.[15] They evaluated the project positively and framed criticisms as suggestions to the company or reassured the audience that the relevant government bodies were confident the company would find adequate solutions to outstanding risks.

This pattern was repeated in the subsequent Q&A session, for which questions and comments had to be submitted in writing at the beginning of the meeting before any information had been provided. In one of his responses, the Minister for Finance, Resources, and Foreign affairs pointed out that it was important not to ask too much of Ironbark (whose director was sitting next to him) in order to maintain a positive investment climate in Greenland. The public had been invited to "come and hear about the Ironbark project." However, the slides presented by the company classified the meeting not as a public meeting intended to provide information, but as a public hearing (*offentlig høring*) and the questions and responses were used to fulfill the consultation requirement of the licensing process.[16]

Extractive companies often emphasize lack of critical questions or opposition at public hearings in their documents. However, the actual practice of these meetings challenges interpretations of nonparticipation or silence as simple disinterest or consent (see also Graugaard et al. 2020), and draws attention to the historical, political, and economic structures in which they are situated. Corporate and political exercises, such as the one described here, suggest that power works in subtle ways to exclude people and voices from and within participatory spaces. In addition, the Ironbark hearing presents an illustration of the ways in which "publics" or "stakeholders" come into being through these processes (Chilvers and Kearnes 2016b). As Braun and Schultz (2010) state, "'the public' [for formal participatory exercises] is never immediately given, but inevitably the outcome of processes of naming and framing, staging, selection and priority setting, attribution, interpellation, categorization and classification" (406).

Greenland's SIA Guidelines stipulate that companies are to carry out a stakeholder analysis to identify "important and less important stakeholders" (Naalakkersuisut 2016a, 24). Half a page is dedicated to this question and a short list is provided of groups in the public sector, nongovernmental organizations (NGOs), and "others"—a category that includes "affected communities," without specifying how a town or settlement qualifies as such (Naalakkersuisut 2016a, 24).[17] In the winter, it takes 20 minutes and an expensive airplane ticket to travel from Aasiaat to Ilulissat. While public meetings for the Ironbark project were held in Sisimiut (south of Aasiaat) and Ilulissat, Aasiaat was bypassed. All three towns are located thousands of miles away from the planned zinc and lead mine, a fact that prompted a member of the administration in Sisimiut to express the following:

> I find it a little bit strange to have public hearings in some towns and not in other towns . . . We have had them here in Sisimiut and in Kangerlussuaq and Ilulissat and Nuuk. And why in those towns and not in other towns? . . . There is no meaning. (May 2016)[18]

The fact that no public meeting was held in Aasiaat about the proposed Ironbark mine in 2015 underscores the sense that, with the departure of Cairn Energy after their second drilling season in 2011, people in Aasiaat had lost their status as key stakeholders on the path toward Greenland's anticipated extractive future. While many had experienced exploration activities in their own town from a distance and a possible return of the oil industry might still loom on the horizon, for the moment, people in Aasiaat are removed both spatially and temporally from extractive operations.

In other parts of the country, some of the mining projects moving through the licensing process are not "new," but have been in development for decades. Others are planned in places of former commercial resource exploitation, in which Greenlanders had no say (see also Aaen 2012, 23; Hansen et al. 2009; Petersen 1995). "Imaginaries that public elites use to frame publics can be powerful," according to Soneryd (2016) "but they are also time and context specific" (152).[19] Around the world (the consequences of), colonial practices continue to shape decision-making and participation in various ways.

Finding the Right Method?

In Greenland's capital Nuuk, members of the administration and NGOs involved in organizing or working to improve public participation processes voiced a sincere concern that they had not "found the right method" for Greenland yet (see also Olsen and Hansen 2014, 77). Some expressed an acute awareness that nonattendance at public meetings was no indication of indifference to planned projects (or of lack of opposition, as has been implied in company documents). Referring to his own experience, one Greenlandic representative of a mining company recounted:

> When I am sitting at some private Kaffemik[20] . . . people are very interested, and people have an opinion. But many of them, they don't come to the public meetings. (Nuuk, September 2015)

Observers of public participation processes in Greenland have noted that direct and confrontational forms of communication have not traditionally formed part of valued social interaction or been a means to resolve disagreement in Greenland (Hansen et al. 2009, 67; Olsen and Hansen 2014, 78). The sociocultural incompatibility of procedures consequently hinders meaningful participation (see also Hansen 2014, 16–17; Hansen and Johnstone 2019, 487). The small size of towns and settlements plays into this (see also Aaen 2012, 59). Several people in Aasiaat were concerned that voicing an unpopular opinion in a public meeting could have negative repercussions for future employment opportunities and one young woman in Nuuk explained:

> There are so many topics that are taboo or that you cannot really talk about, or which are difficult to talk about . . . Oh, we have to think about this and maybe Kim Kielsen[110] is my cousin's husband or whatever and then we have to meet at the next family get-together. So, I have to be able to face him without being embarrassed or without embarrassing his mother or my mother or whatever . . . So, we are very, very careful as a people. (May 2016)

The cultural inappropriateness of some participatory procedures was also highlighted by people in Aasiaat. Interviewees suggested that decision-makers should spend longer than just a few hours in a community to gain a better understanding of the local context, to allow for conversations to take place spontaneously and outdoors, and to give people time to think before providing a response, which would better correspond with Inuit ways of communicating (see also Graugaard et al. 2020, 15).[21]

Colonial Logics of Extraction and Participation

Current administrative—and participatory—procedures in Greenland are the result of the way in which the relationship between Denmark and Greenland has played out over the past two decades. Greenland has gained greater autonomy and taken home an increasing number of policy areas from Denmark, yet a significant number of Danish administrators and experts continue to work in Greenland and staff turnover remains high. While the government of Greenland has used anticipated future extraction to gain greater distance from Denmark (Gad 2017), the set-up and functioning of its administrative structures have continued to position Denmark as a reference point, including in some efforts to improve participation processes. This may find explicit expression, as in the statement of one Danish administrator in Nuuk:

> We compare ourselves to Denmark, for instance, and . . . they know how to do it. Up here, it is relatively new. And we kind of expect both the public and us . . . to be the same as in Denmark. And it is not. (Member of the Environmental Agency for Mineral Resources, Nuuk, September 2015).

As argued by Soneryd (2016), such processes "reproduce ideas of centre and periphery. It is from the 'centres' that 'best practices' can be exported to the 'peripheries' that are not as well equipped with democratic traditions and experiences" (156). In the context of Greenland, "success" defined in terms of measuring up to Denmark is deeply rooted within colonial ideology and practice (Lynge 2011). More often, these dynamics work in subtle ways. In the fall of 2015, the Greenland chapter of Transparency International organized a workshop on public participation in Nuuk. A Danish expert flew in to present a public participation method, which had been tested successfully in Denmark and, it was suggested, might work well in Greenland, too. As a result of such processes not only are methods and technologies transferred; by placing a focus on improving existing practice, the status quo may be reinforced and underlying structures remain unchallenged. In this way, efforts for greater inclusion may lead to reinforced exclusion.

Some people in Aasiaat also referred to colonial experiences to explain why they had not participated actively in public meetings held on various issues in Aasiaat. Aviâja Egede Lynge and Robert Petersen have described such dynamics in Greenland as "mental colonization" (Lynge 2006; 2011) and "internal colonialism" (Petersen 1995, 118).

> Greenlanders are used to the fact that they didn't have anything to say about anything in the government . . . Sometimes I hesitate to say my opinion because I am used to this kind of thinking. (March 2016, translated)

> We were raised not to have a critical opinion, we should not be critical. I think it comes from the colonial times, where you had to say yes to anything the authority said. (April 2016)

> We didn't use to be critical, so people are holding back. Only over the last few years, this has started to change. (December 2015)

There was general agreement that young Greenlanders had become more outspoken and ready to engage in public debates. However, young people have continued to be underrepresented in public hearings organized by extractive companies and the administration (see also Olsen and Hansen

2014, 78), which often represent culturally or socially insecure spaces for communication.[22]

Distanced Stakeholders

Developments in Greenland have highlighted the fact that particular resources gain relevance as a consequence of specific political and economic forces—be they colonial exploitation, geopolitical objectives, nation-building, regional and local needs, or environmental or economic interests—as well as the uncertain outcomes of any efforts to explore, promote, and exploit the subsurface. After decades of costly offshore exploration, it remains unknown whether commercially viable deposits of oil and gas exist in Greenlandic waters. Similarly, the rare earth elements and uranium mine in southern Greenland has been put on hold after a ban on uranium mining in 2021 (McGwin 2022). Actors involved in these processes construct Greenland variously as empty wilderness full of resources to be "found" and exploited, a vulnerable ecosystem in need of protection for all of humankind, or a modern, stable democracy with decision-making based firmly on internationally accepted best practice procedures (see also Nuttall 2012b and this volume; Hastrup and Lien 2020). Whether or not Greenland's population appears in these narratives depends on the broader objectives for which they are produced.

People in places such as Aasiaat resist being made into stakeholders of natural resource extraction. As previous sections of this chapter have shown, reasons for nonattendance at public meetings are diverse and complex. Among them was a claim by individuals that they were not interested in the oil and gas activities that had been carried out. A lack of interest in oil and gas exploration was not expressed in terms of protest against decision-making procedures, but may be a sign of people distancing themselves from extractive activities (see also Hansen and Tejsner 2016). At the same time, people in Aasiaat also have resisted *not* being recognized as stakeholders: despite the great distance between Aasiaat and the planned rare earth elements and uranium mine in the south of Greenland, several dozen Aasiammiut felt compelled to take to the streets to join countrywide anti-uranium demonstrations in early 2016. One of the main objectives of protesters was to pressure the government to hold a public referendum on the issue. Not all wanted to have a say personally, but they expressed deep concern and disagreed with the way in which decisions were being made. Some felt those living in the area would know better—but people living in the vicinity of the planned mine themselves experienced a lack of information about the project and decision-making procedures prior to the ban on uranium mining in 2021 (Hansen and Johnstone 2019). At the time of research for this chapter, a growing number of discussions were also taking place on social media or in the comments' sections beneath news articles on the shared website of Greenland's two national newspapers *Sermitsiaq.AG*, and the website of national broadcaster *KNR.gl*. This revealed a mismatch between stakeholders identified in official documents and the selection of locations for public meetings, and peoples' own interpretations of whether or not they were affected by or felt a connection with planned projects. In Aasiaat, highly informed and opinionated individuals had, for different reasons, never attended a public meeting, but continued to follow developments closely. Opposition to uranium mining and oil exploration in Greenland has also been expressed through paintings and songs. Examples are Bolatta Silis-Høegh's (2015) Exhibition "Lights On Lights Off" or the song "Uulia" by Nive Nielsen & The Deer Children. Such engagement with extractive projects in spaces other than public meetings emphasize that nonparticipation is not equivalent to disengagement or a general lack of interest.[23]

CONCLUSION

Hydrocarbon exploration has created hopes, concerns, and expectations in Greenland, which are tied up with past experiences and current needs at the communal level. However, despite the stated aim of public consultation requirements to "involve relevant stakeholders" (Naalakkersuisut 2016a, 15) in the licensing process, many people and their views and insights have remained excluded from formal decision-making and the public meetings organized by companies and the administration. Developments in Greenland illustrate how certain methods and procedures are in place as a result of colonial practices and policies, which can similarly shape people's reactions to them. Individuals in Aasiaat have at times resisted being drawn into natural resource exploration and extraction processes as stakeholders, and at other times have also resisted *not* being viewed as stakeholders in extraction projects taking place in Greenland. Their engagement with on-going exploration and planning processes has not been restricted to the formal spaces of public participation. These dynamics simultaneously challenge definitions of "stakeholders" and binary narratives of participation/nonparticipation and illustrate how certain efforts for greater inclusion may result in continued exclusion from formal decision-making by reinforcing the status quo.

Postscriptum

This chapter was written based on interviews and observation carried out in the context of a Master's thesis and PhD dissertation, which, upon critical self-reflection, I have recognized as forming part of an extractive, colonial research practice. While research results were shared in Aasiaat and relationships built during the research process have lasted, the research design was set up and carried out in a nonparticipatory, noncollaborative format, which—as has been pointed out in the introduction of this edited collection—are rooted deeply within the exclusionary foundation of Western academic institutions and contribute to epistemic exclusions also in critical resource scholarship. Since then, I have been involved in several projects engaging with this issue and am immensely grateful to Indigenous scholars and friends who have been patient and generous in my ongoing learning journey.

ACKNOWLEDGMENTS

I would like to thank everyone in Kalaallit Nunaat, who so generously shared their time, knowledge, and experiences with me! This work was supported by an ESRC doctoral grant (ES/J500112/1). Thanks to Emma Wilson for proofreading!

NOTES

1. Jørgensen (2019) illustrates how narratives of this past have changed among those affected by relocation in connection with efforts to actively shape the future.
2. Most parts of this paragraph are taken from Doering (2019, 242–243).
3. Public meetings can be public hearings (to be held by the authorities) and information meetings. Distinction is often difficult at the community-level and diverse understandings of their purposes have been noted (Olsen and Hansen 2014).
4. Interest by the hydrocarbon industry has decreased (Poppel 2018) and in 2021, the government announced the planned ratification of the Paris Agreement and suspension of hydrocarbon exploration (Naalakkersuisut 2021).
5. Parts of this paragraph are taken word-for-word from Doering (2019, 19).
6. As noted by the ICC (2020), Indigenous peoples are rights holders, not stakeholders. In this chapter, I engage with the concept of "stakeholders" and follow the terminology used in official English-language documents.
7. The chapter is in parts a revised version of a chapter of my PhD dissertation (Doering 2019, 135–159) with elements taken from the introduction (11–36) and conclusion (232–243) and my Master's thesis (Doering 2013). Where I have taken texts word-for-word, they are marked in footnotes.
8. Cornwall (2006) analyses the historical development of "participation", traces its colonial roots, and provides a brief overview of the emergence of the stakeholder concept.
9. But see also the postscriptum.
10. Cairn Energy operated through its subsidiary Capricorn in Greenland.
11. This section and the next section closely follow Doering (2013). Sentences also taken from Doering (2019, 14).

12. Taken from Doering (2019, 14).

13. Hansen and Johnstone (2019) provide a similar account from the south of Greenland.

14. Chilvers and Kearnes (2016b, 32) write of the making and re-making of publics and democracy.

15. The former department was responsible for approving SIAs, the latter worked closely with the Environmental Agency for Mineral Resources Activities, which approved EIAs.

16. This section to this point is a revised version of a paragraph in Doering (2019, 145–146).

17. The 2009 Guidelines noted that "[i]t might [. . .] be necessary to include the entire population of Greenland and not only the local community, because e.g. the project is situated far away from inhabited areas or the potential labour force living in close proximity of the mining project might be too limited" (BMP 2009, 5; see also Aaen 2012, 58).

18. Parts of this paragraph are taken word-for-word from Doering (2019, 148).

19. Barnes et al. (2003) similarly engage with the construction of publics for participation in the United Kingdom and note the influence of ideologies and intentions.

20. Kaffemiks are a way to celebrate important social events in Greenland, such as birthdays, baptisms, or confirmations.

21. Graugaard et al. (2020) provide recommendations for improved SIA/EIA practice based on a workshop held in 2019 in Nuuk, with a focus on community-led impact assessment. Note that parts of this section are taken from Doering (2019, 152–154, 216).

22. Parts of this section taken word-for-word from Doering (2019, 153–154, 214).

23. For participation outside formally created spaces, see e.g., Chilvers and Kearnes (2016a), Cornwall (2002). An example is Indigenous self-developed FPIC protocols (Lightfoot 2020).

REFERENCES

Aaen, Sara Bjørn. 2012. *Demokratisk legitimitet i høringsprocesser i forbindelse med storskala-projekter i Grønland.* Accessed January 21, 2019. http://www.ga.gl/LinkClick.aspx?fileticket=u3Dqm159SjY%3D&tabid=36&language=da-DK.

Aktion, Kuratorisk, ed. 2006. *Rethinking Nordic Colonialism: A Postcolonial Exhibition Project in Five Acts.* Accessed December 19, 2022. https://www.rethinking-nordic-colonialism.org.

Barnes, Marian, Janet Newman, Andrew Knops, and Helen Sullivan. 2003. "Constituting 'the Public' in Public Participation." *Public Administration* 81 (2): 379–399.

Berry, Dawn Alexandrea. 2012. "Cryolite, the Canadian Aluminium Industry and the American Occupation of Greenland During the Second World War." *The Polar Journal* 2 (2): 219–235. https://doi.org/10.1080/2154896X.2012.735037.

Bishop, Patrick, and Glyn Davis. 2002. "Mapping Public Participation in Policy Choices." *Australian Journal of Public Administration* 61 (1): 14–29. https://doi.org/10.1111/1467-8500.00255.

Bjørst, Lill Rastad. 2020. "Stories, Emotions, Partnerships and the Quest for Stable Relationships in the Greenlandic Mining Sector." *Polar Record* 56 (E23): 1–13. https://doi.org/10.1017/S0032247420000261.

Boassen, Paninnguaq, Èva Cossette-Laneville, Naja Dyrendom Graugaard, Lars Jensen, Björn Lingner, and Regine-Ellen Dora Møller. 2022 "Introduction." In *Greenland - 301 Years Later: Kalaallit Nunaat ukiut 301-nngornerat.* Accessed December 19, 2022. http://postkolonial.dk/wp-content/uploads/2022/06/2-Introduction-Lars-Regine-Pani-BL-ECL-Naja-final-draft.docx.pdf.

———, eds. 2022b. *Greenland—301 Years Later: Kalaallit Nunaat ukiut 301-nngornerat.* Accessed December 19, 2022. http://postkolonial.dk/wp-content/uploads/2022/06/2-Introduction-Lars-Regine-Pani-BL-ECL-Naja-final-draft.docx.pdf.

Braun, Kathrin, and Susanne Schultz. 2010. "'…a Certain Amount of Engineering Involved': Constructing the Public in Participatory Governance Arrangements." *Public Understanding of Science* 19 (4): 403–419. DOI: 10.1177/0963662509347814.

Buchy, M., and S. Hoverman. 2000. "Understanding Public Participation in Forest Planning: A Review." *Forest Policy and Economics* 1 (1): 15–25. https://doi: 10.1016/S1389- 9341(00)00006-X.

Bureau of Minerals and Petroleum [BMP]. 2009. *Guidelines for Social Impact Assessments for Mining Projects in Greenland.* Bureau of Minerals and Petroleum, Greenland. Accessed January 2, 2020. https://www.socialimpactassessment.com/documents/sia_guidelines.pdf.

Chilvers, Jason, and Matthew Kearnes. 2016a. "Science, Democracy, and Emergent Publics." In *Remaking Participation: Science, Environment and Emergent Publics*, edited by Jason Chilvers and Matthew Kearnes, 1–28. London and New York: Routledge.

———. 2016b. "Participation in the Making: Rethinking Public Engagement in Co-Productionist Terms." In *Remaking Participation: Science, Environment and Emergent Publics*, edited by Jason Chilvers and Matthew Kearnes, 1–28. London and New York: Routledge.

Cooke, Bill, and Uma Kothari, eds. 2001. *Participation: The New Tyranny?* London and New York: Zed Books.

Cornwall, Andrea. 2002. "Locating Citizen Participation." *IDS Bulletin* 33 (2): i–x. https://doi: 10.1111/j.1759-5436.2002.tb00016.x.

———. 2006. "Historical Perspectives on Participation in Development." *Commonwealth & Comparative Politics* 44 (1): 62–83. https://doi: 10.1080/14662040600624460.

Dahl, Parnuna Petrina Egede, and Anne Merrild Hansen. 2019. "Does Indigenous Knowledge Occur in and Influence Impact Assessment Reports? Exploring Consultation Remarks in Three Cases of Mining Projects in Greenland." *Arctic Review on Law and Politics* 10: 165–189. https://doi.org/10.23865/arctic.v10.1344.

Doering, Nina. 2013. "Hopes, Concerns and Expectations in Greenland's Emerging Oil Economy." MPhil Thesis, unpublished University of Oxford.

Doering, Nina. 2019. "Public Participation, Democratic Decision-Making and Extractive Resource Management in Greenland." DPhil Thesis, University of Oxford.

Ferguson, James. 2005. "Seeing like an Oil Company: Space, Security, and Global Capital in Neoliberal Africa." *American Anthropologist* 107 (3): 377–382.

Fjellheim, Eva Maria. 2023. "'You Can Kill Us with Dialogue:' Critical Perspectives on Wind Energy Development in a Nordic-Saami Green Colonial Context." *Human Rights Review* 24: 25–51. https://doi.org/10.1007/s12142-023-00678-4

Gad, Ulrik Pram. 2017. *National Identity Politics and Postcolonial Sovereignty Games: Greenland, Denmark, and the European Union.* Copenhagen: Museum Tusculanum Press.

Graugaard, Naja Dyrendom, Anne Merrild Hansen, Parnuna Egede Dahl, and Rasmus Kløcker Larsen. 2020. *LOVISA Workshop Report: Local Knowledge in Environmental and Social Impact Assessment* . AAU Arctic, Aalborg University. Accessed January 20, 2023. https://www.arctic.aau.dk/news/news/lovisa-workshop-report-local-knowledge-in-environmental-and-social-impact-assessment.cid468991

Hansen, Anne Merrild. 2014. *Community Impacts: Public Participation, Culture and Democracy*. University of Copenhagen. Accessed February 3, 2017. http://vbn.aau.dk/files/186256309/Community_Impacts.pdf.

Hansen, Anne Merrild, and Rachael Lorna Johnstone. 2019. "In the Shadow of the Mountain: Assessing Early Impacts on Community Development from Two Mining Prospects in South Greenland." *The Extractive Industries and Society* 6 (2): 480–488. https://doi.org/10.1016/j.exis.2019.01.012.

Hansen, Anne Merrild, Frank Vanclay, Peter Coral, and Anna-Sofie Hurup Skjervedal. 2016. "Managing the Social Impacts of the Rapidly-Expanding Extractive Industries in Greenland." *The Extractive Industries and Society* 3 (1): 25–33. https://DOI: 10.1016/j.exis.2015.11.013.

Hansen, Klaus Georg, Freia Lund Sørensen, and Steen Jeppson, 2009. "Decision Processes, Communication and Democracy: The Aluminium Smelter Project in Greenland." In *Knowledge-Based Tools for Sustainable Governance of Energy and Climate Adaptation in the Nordic Periphery (K-Base)*, edited by Janne Hukkinen, Klaus Georg Hansen, Richard Langlais, Rasmus Ole Rasmussen, Steen Jeppson, Jarkko Levänen, Freia Lund Sørensen, Peter Schmitt, and Stefanie Lange, 55–84. Stockholm: Nordregio. http://www.diva-portal.se/smash/get/diva2:700378/FULLTEXT01.pdf.

Hastrup, Frida, and Marianne Elisabeth Lien. 2020. "Welfare Frontiers? Resource Practices in the Nordic Arctic Anthropocene." *Anthropological Journal of European Cultures* 29 (1): v-xxi. https://doi: 10.3167/ajec.2020.290101.

Inatsisartut. 2009. *Greenland Parliament Act no. 7 of 7 December 2009 on Mineral Resources and Mineral Resource Activities (the "Mineral Resources Act") (Unofficial Consolidation).* Accessed January 2, 2019. https://www.govmin.gl/images/Documents/Mineral_resource_act/Unofficial_translation_of_unofficial_consolidation_of_the_Mineral_Resources_Act.pdf.

———. 2012. *Greenland Parliament Act No. 26 of 18 December 2012 (Unofficial translation).* Accessed January 2, 2019. https://www.govmin.gl/images/Documents/Mineral_resource_act/Unofficial_translation_of_unofficial_consolidation_of_the_Mineral_Resources_Act.pdf.

International Labour Organization [ILO]. 1989. *Indigenous and Tribal Peoples Convention, C169.* Accessed February 14, 2022. https://www.refworld.org/docid/3ddb6d514.html.

Inuit Circumpolar Council [ICC]. 2010. *ICC Charter.* Accessed January 2, 2019. https://inuit.org/?page_id=252.

———. 2020. *Policy Paper on the Matter of "Local Communities."* Accessed January 2, 2021. https://secureservercdn.net/45.40.145.201/hh3.0e7.myftpupload.com/wp-content/uploads/FINAL-ICC-Policy-Paper-on-matter-of-local-communities-2.pdf

———. 2021. *Ethical and Equitable Engagement Synthesis Report: A Collection of Inuit Rules, Guidelines, Protocols, and Values for the Engagement of Inuit Communities and Indigenous Knowledge from Across Inuit Nunaat.* Accessed September 2, 2022. https://hh30e7.p3cdn1.secureserver.net/wp-content/uploads/ICC-EEE-Synthesis-report-WEB.pdf

Inuit Tapiriit Kanatami. 2018. *National Inuit Strategy on Research.* Accessed July 2, 2022. https://www.itk.ca/national-inuit-strategy-on-research/.

Jørgensen, Anne Mette. 2019. "Communitification and Emotional Capital: Producing, Shaping and Re-Shaping Communities Before and After Mining in Norbotten and Disko Bay." *Polar Record* 56 (e7). https://doi.org/10.1017/ S0032247419000548.

Kangasluoma, Sohvi, and Hanna Lempinen. 2022. "Making of the Arctic Dream – Affective Resources in the Strategies of Arctic Coastal States." *Globalizations.* https://doi.org/10.1080/14747731.2022.2091869.

Langhoff, Rune. 2013. *Med folkets mandat? Høringsprocesser og borgerinddragelse pa° råstofområdet.* ICC and WWF. Accessed January 16, 2022. http://awsassets.wwfdk.panda.org/downloads/med_folkets_mandat___dansk.pdf.

Lawrence, Rebecca, and Rasmus Kløcker Larsen. 2017. "The Politics of Planning: Assessing the Impacts of Mining on Sami Lands." *Third World Quarterly* 38 (5): 1164–1180. DOI: 10.1080/01436597.2016.1257909.

Lempinen, Hanna, and Marjo Lindroth. 2021. "Fear and Hoping in the Arctic. Charting the Emotional Fabric of Resource Extraction." *The Extractive Industries and Society* 8 (2): 100872. https://doi.org/10.1016/j.exis.2021.01.007.

Lightfoot, Sherly. 2020. "Indigenous Laws and Governance in Indigenous Self-Developed FPIC Protocols." https://fnemc.ca/wp-content/uploads/2015/07/Sheryl-Lightfoot-Expert-Paper-Indigenous-Legal-Framework.pdf

Lindroth, Marjo. 2020. "Merging of Resources and Independence." In *Resources, Social and Cultural Sustainabilities in the Arctic*, edited by Monica Tennberg, Hanna Lempinen, and Susanna Pirnes, 15–26. London and New York: Routledge.

Lynge, Aviâja Egede. 2006. "The Best Colony in the World. Act 2: Greenland, April 21 – May 14, 2006." In *Rethinking Nordic Colonialism: A Postcolonial Exhibition Project in Five Acts*, edited by Kuratorisk Aktion. Accessed December 19, 2022. http://rethinking-nordic-colonialism.org/files/pdf/ACT2/ESSAYS/Lynge.pdf.

———. 2011. "Mental Decolonization in Greenland." *Inter-Nord* 21: 273–276

McGwin, Kevin. 2022. "Greenland Plans to Reject Rare-Earth Mining Permit." *Polar Journal*, August 5, 2022. https://polarjournal.ch/en/2022/08/05/greenland-plans-to-reject-rare-earth-mining-permit/.

Mitchell, Terry, Courtney Arsenau, Darren Thomas, and Peggy Smith. 2019. "Towards an Indigenous-Informed Relational Approach to Free, Prior, and Informed Consent (FPIC)." *International Indigenous Policy Journal* 10 (4). https://doi.org/10.18584/iipj.2019.10.4.8372.

Naalakkersuisut. 2014. *Greenland's Oil and Mineral Strategy 2014-2018*. Accessed March 19, 2019. https://naalakkersuisut.gl/~/media/Nanoq/Files/Publications/Raastof/ENG/Greenland oil and mineral strategy 2014-2018_ENG.pdf.

———. 2016a. *Social Impact Assessment (SIA) Guidelines on the Process and Preparation of the SIA Report for Mineral Projects*. Accessed September 20, 2022. https://govmin.gl/wp-content/uploads/2019/09/SIA_guideline.pdf.

———. 2016b. *Guidelines for Preparing an Environmental Impact Assessment (EIA) Report for Mineral Exploitation in Greenland.* Accessed September 20, 2022. https://govmin.gl/wp-content/uploads/2022/05/VVMEIA_UK.pdf.

———. 2020. *Greenland's Mineral Strategy 2020-2024*. Accessed September 20, 2022. https://govmin.gl/wp-content/uploads/2020/03/Greenlands_Mineral_Strategy_2020-2024.pdf.

———. 2021. *Greenland Joins the Paris Agreement.* Accessed September 22, 2022. Naalakkersuisut. https://perma.cc/CJJ5-GMWL.

Naalakkersuisut and Government of Denmark. 2009. *Act on Greenland Self-Government*. Accessed December 29, 2018. https://naalakkersuisut.gl//~/media/Nanoq/Files/Attached Files/Engelske-tekster/Act on Greenland.pdf.

Nive Nielsen & The Deer Children. 2012. "Uulia." In *Nive Sings*. Glitterhouse (Indigo), CD.

Nunaoil. n.d. *Who We Are*. Accessed September 20, 2022. https://nunaoil.gl/en/about-nunaoil-a/s/who-are-we.

Nuttall, Mark. 2012a. "Imagining and Governing the Greenlandic Resource Frontier." *The Polar Journal* 2 (1): 113–124. https://doi.org/10.1080/2154896X.2012.679563.

———. 2012b. "The Isukasia Iron Ore Mine Controversy: Extractive Industries and Public Consultation in Greenland." *Nordia Geographical Publications* 41 (5): 23–34.

———. 2013. "Zero-Tolerance, Uranium and Greenland's Mining Future." *The Polar Journal* 3 (2): 368–383. https://doi: 10.1080/2154896X.2013.868089.

———. 2016. "The Making of Resource Spaces in Greenland." *Hotspots* (blog), *Fieldsites*, July 29, 2016. https://culanth.org/fieldsights/the-making-of-resource-spaces-in-greenland.

Olsen, Anna-Sofie Hurup, and Anne Merrild Hansen. 2014. "Perceptions of Public Participation in Impact Assessment: A Study of Offshore Oil Exploration in Greenland." *Impact Assessment and Project Appraisal* 32 (1): 72–80. https://doi.org/10.1080/14615517.2013.872842.

Petersen, Robert. 1995. "Colonialism as Seen from a Former Colonized Area." *Arctic Anthropology* 32 (2): 118–126.

Poppel, Birger. 2018. "Arctic Oil & Gas Development: The Case of Greenland." *Arctic Yearbook 2018*: 1–32. https://arcticyearbook.com/images/yearbook/2018/Scholarly_Papers/19_AY2018_Poppel.pdf.

Qaasuitsup Kommunia. 2011. *Offshore Onshore. Selection of Areas for Offshore-Related Onshore Activities in Qaasuitsup Municipality.* Accessed March 8, 2013. http://www.qaasuitsup.gl/~/media/Files/Erhverv/Olie-og-mineraler/Offshore-onshore/Offshore_Onshore_ENG.ashx.

Reed, Mark S. 2008. "Stakeholder Participation for Environmental Management: A Literature Review." *Biological Conservation* 141 (10): 2417–2431. https://doi.org/10.1016/j.biocon.2008.07.014.

Rowe, Gene, and Lynn Jayne Frewer. 2000. "Public Participation Methods: A Framework for Evaluation." *Science, Technology, & Human Values* 25 (1): 3–29. https://doi:10.1177/016224390002500101.

Sámiráđđi. 2019. *Sámi Arctic Strategy // Samisk Strategi for Arktiske saker // Sámi Árktalaš Áigumušat – Securing Enduring Influence for the Sámi People in the Arctic Through Partnerships, Education and Advocacy.* Accessed June 2, 2022. https://static1.squarespace.com/static/5dfb35a66f00d54ab0729b75/t/5e71ebcc8554fb16d61009bc/1584524293838/Samerådet+arctic+strategies+Web.pdf .

Secretariat of the Convention on Biological Diversity. 2004. "Akwé: Kon Guidelines." https://www.cbd.int/doc/publications/akwe-brochure-en.pdf.

Sejersen, Frank. 2014. *Efterforskning og udnyttelse af råstoffer i Grønland i historisk perspektiv.* University of Copenhagen. Accessed June 2, 2018. https://greenlandperspective.ku.dk/this_is_greenland_perspective/background/ report-papers/Efterforskning_og_udnyttelse_af_r_stoffer_i_Gr_nland_i_ historisk_perspektiv.pdf.

Siegstad, Mia Olsen, and Mads Fægteborg. 2015. *'Ajorpoq' - vi får ingen svar!* Accessed January 18, 2019. https://naalakkersuisut.gl/~/media/Nanoq/Files/Attached Files/Formandens Departement/Puljemidler/Ajorpoq Kuannersuit.pdf.

Silis-Høegh, Bolatta. 2015. *Lights On Lights Off.* [Exhibition]. Accessed May 2, 2019. https://www.bolatta.com/storm#/lightsonlightsoff.

Simprentis. 2013. *PetroChallenge Greenland.* Accessed March 9, 2013. http://www.simprentis.com/petrochallenge/greenland/.

Soneryd, Linda. 2015. "Technologies of Participation and the Making of Technologized Futures." In *Remaking Participation: Science, Environment and Emergent Publics*, edited by Jason Chilvers and Matthew Kearnes, 144–161. London and New York: Routledge.

Statbank Greenland. 2017. *Greenland in 2017.* Statistics Greenland. Accessed December 10, 2020, https://stat.gl/publ/en/GF/2017/pdf/Greenland%20in%20Figures%202017.pdf.

Tejsner, Pelle. 2014. *Historical Experiences and Contemporary Lessons from Quallissat.* Aarhus University. Accessed December 4, 2019 https://arctic.au.dk/news-and-events/news/show/artikel/historical-experiences-and-contemporary-lessons-from-qullissat/.

Uldum, Simon. 2017. "Evakueret: Vi skylder en stor tak til Aasiammiut." *KNR*, June 20, 2017. https://knr.gl/da/nyheder/vi-skylder-en-stor-tak-til-aasiammiut.

United Nations General Assembly. 2007. *United Nations Declaration on the Rights of Indigenous Peoples.* https://www.refworld.org/docid/471355a82.html

Weszkalnys, Gisa. 2008. "Hope & Oil: Expectations in São Tomé e Príncipe." *Review of African Political Economy* 35 (117): 473–482. https://doi.org/10.1080/03056240802411156.

Weszklanys, Gisa. 2011. "Cursed Resources, or Articulations of Economic Theory in the Gulf of Guinea." *Economy and Society* 40 (3): 345–372.

Wilson, Emma. 2015. *Energy and Minerals in Greenland: Governance, Corporate Responsibility and Social Resilience.* London: IEED. http://pubs.iied.org/16561IIED

Wilson, Emma, Anne Merrild Hansen, and Elana Wilson Rowe. 2017. "Imagining the Future: Local Perceptions of Arctic Extractive Projects That Didn't Happen." In *Arcticness: Power and Voice from the North*, edited by Ian Kelman, 130–149. London: UCL Press. https://www.jstor.org/stable/j.ctt1tm7jp1.1.

World Bank. 1996. *The World Bank Participation Sourcebook.* https://doi.org/10.1596/0-8213-3558-8.

Conclusion

Extractive Orientations

Gavin Bridge

Is extraction exhausted? Has the potential of "extraction," as a diagnostic tool or speculative concept, been all mined out? Research on extraction has proliferated over the past decade, diversifying its application and expanding the concept's meaning. Planetary infrastructure and finance; artificial intelligence and the algorithmic processing of digital data; speculative plays on urban real estate; Indigenous critiques of extractive epistemologies and research methods: across these and other thematic areas of inquiry, extraction has slipped its historic moorings to the primary sector and no longer specifies a (niche) concern with the material processes and geographies of mining. Extraction has become a capacious critique of control, exploitation, and exposure to violence and a synonym, even, for capitalism, colonialism, or modernity. What's at stake in this "journey" of extraction, from sectoral specification to an organizing concept?

My argument in this concluding chapter moves in two directions. Extraction's proliferation risks hollowing it out: applied to everything, it becomes so encompassing that it loses the capacity to specify and differentiate. Yet a strategy of simultaneously *extensifying* and *intensifying* the notion of extraction can enhance its analytical utility. Extensification here references the significant speculative potential of concepts like extraction and extractivism. It is an orientation toward an expansive mode of enquiry—already well underway—that makes connections across differences and relies, in part, on extraction's affective charge for political mobilization. More specifically, I argue that much of extraction's capacity to connect, galvanize, and enroll is linked to the "ex" of extraction—that is, to the moral and ethical claims associated with removal and involuntary displacement. Intensification, on the other hand, is an orientation toward the diagnostic value of extraction in relation to specificities of primary sector economies. Extraction's significant potential in this intensive mode can be harnessed, I argue, by focusing less on the "ex" and more in the "tract"—that is, by paying attention to the technoeconomic practices and sociomaterial configurations through which extraction attains its grip on the world, and how primary sector economies take form and endure. I illustrate this briefly by outlining four modes of economy that can take hold around (the potential for) physical displacement.

Overall, the chapter affirms the multiple possibilities of extraction as a productive concept and its capacity to do different things. It acknowledges the speculative possibilities of thinking with the "ex" of extraction for connecting across differences. Unpacking the "tract" part of the equation, however, can diagnose the geographical, material, and social specificities of primary sector processes, ensuring these are not obscured by speculative moves to identify connections across space and time. Together these two orientations—extensive and intensive, simultaneous yet counterposed—allow the concept of extraction to emerge and evolve, sustaining its generative capacity for thinking and doing.

EXTRACTION: CLASSIC EDITIONS

By convention, references to extraction perform a two-step parsing of economy: extraction distinguishes the primary sector from manufacturing and services; and then, within the primary sector, it separates extractive activities from those of agriculture. Deployed this way, extraction still names an enormous empirical diversity of practices and materials with important differences in labor regimes, forms of economic organization, political context, and ecological implications. Yet, amid this diversity of materials, practices, and implications, extraction remains a useful category because it names a key commonality: removal and mobilization. Classically, then, extraction refers to materials and bodies that have been made to move—an enforced geography of materials and bodies moving across space (vertically and horizontally) that is, at the same time, a destructive-creative process of remaking socio-ecological and spatial relations. Foremost, then, extraction references flow and involuntary displacement—a metabolic process of material mobilization understood, for the most part, from the perspective of the tributary locale. In commodity chain analysis, for example, extraction occupies the furthest left-hand station in a chain of material movement and transformation that stretches rightwards across the page. Extraction here is the originating point for a linear flow that supplies manufacturers with energy and raw materials which, transformed by labor into commodities for exchange, then become decommodified in use (or recommodified as post-consumer wastes via recovery and recycling). Similarly, classic accounts of the Anthropocene's Great Acceleration (Steffen et al. 2015) reference extraction as an anthropogenic planetary flux—a lithospheric displacement of fossil fuels, sulfide ores, mercury, phosphates, uranium, or rare earths (for example) that has surged in scale over time to now assume geological significance.

Classic accounts of extraction do more than simply reference flow, however. Work on extraction in political ecology (and related fields) adopts the same primary sector focus as technical and liberal accounts of extraction (as material mobilization) but departs from their largely apolitical analysis. Adopting a critical perspective attuned to revealing and challenging inequalities in economic and political power, this work embeds material displacement within historical geographies of colonialism and capitalism to show how extractive primary sector activities are "patterned in a particular way"–that is, how they dissolve and reconstitute forms of social and territorial (dis)order (Ye et al. 2020, 156; Rasmussen and Lund 2018; Postar et al. 2023). Contributors to this book show how these forms of order are predicated on exclusions and inclusions that reproduce violent exploitative processes on the ground. They unpack extraction into a series of binaries, revealing how they are constitutive of extraction rather than incidental or unfortunate side effects. They document how extraction's uneven power geometries are being produced, experienced, and contested, even as resource-owning states and international investors (highly concentrated forms of economic power with the capacity to control exclusive access rights) champion the primary sector—without a hint of irony—as a means to deliver "inclusive growth" (Postar et al. 2023).

Understanding how binary geographies of inclusion "and exclusion are . . . part and parcel of . . . the remaking of uneven development" (Werner 2016, 460) in the context of extraction is an urgent task. The social and spatial ordering effects consequent to extraction are central to the experiences and fantasies of extractivism, a mode of accumulation and an ideology of development premised on the large-scale (mono)production of natural resources and their export in a largely unprocessed state (Acosta 2013; Gómez-Barris 2017). A growing body of work on green colonialism and climate extractivism shows how the capacity of developmentalism to absorb critique and promote extractivism (yet again) should not be underestimated (Fjellheim 2023; Zografos and Robbins 2022; Voskoboynik and Andreucci 2022). Decarbonization strategies premised on reducing fossil fuel consumption are driving horizontal (i.e., land-hungry) energy regimes that valorize, claim, and convert land for its renewable energy generation potential; at the same time, decarbonization is mineral intensive, accelerating extraction of many (nonfuel) materials such as copper, nickel, lithium, and rare earths (Bazilian 2018; Huber and McCarthy 2017; Dunlap 2020).

Extractivism, then, is arguably stronger now than ever, whether in the guise of "pro-poor" growth, neoextractivist models of "inclusive development" that incorporate participatory processes and elements of redistribution, or climate change mitigation. Each modality of extractivism rests on specific exclusions and inclusions, yet they tend to amplify one another to reproduce and consolidate extractivism as a system of domination, violence, and dispossession. Isla's (2021, 70) notion of greening as the "highest stage" of extractivism captures the interlocking and mutually reinforcing character of exclusion, manifesting as the "massive expropriation of territories, depredation and contamination of the soil, and dispossession of the workers, through the monetization of nature that requires the devaluing of all other forms of social existence, transforming skills into deficiencies, commons into resources, knowledge into ignorance, autonomy into dependency, and men and women into commodified labour power whose needs require the mediation of the markets." Critical social theory has long acknowledged extraction's intimate connection with other forces of domination. Brechin's (1999) "pyramid of mining," for example, expresses the durable power geometries of extraction, drawing its inspiration from Mumford's (1934) description of the mechanisms of wealth and power that link mining, mechanization, metallurgy, militarism, and moneymaking. At the apex of the pyramid is mining as this extractive activity, writes Brechin (1999, 19), "retains a seminal and dominant role over the other four activities . . . from that most fundamental of industries issue the others, and from the union of all five, joined in a crystalline lattice of enduring stability and hierarchical organisation, the pyramid derives its accelerating power to transform both human societies and the organic world."

The concept of *extractivismo*—originating in analyses of neoliberal policies promoting raw material production and export from Latin America and contextualizing them by reference to centuries of land appropriation and dependent forms of development—captures the primary sector's interdependence with (and reproduction of) wider forms of social and territorial order (Gudynas 2018; Alimonda 2012; Domínguez Martín 2021). It links extraction-as-material mobilization to primitive accumulation, a mode of appropriation synonymous with colonial plunder and producing forms of territorial order and racial stratification that transcend extraction as a specific sector or economic activity. Extensive analyses of *neoextractivismo* in Latin America highlight both the continuity of raw material extraction and primary exports under left-Indigenous governments, and also a "deepening of the dynamic of dispossession . . . (around) land, resources, and territories, principally by large corporations, in multiscalar alliances with different governments" (Svampa 2015, 66). It also relates contemporary territorial struggles to "the exclusions and violence created by sedimented histories of colonial extractivism," while showing how extractivism can also generate novel territorialities, institutions, and identities as existing inclusions and exclusions are recombined and/or reinterpreted (Postar et al. 2023, 11; Rasmussen and Lund 2018; Anthias 2018). In sum, classic work on extraction centers on material displacement—a geographical flux or "dragging out" (Latin: *extrahere*) of materials from one location to another—with critical analyses showing how displacement is, at the same time, an assembling and disassembling of social and territorial order that exceeds the sites of extraction.

EX-TRACT: HARNESSING THE SPECULATIVE POTENTIAL OF EXTRACTION

The capacity of extraction/extractivism to index geographical displacement and accompanying processes of spatial and social ordering means increasingly these terms function as general signifiers. They reference the kind of exploitative relations and asymmetric structures constituted through difference documented by many of the chapters in this book, but gesture toward a wider horizon beyond the primary sector. No longer tethered to its conventional role as a sectoral identifier, extraction is experiencing a form of "conceptual creep" to reference wider organizational logics, sites of violence, and modalities of power—an "ever more complex web of extractivisms" (Chagnon et al. 2021, 177). The capacity for this expansion pivots, to a large degree, on the "ex" of extraction—the asymmetric practices and geographies of appropriation, removal, and transplantation. The "ex" connects to ontologies of separation (for which the separability of minerals from their ecological and socio-historical context offers a grounding abstraction); to experiences of severance as violence and (dis)possession (directed at lands, bodies and knowledges); to processes of valorization/devalorization that occur through the making and unmaking of external relations; and to political asymmetries and epistemological hierarchies.

This "ex" of extraction has significant speculative potential as it enables connections to be forged across substantial differences (materials, bodies, territories, and temporalities), linking displacement of materials to other flows involved in colonial and capitalist economies and, in the process, stripping extraction of some of its liberal conceptual baggage. For example, displacement emerges from this process as more than a physical transfer of materials and, instead, as an appropriation and transfer of value (a process of seizing and plunder, in a continuation of colonial looting) that, at the same time, also allocates exposure to harm and vulnerability. Associated initially with A. N. Whitehead (1929), and more recently revived in fields like human geography, speculative potential refers to "a style of thinking that prioritises an openness to what thought might become, and which therefore reconfigures empirical problems beyond what seems given in an immediate experience" (Williams and Keating 2022, 2). It is, in Whitehead's useful phrase, about "mak(ing) thought creative of the future" (ibid.). This style of conceptual work considers how concepts and abstractions condition thought and practice, while also recognizing the possibilities of their "relative emptiness (which allows them to assume different meanings and to relate to many other, different, and conflicting terms)" (Ophir 2018). To think about extraction through the lens of speculative potential, then, is to recognize how concepts emerge through material processes (so that their meaning and utility evolves through unconventional application) and, more broadly, that concepts like extraction can do several different things.

The speculative possibilities for extrapolating from specific experiences of the extractive sector to economies at large has been recognized for a while. Ferguson's (2006) work on the governance of mineral extraction in Africa, for example, considers how the spatializations of order and disorder associated with extraction might presage the emergence of broader patterns of political-economic governance. His account of the socially thin relations around offshore oil in Angola—in comparison to the historically thick social relations (extending to housing, education, and social service provision) associated with the Zambian Copperbelt—led him to speculate on a mode of "extractive neoliberalism" that is "oil-like" (ibid.; 2010, 204) in its patchwork territoriality, reliance on private contractors, and abdication of responsibility by (foreign) private capital for the public realm (see also Appel 2012a).

Recent work with extraction moves further in this direction, positioning extraction as a systemic or structural logic working across multiple domains conventionally considered as separate. Dunlap and Jakobsen's (2020, 1) notion of "total extractivism" is arguably its imaginative apogee, a brute scaling of extraction into a "totalizing imperative and tension" driving the global capitalist economy and "at the heart of the present catastrophic trajectory." Here extractivism works to connect a suite of materials and technologies—minerals, hydrocarbons, intensive agriculture, renewable energy, digital technologies—that are often positioned discretely (and frequently in opposition) through analytical lenses like sustainability or energy transition. It speculates on their intersections, amalgamation, and their aggregate colonizing, converting, and consuming effects, making it possible to reframe development as a giant alimentary/colonic process of material displacement. The authors harness the speculative potential of extractivism to create a dissident agenda, located outside the technocratic registers through which global environmental change is normally narrated, which frames technocapitalist development as "progress of the Worldeater" (7).

A similar if more specific move is at work in Mezzadra and Neilson's (2015, 1) elevation of extraction as a diagnostic tool for "really tak(ing) stock of the underlying transformations of capitalism" in a way that can push beyond "neoliberalism" as explanation. Here extraction—together with finance and logistics—are no longer sectors or sites, but pervasive and defining logics. The effect is to extend extraction outwards as an "operation of capital," a logic manifest in spaces other than raw material production such as the subprime mortgage crisis in the United States and the extraction of value from *favelas* in Brazil. Ye et al. (2020, 163) similarly expand extractivism from an economic niche to a systemwide phenomenon, arguing that "extractivism has travelled from de-regulated spaces at the margins . . . to become a main feature of global capitalism as a whole (so that) more and more economic sectors are (re-)constructed in extractivist ways." They highlight the appropriation and draining of value produced elsewhere, oligopolistic networks that centralize wealth, and the control of circulatory flows rather than ownership of production.

Equally expansive approaches to extraction can be found in decolonial feminist scholarship, and in work on the aesthetics of the Anthropocene, where extraction names a field of racialized and gendered power relations that works by categorizing minerals, bodies, and multiple other forms of matter as active or inert (Murrey and Mollett 2023; Yusoff 2019). Extraction here references categorical distributions of matter (i.e., ontological inclusions and exclusions), together with their systemic effects that materialize as forms of "embodied-landed violence" reaching well "beyond subsoil resources, environmental contamination and displaced communities" (Murrey and Mollett 2023, 99). Yusoff's notion of an "extractive grammar"—"inhuman, property, value, possession . . . categories (that) move across territory, relation, and flesh" (2019, 4)—articulates this expansive understanding of extraction as a system of racialized signification. Extraction here is generalized and systematized: not only "the negation of racialised peoples and the geographies they inhabit" at sites of material displacement, it is also "more fundamentally . . . a symbolic and systemic structure that reduces the worth of racialised peoples and geographies to the marketable value of an extractable resource" (Çaylı 2021, 1392). For Yusoff, for example, extraction is both the moment of colonial racial categorization (positioning extraction in a political–historical relationship to racial difference) *and* the occlusion of racialized difference that is inherent to the "*Anthro*pocene."

The capacity of extraction and extractivism for forging conceptual connections has encouraged speculation on their potential as "organizing concepts" for the critical social sciences. Chagnon et al. (2022, 762), for example, consider how extractivism references "socio-ecologically destructive processes of subjugation, depletion, and non-reciprocal relations, occurring at all levels of practice" in ways that make it an "effective tool for sharpening critiques of what constitutes the 'sustainable' in development practices, while simultaneously opening the possibility for transformational practices, policies, and designs." Significantly, it is the generality of the "ex"—a lack of diagnostic specificity about techniques of separation and removal—that is key to extraction's speculative potential as it affords an openness and creative possibility, allowing extraction to proliferate semantically, producing what Ophir (2018) describes as "dense semantic intersections that connect (it) with many other terms." When mobilized for its speculative potential, the concept of extraction "construct(s) plural rather than singular narratives, recuperate(s) multiple rather than complete forms of knowledge, value(s) holding open what is at stake and can be brought into purview and, in doing so, intensif(ies) alternative possibilities" (Williams and Keating 2022, 1–2). It points—as highlighted by chapters in this book—to the possibility of thinking outside binaries of inclusion/exclusion and developing alternative registers to those of action/resistance.

EX-TRACT: DIAGNOSING HOW RAW MATERIAL ECONOMIES TAKE HOLD

An alternative orientation to extraction, I suggest, is to refocus on the specificities of primary sector economies. The impulse here is to double-down on the diagnostic capacity of extraction—its potential for parsing difference among modes of production and for identifying the particularities, contradictions, and evolving practices of raw material extraction. The value of extraction here, I suggest, lies not so much in the "ex" but in the "tract"—the technoeconomic practices and sociomaterial orders through which extraction (as a primary sector activity centered on the mobilization of materials) attains its grip on the world. The focus is less on the geographical mobilization of materials and the spatial transfer of value, and more on the inclusions and exclusions through which extractive, primary sector activities come to take hold and endure. Focusing on the "tract" of extraction highlights material appropriation and mobilization as a project—a technoeconomic and material "confrontation" with space and nature as both an obstacle and source of wealth and power—and how these are made conducive to accumulation and/or control (Banoub et al. 2021; Bustos-Gallardo et al. 2021; Moore 2015; Rasmussen and Lund 2018). Tsing's (2005; 2012) account of resource-making through "patches"—and her master-narratives of capitalist development occurring through "friction" and "zones of awkward engagement"—point the way here. They suggest how materials cannot be merely drawn from locations but must be "coaxed or coerced," and how scalable capital need always articulate with the specificity of each extraction site and with nonscalable geological and biological systems. In other words, a "messy engagement with difference is the assumed starting point" (Appel 2012b, 697).

Rather than generalizing extraction to establish connections across domains of difference, the impulse here is to focus inward on the extractive sector (classically understood) to understand its specificity and disclose internal differences. Exploring the tractability of raw material extraction and attending to difference (rather than generalizing across it) has the effective of multiplying extraction: it treats extraction not as a unitary logic or imperative but as grounded projects centered on claiming, severing, and mobilizing materials that, in practice, amalgamate and combine several forms of economy, each with different consequences for socio-political ordering. Significantly, these plural economies of extraction are not all of a piece—they differ, for example, in their economic objectives (ownership, production), economic benchmarks (cash flow, capital gain), and material geographies—suggesting how extractive projects can be subject to intercapitalist tensions (and how these differences might work as cracks, offering ways to loosen the grip of extractive projects outside the binary of "resistance"). I briefly outline here four modes of economy around extraction. They are not intended to be comprehensive and are necessarily stylized accounts.

Treasure

"Treasure" here articulates the classical political economy understanding that resources are "gifts of nature," materials for which the work of formation occurs prior to human labor and which can be captured and claimed. Gold nuggets, oil accumulations, and coal seams all owe their existence, as physical concentrations of materials, to the deep-time processes of astrophysics (in the case of metals) or geological time (in the case of oil and coal). They serve, therefore, as an economic "outside" so that extraction (as treasure taking) is characterized by "production without reproduction"—"the resources needed are not reproduced—neither within the enterprise that consumes them nor obtained, through market exchange, from other enterprises that produced them" (Ye et al. 2020, 157). Subject to epistemological work to render them legible and actionable, resources need only to be captured and removed for their value to be appropriated—something the popular idea of "bounty" or "treasure" readily conveys. Capturing is a primary form of economy in relation to extraction: an act of appropriation directed toward materials that have social utility and value through exchange. Rarely is this a wholesale taking of materials from the earth's crust: more often, it involves selective appropriation, a process of identifying, sorting, and sifting the most valuable forms (e.g., concentrated minerals, productive land). We can think of these concentrations of materials as a form of time-space appropriation, a condensation of deep-time and wide spatial horizons into a specific material form (Hornborg 2006). This is perhaps clearest with something like oil, which, as the rendered bodies of plants and animals, concentrates into individual oil fields the solar energy of photosynthesis from across the space of continents and the time of millions of years. Similarly in the case of land, productivity also relates to concentrations and distributions (of nutrients, organic matter, water, among others), although these are often, at least in part, the work of previous generations.

Extraction then is premised on opportunities for primitive accumulation, for laying claim to sources of value in whose production the capitalist has not invested (the fruits of other people's—or nature's—labor) and for enforcing that claim over claims of others. It is simultaneously a profound act of inclusion—of binding the extra-economic (the constitutive outside of the economy) into the economic process—while also being violently exclusionary since it rests on determining (via a claim to property) who has rights to remove materials (and who does not), what kinds of materials can be extracted, whose knowledge systems frame the space-times through which extraction emerges as a possibility, and who must be denied access, expelled, or dispossessed for extraction to take place. Chapters in this book illustrate empirically these forms of exclusion underpinning extraction, and how they are simultaneously practical, epistemological, and ontological. Indigenous and decolonial scholarship shows how extractive activities rely on the exclusion, marginalization, or miscommunication of knowledge systems that engage with the more-than-human in relational ways (Fjellheim 2023; Cameron 2015; Çaylı 2021). It unpicks the way recognition and the liberal politics of inclusion can work to legitimize extraction, operating as a form of "inclusionary control" that intensifies dispossession (Coulthard 2014; Tornel 2023).

Extraction as primitive accumulation takes place in multiple different technological forms. Artisanal gold panning and offshore oil production both rest on appropriation, but show how exclusion works through property, and yet is not limited to it: technological requirements can create high barriers to entry around, for example, the specialized equipment and/or capital costs necessary for project development, excluding all but the world's largest firms which are able to access global sources of finance (undermining efforts to "Indigenize" extraction by promoting domestic and "national" extractive capital). And because property is always embedded in existing social formations, there are formal and customary exclusions—as well as manipulations of property law by the powerful—that prevent marginalized peoples from claiming land, exclude them from consultations on land use, and reproduce relations of domination by, for example, disregarding non-Western knowledge or displacing root causes of conflict (Fjellheim 2023). Physical violence, meted out against other land users and against other claimants, frequently involves factions of the state. In short, the exclusions associated with primitive accumulation revolve around property, but are not limited to it.

Rent

The second distinctive opportunity extraction affords, as a mode of economy, is the opportunity for rent. Qualitative differences in resources (ease of accessibility, grade, purity, among others) and in social and environmental protections, mean some sites have lower extraction costs than others. These production costs are locally embedded to the extent they depend on the quality of the resource and on socio-politically negotiated settlements in relation to social and environmental protections. Yet prices for many extracted commodities are international as they reflect the costs of the highest-cost producer whose production fulfils demand. Because prices arise "globally" but production costs arise "locally" there can be big differences in profitability between the highest and lowest cost producers (Bomsel 1992). Very large rents are possible in oil and gas, for example, as by far the largest driver of (Ricardian) rent is the quality of the resource: once oil or gas wells are in production, labor productivity in oil and gas extraction is more or less the same around the world. Rents tend to be lower overall in mining than in oil and gas, as significant variations in labor productivity in mining means levels of differential rent are related to both quality of resource and productivity of labor (Bomsel 1992).

The opportunity for differential rent gives rise to competition to control the best sites. A good deal of the geographical strategy adopted by large resource extraction firms is about precisely this dynamic—seeking to monopolize ownership of (or access to) resource patches that afford the greatest differential rent. This materializes in large international mining firms, for example, as a strategy of occupying the "bottom quartile" of projects worldwide—that is, the 25 percent of all operating mines that have the lowest production costs. Focusing on rent highlights the capacity of landlords—"the person who by virtue of title to a portion of the globe has become the proprietor of these natural objects"—to lay claim to some of these differential rents (Marx 1894). A focus on distribution (between landlord and capitalist) shifts attention from differential rent based on resource quality to a Marxian rent analysis based on property, and how the landlord can "wrest surplus profit from functioning capital . . . thanks to his title to this piece of the globe endowed with singular properties" (Marx 1894).

From this perspective, the opportunity for rent revolves around the landlord/capitalist relation, although for historically and geographically specific reasons, the landlord in relation to natural resources is frequently the state. The history of state ownership of resources is rich and complex, arising often in the twentieth century as a consequence of specific contradictions associated with private resource ownership. In the United Kingdom, for example, the state became the steward of underground coal resources in the 1930s as a way of overcoming fragmented private ownership which had become an obstacle to accumulation (Fine 1990). In the post-war period, the doctrine of permanent sovereignty over natural resources secured ownership of resources to the state in many newly independent states in Asia and Africa, as well as in countries in Latin America independent since the 1820s, in a context where the developmental state was seen as both a bulwark against imperialism and a modernizing force. The state, then, has become a significant actor in the political economy of resource extraction, so that the "variable geometry" of the state, in terms of its porosity and relation to multinational capital and social movements, has a dominant role in shaping inclusions and exclusions that materialize as, for example, governance of risks, distribution of royalties and rents, and the epistemologies through which extraction and its consequences are disclosed (Svampa 2015).

Scale Economies

Extraction holds significant opportunities for capturing economies of scale, in part because other strategies for improving profitability are limited: there is not much potential for achieving economies of scope (using the fixed stock of extractive equipment to produce a range of different goods); and, with some notable exceptions (such as in the case of oil or tin) where cartelization or market-cornering is achieved, limited opportunities to drive accumulation by pushing up price. Economies of scale take the form of scaling up of extractive machinery and infrastructure so that individual unit costs for handling and processing material falls. By reducing costs of production at a particular facility, individual extractors can temporarily capture more differential rent, although this is only temporary because declining costs of production draw down price over time.

This dynamic, the "technological treadmill," is a conspicuous feature of extraction (Hanink 2005) and has several implications. First, it drives an innovation trajectory in extraction toward ever greater economies of scale, which is particularly visible around extractive and motive power (digging and hauling machinery), material processing, and transportation (such as conveyors, pipelines, and port facilities). Second, larger trucks and higher capacity machinery represent an increase in the technical composition of capital, the ratio of capital to labor. These growing capital costs mean barriers to entry rise: smaller (domestic) firms become squeezed out, as extraction increasingly requires firms with access to finance, a process that drives consolidation and, along with it, the development of new sources of financing. Third, growing capital requirements lead to expanding debt and a structural tendency to over-invest in capacity that, ultimately, powers the boom-bust cycle of extraction (experienced as squeezes on labor, crushing write downs, abandonment, and closure). Finally, economies of scale in processing and transportation require increased inputs that, as a general rule, cannot be generated by intensification in the way it is possible to increase yields from the same area with agriculture (Bunker and Ciccantell 2005). Economies of scale in transportation, then, drive a logic of extensification in extraction, materializing as deepening mines, expanding pits, and larger farms and plantations (see Fine 1994, 285). Here the scalability of capital confronts the limited scalability of the geological, biophysical, and labor systems that subtend the primary sector, as opportunities for formal subsumption of labor and nature (by merely expanding throughput) falter and extractive projects search for ways to reconfigure materials, ecologies, and bodies in an effort to scale output from nonscalable inputs (i.e., achieve a real subsumption of nature). The differential opportunities of biologically and geologically based systems generate the plantation and the mine as distinctive forms of extractive land and labor relations (Peluso 2017). The inclusions and exclusions that subtend mines and plantations are not only technical and economic but fundamentally ecological, as extractive capital (and communities) navigate "the dialectic between the ever-mounting material-throughput demands of an ever-growing mass of capital and the ever-mounting biophysical degradation that ensues through the endless accumulation of this capital" (Moore 2010, 38).

Financial Speculation

The treasure like character of extraction offers the speculative possibility of an 'upside': the possibility of striking it rich, particularly in metal mining and in oil (less so in coal or bulk minerals like bauxite). Histories of extraction have enough of these bonanza type finds to offer a compelling narrative to attract certain forms of speculative finance. At the core of speculation is an inverse relation between value and certainty: speculation thrives on possibility, while certainty transforms it—either by killing the possibility outright, or by converting what is ultimately a gamble into a sober projection of economic return. The constitutive uncertainty that makes possible extractive economies of speculation derives from multiple sources (commodity price risk, geopolitical risk, and others). At the development phase of extractive projects—where, in mining, for example, a lot of public capital is raised—there is a fundamental spatial uncertainty. Material heterogeneity offers affordances for speculation, enabling land and unextracted materials to circulate as possibilities for future revenues. Such speculative dreams—"conjuring the prospect of wealth from the spaces in between the drill holes" (Majury 2014, 548)—are fundamentally volumetric in character as they relate to the qualities of specific territory, such as the scale and extent of mineralization, its consistency/uniformity across horizontal and vertical dimensions, and the proximity and density of infrastructure.

These material qualities of territory offer different possibilities for capital circulation. Classic commodity production, via the activity of mining for example, is one option but sinking capital in mines, producing materials, and then realizing their value through sale is a slow and risky means to accumulation. The uncertain qualities of the underground enable "economies of appearances" to take hold, with mining share booms prime examples of this form of spectacular accumulation in which assumed future value of extracted materials is captured in the present (Tsing 2000, 2005). This distinction between extraction-as-commodity production and extraction-as-speculation is a site of discipline. It performs materially significant inclusions and exclusions that, together, constitute financial markets, including the apparatus of public warrant and systems of disclosure related to company registration and public equities, such as those of the Securities and Exchange Commission. Extraction's capacity for "rank speculation," for example, saw mining physically excluded from the London Stock Exchange until the mid-nineteenth century; while new standards of disclosure introduced to the Toronto Stock Exchange in the 1990s, following the Bre-X scandal, limited authorship of technical disclosure, mandated more inclusive access to geoscientific data via transparency, and rescaled the warranting process to include reference to non-Canadian reporting and valuation regimes (Majury 2014). To be clear, primary sector extractive activity is more than a set of composite economies: it fashions gendered, racialized and sexed subjects, buttresses (exclusionary) national histories, provokes alternative visions for the future, and much more besides. Outlining some of the multiple forms of economy folded within conventional extractive activity, however, serves to illustrate a diagnostic orientation toward extraction—that is, understanding how extractive activities operate—centered on the geographical, technoeconomic, and material specificities of primary sector processes.

CONCLUSIONS

Extraction is not exhausted as a concept—as the creative work unpacking power dynamics of resource exploitation in this collection demonstrates—although that is in part because a concept can be and do multiple things. I have argued here that working critically with the concept of extraction requires a dual orientation. On the one hand, a (centripetal) orientation toward greater analytic specificity by, for example, paying attention to the distinctiveness of extractive, primary sector economies and the processes through space and nature are made tractable, and are held together. And on the other, an expansive (centrifugal) orientation in which extraction casts off its sectoral roots and spins outward, encountering novel sites, registers, and concerns. This speculative move harnesses extraction's classic concerns with displacement and social ordering, using it to offer speculative connections and multiply extraction's "spaces of denotation and connotation" (Ophir 2018). This kind of approach acknowledges that a concept like extraction is neither fixed nor given but acquires its capacities through where and how it is performed. From this perspective, extraction and extractivism are not mined out, and there is productive work with them still to be done. I have argued for a kind of kinetic binary—a double movement—through which to simultaneously intensify and extensify the notion of extraction. Kinetic here speaks to a sense of continuing conceptual travel and emergence, a "becomingness" of extraction as it evolves through speculative gestures, diagnostic introspection, and other orientations it may be tasked to perform.

ACKNOWLEDGMENTS

I would like to thank Negar Elodie Behzadi, Nina Döring, and Stephanie Postar for helpful editorial comments on an earlier draft of this chapter; and Penelope Anthias, Eric Boyd, and Carlos Tornel for generous collegial engagement with its themes. I am grateful for the generative environment created by hosts (Paul Langley and Léonie Newhouse) and participants in the *Extraction & Future Fantasies Workshop* at Durham (January 2023) where some of the arguments here were first aired. Thanking the above does not imply their responsibility for the content of the chapter: that is mine alone.

REFERENCES

Acosta, Alberto. 2013. "Extractivism and neo-extractivism: Two sides of the same curse." In *Beyond Development: Alternative Visions from Latin America*, edited by Imre Szűcs, Quito and Amsterdam: Permanent Working Group on Alternatives to Development.

Alimonda, Héctor. 2012. "Debating development in Latin America: From ECLAC to the Brazilian workers' party." In *Inside a Champion: An Analysis of the Brazilian Development Model.* Berlin and Washington: Heinrich Böll Foundation.

Anthias, Penelope. 2018. "Indigenous peoples and the new extraction: From territorial rights to hydrocarbon citizenship in the Bolivian Chaco." *Latin American Perspectives* 45(5): 136–153.

Appel, Hannah C. 2012a. "Walls and white elephants: Oil extraction, responsibility, and infrastructural violence in Equatorial Guinea." *Ethnography* 13(4): 439-465.

Appel, Hannah C. 2012b. "Offshore work: Oil, modularity, and the how of capitalism in Equatorial Guinea." *American Ethnologist* 39(4): 692–709.

Banoub, Daniel, Gavin Bridge, Beatriz Bustos, Irmak Ertör, Marien González-Hidalgo, and Julie Ann de los Reyes. 2021. "Industrial dynamics on the commodity frontier: Managing time, space and form in mining, tree plantations and intensive aquaculture." *Environment and Planning E: Nature and Space 4*(4): 1533–1559.

Barandiarán, Javiera. 2019. "Lithium and development imaginaries in Chile, Argentina and Bolivia." *World Development* 113: 381–391.

Bomsel, Olivier. 1992. "The political economy of rent in mining countries." In *Mineral Wealth and Economic Development*, edited by John Tilton, 67–129. Washington: Resources for the Future Press.

Brechin, Gray. 1999. *Imperial San Francisco: Urban Power, Earthly Ruin*. Berkeley: University of California Press.

Bridge, Gavin. 2007. "Acts of enclosure: Claim staking and land conversion in Guyana's gold fields." In *Neoliberal Environments: False promises and Unnatural Consequences*, edited by Nik Heynen, James McCarthy, Scott Prudham, and Paul Robbins, 74–86. London: Routledge.

Bunker, Stephen G., and Paul S. Ciccantell. 2005. *Globalization and the Race for Resources*. Baltimore: Johns Hopkins University Press.

Bustos-Gallardo, Beatriz, Gavin Bridge, and Manuel Prieto. 2021. "Harvesting lithium: Water, brine and the industrial dynamics of production in the Salar de Atacama." *Geoforum* 119: 177–189.

Cameron, Emilie. 2015. *Far Off Metal River: Inuit Lands, Settler Stories, and the Marking of the Contemporary Arctic*. Vancouver and Toronto: UBC Press.

Çaylı, Eray. 2021. "The aesthetics of extractivism: Violence, ecology, and sensibility in Turkey's Kurdistan." *Antipode* 53(5): 1377–1399.

Chagnon, Christopher W., Sophia E. Hagolani-Albov, and Saana Hokkanen. 2021. "Extractivism at your fingertips." In *Our Extractive Age: Expressions of Violence and Resistance*, edited by Judith Shapiro and John-Andrew McNeish, 176–188. London: Routledge.

Chagnon, Christopher W., Francesco Durante, Barry K. Gills, Sophia E. Hagolani-Albov, Saana Hokkanen, Sohvi M. Kangasluoma, Heidi Konttinen, Markus Kröger, William LaFleur, Ossi Ollinaho, and Marketta P.S. Vuola. 2022. "From extractivism to global extractivism: The evolution of an organizing concept." *The Journal of Peasant Studies 49*(4): 760–792.

Coronil, Fernando, 1997. *The Magical State: Nature, Money, and Modernity in Venezuela*. Chicago: University of Chicago Press.

Coulthard, Glen Sean. 2014. *Red Skin, White Masks: Rejecting the Colonial Politics of Recognition*. Minneapolis: University of Minnesota Press.

Domínguez Martín, Rafael. 2021. "El extractivismo y sus despliegues conceptuales." *Revista Territorios y Regionalismos* 4: 1–26.

Dunlap, Alexander. 2020. "Wind, coal, and copper: The politics of land grabbing, counterinsurgency, and the social engineering of extraction." *Globalizations* 17(4): 661–682.

Dunlap, Alexander, and Jostein Jakobsen. 2020. *The Violent Technologies of Extraction: Political Ecology, Critical Agrarian Studies and the Capitalist Worldeater*. Basingstoke: Palgrave Pivot.

Ferguson, James. 2006. *Global Shadows: Africa in the Neoliberal World Order*. Durham: Duke University Press.

Fine, Ben. 1990. *The Coal Question: Political Economy and Industrial Change from the Nineteenth Century to the Present Day*. Abingdon: Routledge.

Fine, Ben. 1994. "Coal, diamonds and oil: Toward a comparative theory of mining?" *Review of Political Economy 6*(3): 279–302.

Fjellheim, Eva Maria. 2023. "'You Can Kill Us with Dialogue:' Critical Perspectives on Wind Energy Development in a Nordic-Saami Green Colonial Context." *Human Rights Review* 24: 25–51.

Fry, Matthew, and Trey Murphy. 2021. "The Geo-Imaginaries of Potential in Mexico's Burgos Basin." *Political Geography* 90. https://www.sciencedirect.com/science/article/abs/pii/S0962629821001220?via%3Dihub.

Gudynas, Eduardo. 2018. "Extractivisms: Tendencies and consequences." In *Reframing Latin American Development*, edited by Ronaldo Munck and Raul Delgado Wise, 61–76. London: Routledge.

Hanink, Dean. 2005. "Resources." In *A Companion to Economic Geography*, edited by Eric Sheppard and Trevor J. Barnes, 227–241. Maldon and Oxford: Blackwell Publishing Ltd.

Hornborg, Alf. 2006. "Footprints in the cotton fields: The Industrial Revolution as time–space appropriation and environmental load displacement." *Ecological Economics* 59(1): 74–81.

Huber, Matt T., and James McCarthy. 2017. "Beyond the subterranean energy regime? Fuel, land use and the production of space." *Transactions of the Institute of British Geographers* 42(4): 655–668.

Isla, Ana. 2021. "'Greening,' the Highest Stage of Extractivism in Latin America." In *The Routledge Handbook on Ecosocialism*, edited by Leigh Brownhill, Salvatore Engel-Di Mauro, Terran Giacomini, Ana Isla, Michael Löwy, and Terisa E. Turner. London: Routledge.

Kuchler, Magdalena, and Gavin Bridge. 2018. "Down the black hole: Sustaining national socio-technical imaginaries of coal in Poland." *Energy Research & Social Science* 41: 136–147.

Li, Tania M. 2014. "What is land? Assembling a resource for global investment." *Transactions of the institute of British Geographers* 39(4): 589–602.

Majury, Niall. 2014. "'Trusting the numbers': Mineral prospecting, raising finance and the governance of knowledge." *Transactions of the Institute of British Geographers* 39(4): 545–558.

Marx, Karl. 1894. "Transformation of surplus-profit into ground-rent." *Capital* III(part VI). Available at https://www.marxists.org/archive/marx/works/1894-c3/ch46.htm#r40

Mezzadra, Sandro, and Brett Neilson. 2015. "Operations of capital." *South Atlantic Quarterly* 114 (1): 1–9.

Moore, Jason. 2015. *Capitalism in the Web of Life: Ecology and the Accumulation of Capital*. London: Verso Books.

Moore, Jason. 2010. "'Amsterdam is standing on Norway'. Part I: The alchemy of capital, empire and nature in the diaspora of silver, 1545–1648." *Journal of Agrarian Change* 10(1): 33–68.

Ophir, Adi. 2018. "Concept." In *Political Concepts: A Critical Lexicon*, edited by. J. M. Bernstein, Adi Ophir, and Ann Laura Stoler. New York: Fordham University Press. https://www.politicalconcepts.org/concept-adi-ophir/

Peluso, Nancy. 2017. "Plantations and mines: Resource frontiers and the politics of the smallholder slot." *Journal of Peasant Studies* 4(4): 834–69.

Rasmussen, Mattias Borg, and Christian Lund. 2018. "Reconfiguring frontier spaces: The territorialization of resource control." *World Development* 101: 388–399.

Richardson, Tanya, and Gisa Weszkalnys. 2014. "Introduction: Resource materialities." *Anthropological Quarterly* 87(1): 5–30.

Steffen, Will, Wendy Broadgate, Lisa Deutsch, Owen Gaffney, and Cornelia Ludwig. 2015. "The trajectory of the Anthropocene: The great acceleration." *The Anthropocene Review* 2(1): 81–98.

Svampa, Maristella. 2015. "Commodities consensus: Neoextractivism and enclosure of the commons in Latin America." *South Atlantic Quarterly* 114(1): 65–82.

Tornel, C., 2023. Energy justice in the context of green extractivism: Perpetuating ontological and epistemological violence in the Yucatan Peninsula. Journal of Political Ecology, 30(1), pp. 1–28. Available online at https://journals.librarypublishing.arizona.edu/jpe/article/5485/galley/5337/view/.

Tsing, Anna Lowenhaupt. 2000. "Inside the economy of appearances." *Public Culture* 12(1): 115–144.

Tsing, Anna Lowenhaupt. 2005. *Friction: An Ethnography of Global Connection.* Princeton: Princeton University Press.

Tsing, Anna Lowenhaupt. 2012. "On Nonscalability: The Living World Is Not Amenable to Precision-Nested Scales." *Common Knowledge* 18(3): 505–524.

Valdivia, Gabriela, Matthew Himley, and Elizabeth Havice. 2021. "Critical resource geography: An introduction." In *The Routledge Handbook of Critical Resource Geograph*y, edited by Matthew Himley, Elizabeth Havice, and Gabriela Valdivia, 1–20. London: Routledge.

Voskoboynik, Daniel Macmillen, and Diego Andreucci. 2022. "Greening extractivism: Environmental discourses and resource governance in the 'Lithium Triangle'." *Environment and Planning E: Nature and Space* 5(2): 787–809.

Watts, Michael J. 2004. "Antinomies of community: Some thoughts on geography, resources and empire." *Transactions of the Institute of British Geographers* 29(2): 195–216.

Werner, Marion. 2016. "Global production networks and uneven development: Exploring geographies of devaluation, disinvestment, and exclusion." *Geography Compass* 10(11): 457–469.

Weszkalnys, Gisa. 2016. "A doubtful hope: Resource affect in a future oil economy." *Journal of the Royal Anthropological Institute* 22(S1): 127–146.

Whitehead, Alfred North. 1929. *The Function of Reason.* Princeton: Princeton University Press.

Williams, Nina, and Thomas Keating. 2022. "From Abstract Thinking to Thinking Abstractions: Introducing Speculative Geographies." In *Speculative Geographies: Ethics, Technologies, Aesthetics*, edited by Nina Williams and Thomas Keating, 1–32. Singapore: Palgrave Macmillan.

Ye, Jingzhong, Jan Douwe van der Ploeg, Sergio Schneider, and Teodor Shanin. 2020. "The incursions of extractivism: Moving from dispersed places to global capitalism." *The Journal of Peasant Studies* 47(1): 155–183.

Zografos, Christos, and Paul Robbins. 2020. "Green sacrifice zones, or why a green new deal cannot ignore the cost shifts of just transitions." *One Earth* 3(5): 543–546.

Index

About the Contributors

Gavin Bridge is Professor of Economic Geography at Durham University and a Fellow of the Durham Energy Institute. He is interested in the spatial and temporal dynamics of extractive industries and the political ecologies of profit, property and possibility associated with oil, gas, and mining. His research has been funded by the US National Science Foundation, UK Research and Innovation, National Geographic Society, Leverhulme Trust, British Academy, Newton Fund, and Horizon 2020. Books include Oil (with Philippe LeBillon), the Routledge Handbook of Political Ecology (with James McCarthy and Tom Perreault) and he is lead author of Energy and Society: A Critical Perspective (Routledge). His current work focuses on the production networks and supply chains associated with high and low carbon economies.

Negar Elodie Behzadi is a lecturer in human geography at the University of Bristol. She holds a doctorate in human geography from the University of Oxford, and a masters in environment and development from King's College London. Negar is a feminist political geographer and political ecologist whose work focuses on how intersectional forms of exclusion and marginalization are produced, reproduced, and contested in stressed environments. Her research brings the insights of feminist geography and the sensibilities of an ethnographer to issues related to resource extraction in Tajikistan (Central Asia) and Iranian migrants in France. Negar has also codirected two ethnographic films, and her work uses creative practice and methods.

Teresa Bornschlegl holds an MSc in human ecology from Lund University (Sweden) and a PhD in geography from Clark University (USA). After a postdoctoral fellowship with the Trans-Andean Network for Sustainability (trAndeS) researching the challenges and opportunities arising from the use of geospatial technologies for the environmental oversight of extractive industries, she started as a research associate and policy advisor at the sector program Extractives and Development at the Federal Institute of Geosciences and Natural Resources (BGR) in Germany. Her current work focuses mainly on environmental risks, environmental due diligence, and remote sensing in the mining sector and the politics thereof.

Emilie Cameron is a human geographer and associate professor in the Department of Geography and Environmental Studies at Carleton University (Canada). Her research focuses broadly on critical northern geographies, including critical approaches to climate change, economic and social development, land claims, environmental assessment, health, and knowledge production in the Canadian Arctic. She is particularly interested in the geographies of resource extraction, empire, and labor in the contemporary North.

Isabel Crowhurst is reader in sociology at the University of Essex, UK. Through the conceptual lenses of intimate and economic citizenship, her work explores how social norms operate as mechanisms of exclusion and as sites of contestation. Recent publications include *Third Sector Organizations in Sex Work and Prostitution* (2021, with Dewey and Izugbara) and *The Tenacity of the Couple Norm* (2020, with Roseneil, Hellesund, Santos, and Stoilova).

Penda Diallo is an academic, researcher, and Environmental, Social, and Governance specialist focusing on the mining sector, working at the intersection between NGOs, academia, industry, government, and rural and local communities. Penda is the Qualitative Research Manager at the Aluminum Stewardship Initiative (ASI). She is the author of *Regime Stability, Social Insecurity and Bauxite Mining in Guinea: Developments Since the Mid-Twentieth Century* (Routledge 2020). She has over ten years of experience working in academia, research, international development, and consultancy on issues at the intersection between sustainable development, mining, conflict, and security. Given her work in development, human rights, resilience, governance, social performance, and sustainability, her work has been in multicultural and multidisciplinary settings where Penda has utilized her multilingual skills. She is a bridge-builder connecting stakeholders from various backgrounds and industries to collaborate on responsible mining and development projects that support local community development in mining regions. Her work extends to mining regions globally.

Nina Nikola Doering holds an MPhil in development studies and a DPhil in geography from the University of Oxford (UK). Her research has engaged with public participation and decision-making in extractive resource management in Greenland. More recently, she has focused on knowledge cocreation, research ethics and decolonial approaches to research. Nina works as research group leader of the Arctic Governance research group at the Research Institute for Sustainability - Helmholtz Centre Potsdam (RIFS) in Potsdam, Germany.

Asel Doolotkeldieva is a senior lecturer at the OSCE Academy in Bishkek, Kyrgyzstan. She earned her PhD from the University of Exeter (UK) in politics and previously worked as a visiting fellow at ZMO (Berlin, Germany) and College Mondial, FMSH (Paris, France). Her academic interests include political economy of resource extraction in Central Asia as well as political mobilizations and protests. She published on various aspects of gold mining, neoliberalism, and authoritarianism in Kyrgyzstan.

Rebecca Elmhirst is professor of human geography in the Centre for Spatial, Environmental, and Cultural Politics at the University of Brighton in the United Kingdom. Her research interests lie in feminist political ecology, and the everyday politics of natural resource governance, livelihoods, and migration in extractive landscapes in the United Kingdom and Southeast Asia. She is part of the Extracting Us curatorial collective (extractingus.org), an ongoing project that brings together creative work and reflections from artists-activists-researchers in relation to diverse extractive contexts.

Gavin Hilson is professor and chair of sustainability in business at the University of Surrey Business School (UK). He is a leading global authority on the social and environmental impacts of artisanal and small-scale mining (ASM) in sub-Saharan Africa. Professor Hilson has published over three hundred articles, book chapters, and reports on the subject. Moreover, in recognition of his specialized knowledge, he has consulted on a number of ASM matters for a range of organizations, including the World Bank, OXFAM America, United Nations, GTZ, and Rio Tinto. Professor Hilson received bachelors and masters degrees from the University of Toronto (Canada), and his PhD from Imperial College London (UK).

Leah S. Horowitz is an associate professor of environmental studies at the University of Wisconsin-Madison (USA). As a critical cultural geographer, she uses ethnographic methods to examine social relationships, cultural specificities, and power dynamics within environmental governance, across a range of scales. At the local scale, she examines community-led efforts to influence environmental decision-making processes. Simultaneously, her research investigates ideologies, policies, and power relations of elites at the national and global scales, so as to better understand structures conditioning community-based engagements. She has studied Indigenous Kanak communities' negotiations with and resistance to mining projects in New Caledonia, and communities' engagements, in New Caledonia, Malaysian Borneo, and the United States, with both rural and urban biodiversity conservation. Currently, her research program focuses on American Indian-led resistance to oil and gas pipelines. She has published in, *inter alia*, *Progress in Human Geography*; *Gender, Place and Culture*; and *Transactions of the Institute of British Geographers*.

Nanang Indra Kurniawan is a lecturer at the Department of Politics and Government, Faculty of Social and Political Sciences, UGM (Indonesia). His research interests include Civil Society, Environmental Politics, Natural Resource Politics, and Politics of Technology. Before becoming a lecturer in 2006, he was involved as a researcher and NGO activist with the Institute for Research and Empowerment (IRE), Yogyakarta. Among his publications is the book *Globalisasi dan Negara Kesejahteraan Norwegia* (*Globalisation and the Norwegian Welfare State* 2009). His journal articles include "Struggle to Gain Representation: Mixed Politics in Democratizing Indonesia" (published in *Power Conflict and Democracy Journal*, Vol. II No.2. 2010) and "Claiming Indigenous Rights through Participatory Mapping and the Making of Citizenship" (cowritten with Ståle Angen Rye, published in *Political Geography*).

Kuntala Lahiri-Dutt is a professor at the Crawford School of Public Policy, College of Asia and the Pacific, The Australian National University (Australia). Kuntala is a feminist activist-researcher who works with local communities on questions of gender, the environment, and natural resources in developing countries (India, Indonesia, Lao PDR, Mongolia, Pakistan, and Papua New Guinea). Her scholarship explores the gendered aspects of artisanal and small-scale mining, water resource management, small-holder agriculture, and extractive industries. Kuntala has also been an expert member or author of various United Nations and World Bank groups and publications on women, gender, sustainability, and resource extraction.

Philippe Le Billon is professor at the University of British Columbia (Canada) with the Department of Geography and the School of Public Policy and Global Affairs. He holds an MBA (Paris 1, France) and PhD (Oxford, UK). Prior to joining UBC, he was a research associate with the Overseas Development Institute (ODI) and the International Institute for Strategic Studies (IISS). He works on linkages between environment, development, and security. His two latest books including *Environmental Defenders: Struggles for*

Life and Territory (2021, with Mary Menton), *Oil* (2017, with Gavin Bridge) and *Wars of Plunder: Conflicts, Profits and the Politics of Resources* (2013).

Éléonore Lèbre is a research fellow at the University of Queensland's Sustainable Minerals Institute, Centre for Social Responsibility in Mining (Australia). She is a recipient of the 2022 Discovery Early Career Researcher Award from the Australian Research Council. Her research centers on using global spatial data to develop a multidisciplinary understanding of social, environmental, and governance contexts around mineral resource extraction. She connects with topical issues such as the low-carbon energy transition, critical metal supply, deep-sea mining, and tailings dam failures. Dr. Lèbre has a PhD in Industrial Ecology from the Sustainable Minerals Institute's Centre for Mined Land Rehabilitation (Australia). Her thesis focuses on circular economy initiatives in the mining industry, following objectives of waste minimization and resource efficiency.

Päivi Lujala is a professor of geography at the University of Oulu, Finland. Her research focuses on two broad topics: management of valuable natural resources in the Global South and adaptation to climate (change)-related natural hazards. She has led several multidisciplinary projects on the links between primary commodity sector, development, and security and has published widely on these topics. Her recent publications include "Transparent for Whom? Dissemination of Information on Ghana's Petroleum and Mining Revenue Management" (with C. Brunnschweiler and I. Edjekumhene, *Journal of Development Studies* 2020), "Transparency in Environmental and Resource Governance: Theories of Change for the EITI" (with P. Le Billon and S. A. Rustad, *Global Environmental Politics* 2021), and "Spatial Crowdsourcing in Natural Resource Revenue Management" (with M. Ogbe, *Resources Policy* 2021). Lujala's current research on climate change focuses on climate migration in the Global South.

Sharlene Mollett is an associate professor and chair in the Department of Global Development Studies with cross-appointments in the Department of Human Geography (UTSC) and the Graduate Program in the Department of Geography and Planning at the University of Toronto (Canada). As cultural geographer, her work interrogates multiple forms of power shaping land conflicts in Latin America. Mollett has published widely in a variety of geography and Latin American regional journals including the *Annals of the American Association of Geographers*, *Antipode*, *Latin American Research Review,* and *Cultural Geographies*. She currently holds the award, Distinguished Professor in Feminist Cultural Geography, Nature and Society at the University of Toronto, 2020–2024, and is an editor at *Social and Cultural Geography*.

Amber Murrey is a decolonial political geographer, ethnographer, and educator. Her research on resistance and social change in Africa is empirically grounded and integrates the political geographies of environmental and socio-political struggles with decolonial thought and resistance studies. She is the editor of *'A Certain Amount of Madness': The Life, Politics and Legacies of Thomas Sankara* (Pluto 2018) and coauthor (with Patricia O. Daley) of *Decolonizing Development Studies: Learning Disobedience* (Pluto 2023). Amber is an associate professor of political geography at the University of Oxford and Tutorial Fellow at Mansfield College (UK). She has taught at the American University in Cairo (Egypt), Clark University and Boston College in Massachusetts (USA), and Jimma University (Ethiopia).

Mark Nuttall is a social anthropologist and has a particular focus on the societies and environments of the circumpolar North. He is interested in understanding human-environment relations, especially within a context of historical and contemporary environmental changes which affect livelihoods, human engagement with the environment, resource use, and development at multiple and interrelated levels. Mark has carried out extensive anthropological research and fieldwork in Greenland, Canada, Finland, Scotland, Alaska, and Wales. He is professor and Henry Marshall Tory Chair of Anthropology at the University of Alberta (Canada), and adjunct professor and Ilisimatusarfik/University of Greenland and the Greenland Climate Research Centre (Greenland).

John Owen's current program of research is concerned with building a foundational knowledge base on resettlement and livelihood displacement in the mining and resource sectors. He is active in promoting the development and use of social data technologies as a means for improving due diligence processes surrounding mining and mine-related activities. John is a professorial research fellow at the Centre for Social Responsibility in Mining at UQ's Sustainable Minerals Institute (SMI) (Australia). He is an editorial board member for *Resources Policy*, *Corporate Social Responsibility and Environmental Management*, and *The Extractives Industries and Society*. He is also a member of the Society of Business Ethics, the Project Management Institute, and the International Association for Business and Society.

Stephanie Postar is an environmental anthropologist specializing in energy and natural resources in the Global South. Her research engages with the politics and practices that shape decisions about and attitudes toward natural resource use and management in Tanzania. Stephanie holds a doctorate in Social Anthropology from the University of Oxford (UK). She is a British Academy Postdoctoral Fellow at the London School of Economics and

Political Science (UK) and previously was a Ciriacy-Wantrup Postdoctoral Fellow in Natural Resource Economics and Political Economy at the University of California, Berkeley (USA).

Ståle Angen Rye is a professor in political geography at the Norwegian University of Science and Technology, NTNU where he is affiliated to the Department of Geography (Norway). He is currently involved in several transdisciplinary research projects related to citizens participation in natural resource governance and sustainable transitions in Indonesia. Rye has extensive field experience from Indonesia which includes research on mining communities, mineral extraction, participatory mapping, and internet use in the environmental movement. Globalization, transnational relations, and citizenship are central dimensions in most of his research.

Titus Sauerwein is a researcher specializing in the governance of the artisanal and small-scale mining (ASM) sector. His research on extractive industries in sub-Saharan Africa examines the dynamics of interactions and engagement between different actors at the fringes of the state, where the boundaries between formal and informal, customary and statutory, and private and public are typically obfuscated. Since 2017, he has carried out ethnographic and archival research on the gold mining frontier in Côte d'Ivoire, where large-scale gold mining companies and ASM communities increasingly interact, and which has often culminated in violence. Dr. Sauerwein holds a PhD at the Surrey Business School (UK), an MSc in International Relations from the University of Amsterdam (Netherlands) and a MA in African Studies from Leiden University (Netherlands).

Massaran Traoré has worked for more than a decade as a specialist on natural resources and local development in the West African region. She is the founder of Alliance for Research and Integrated Development (ARDI), an NGO dedicated to developing strategies for good governance and supporting gender-sensitive capacity building in the extractive industries. Ms Traoré holds an MSc in international development from the University of Reading (UK) and is currently a doctoral candidate at the University of Surrey Business School (UK), where she is carrying out research on women's livelihoods in Mali's artisanal gold mining sector.

Robin West is senior lecturer in sociology at London Metropolitan University (UK). His research focuses on cultural memory and questions of contested representations. He retains an active interest in the construction of meaningful relationships between cultural landscapes and the normative frameworks that underpin social and environmental transformation. He is currently working on a project exploring affect and the performance of collective memory related to the intersections of dark tourism, historic state violence and reparation, and the afterlife of victimization in South Korea.

Gisa Weszkalnys is associate professor in anthropology at the London School of Economics (LSE) (UK). Her research examines future making as a political, material, and affective practice critical to contemporary capitalism. She has carried out intensive research on the politics of urban planning, resulting in the monograph *Berlin, Alexanderplatz: Transforming Place in a Unified Germany* (Berghahn 2010) and an edited volume *Elusive Promises: Planning in the Contemporary World* (ed. with Simone Abram, Berghahn 2013). More recently, her research has focused on the inconsistent temporalities of hydrocarbon extraction against the backdrop of climate change and the transformation of the industry, both in the United Kingdom and in Africa. Her current book project tracks the speculative logics at play in ongoing offshore hydrocarbon exploration in São Tomé and Príncipe building on both ethnographic and archival work.

Anna J. Willow is a professor of anthropology at the Ohio State University (USA). An environmental anthropologist who studies how individuals and communities experience and respond to externally imposed resource extractive development, she is the author of *Strong Hearts, Native Lands: The Cultural and Political Landscape of Anishinaabe Anti-Clearcutting Activism* (2012) and *Understanding ExtrACTIVISM: Culture and Power in Natural Resource Disputes* (2018). Willow received her PhD in cultural anthropology from the University of Wisconsin-Madison (USA) as well as an MS in natural resources and environment from the University of Michigan (USA).